E&M Endocrinology and Metabolism

Progress in Research and Clinical Practice

Piero P. Foà
Series Editor

Endocrinology and Metabolism
Progress in Research and Clinical Practice

Piero P. Foà
  Series Editor

Piero P. Foà
Editor

Humoral Factors in the Regulation of Tissue Growth

Blood, Blood Vessels,
Skeletal System, and Teeth

With 43 Figures

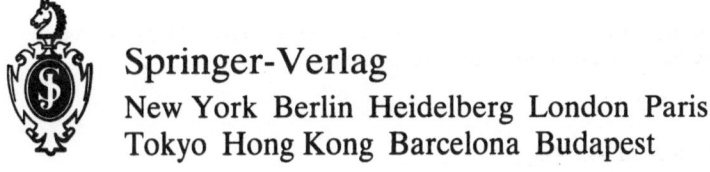
Springer-Verlag
New York Berlin Heidelberg London Paris
Tokyo Hong Kong Barcelona Budapest

Piero P. Foà, M.D., Sc.D.
Professor Emeritus
Department of Physiology
Wayne State University
Detroit, MI
Mailing address:
  2104 Rhine Road
  West Bloomfield, MI 48323 USA

Library of Congress Cataloging-in-Publication Data
Humoral factors in the regulation of tissue growth: blood, blood vessels, skeletal system,
  and teeth/Piero P. Foà, editor.
    p. cm.—(Endocrinology and metabolism; 5)
  Includes bibliographical references and index.
  ISBN-13:978-1-4613-9274-3   e-ISBN-13:978-1-4613-9272-9
  DOI: 10.1007/978-1-4613-9272-9
  1. Growth factors. 2. Neovascularization. 3. Erythropoiesis.
4. Musculoskeletal system—Growth—Regulation. I. Foà, Piero P.
(Piero Pio), 1911– . II. Series: Endocrinology and metabolism
(New York, N.Y.); 5.
    [DNLM: 1. Growth Substances—physiology. 2. Muscles—growth &
development. W1 EN396SN v.5/WE 500 H925]
QP552.G76H86 1992
612.4—dc20
DNLM/DLC
for Library of Congress

                                                                    92-20284

Printed on acid-free paper.

Production managed by Henry Krell; manufacturing supervised by Jacqui Ashri.
Typeset by Best-set Typesetter Ltd., Hong Kong.

9 8 7 6 5 4 3 2 1

ISBN-13:978-1-4613-9274-3

Preface

More than 70 years have elapsed since H.M. Evans and J.A. Long discovered the growth-promoting effect of crude saline extracts of the anterior pituitary and almost exactly 40 years from Rita Levi-Montalcini's discovery that the growth and differentiation of neurons is stimulated in the presence of mouse S-180 tumor transplants. Since then, the pace setting observations of Evans and Long were followed by the preparation of ever more potent pituitary extracts, by the isolation of pure growth hormone (GH), by its molecular identification and, finally, by its biosynthesis by recombinant DNA technology: a wondrous, even if now familiar sequence of events. Similarly, Levi-Montalcini's work and the subsequent observation that snake venom and extracts of mouse submaxillary gland shared the growth-promoting powers of sarcomatous cells led her and Stanley Cohen to the discovery of the nerve growth factor (NGF). GH and NGF are but two representatives of a rapidly increasing number of organ and tissue-specific growth-stimulating substances whose discovery created a revolution in medicine and blurred the frontiers of endocrinology. Indeed, any distinction between growth factors and hormones was rendered obsolete by the observation that the insulin-like growth factor 1 (IGF-1) is identical to the GH-dependent somatomedin C and by the realization that circulating hormones often exert their effect by releasing local target-specific factors which in an autocrine or paracrine fashion activate receptor-mediated mechanisms of signal transduction.

The chapters of this book were selected not to cover the waterfront, as only a series of monographs could do justice to that purpose, but rather to discuss the nature and mechanism of action of growth-stimulating substances whose biologic and clinical activities in man have been demonstrated or are being investigated. Thus, the book starts with a chapter on angiogenesis in health and disease, discussing the various factors that regulate blood vessel growth during embryologic development, follicle maturation, menstrual cycle, implantation of the embryo and formation of the placenta, and such pathologic events as joint inflammation and inflammatory processes associated with wound healing, tumor growth and diabetic retinopathy. The normal vascularization and pathologic neovascularization of the retina is dealt with in greater detail in a separate chapter that discusses the identity, mechanism of action and clinical significance of angiogenic factors and their relationship to the pathogenesis of potentially blinding retinal disease. A chapter on the humoral regulation of erythropoiesis discusses the nature, secretion, mode of action and clinical usefulness of erythropoietin, interleukin-3, erythroblast-enhancing factor, and other stimulatory and inhibiting substances. This is followed by a chapter on the role of the colony-stimulating factors and the interleukins in leukopoiesis, which includes a discussion of pilot studies carried out at the Sloan-Kettering Cancer Center and elsewhere attempting to boost blood cell production in patients with AIDS, with lymphoid malignancies, or undergoing bone marrow transplantation and suffering the hematologic consequences of chemotherapy. The normal and pathologic roles of the lymphokinins are discussed separately in greater detail, while another chapter discusses inhibitory rather than stimulatory effects on cell proliferation, emphasizing the role of transforming growth factor B, interferon and the cytokines and their possible role as antineoplastic agents. Another chapter discusses how the cytokines regulate hypothalamic and pituitary hormone secretion, an example of interaction between growth factors and hormones that reaches the highest degree of complexity in the case of the humoral agents (parathyroid hormone, calcitonin, vitamin D, growth factors and prostaglandins) that regulate bone growth, remodeling and repair. This discussion provides the necessary information required to understand the numerous endocrine, nutritional, and iatrogenic factors involved in the pathogenesis of osteopenia and the special problems of bone mineral metabolism associated with the menopause. Physical inactivity causes loss of bone mass, a problem extensively investigated during space flight in a gravity-free environment, an interesting model of unloading and immobilization described in a chapter based on the experience of the Russian space program. The rapidly growing body of knowledge on the role of growth factors in the physiology of joints and in the pathogenesis of arthritis is the topic of another chapter, and the monograph ends with a dissertation on the control of tooth and periodontal tissue growth and repair.

Considering the scope and variety of the subject matter, I believe that this book represents a good sampler of current basic and applied knowledge and a tantalizing inducement to further reading. A rich bibliography will guide the reader in this effort.

PIERO P. FOÀ

References

1. Evans HM, Long JA. The effect of the anterior lobe of the hypophysis administered intraperinoteneally upon growth, maturity, and oestrus cycles of the rat (Abstr). *Anat Rec.* 1921;21:61.
2. Levi-Montalcini R. Effect of mouse tumor transplants on the nervous system. *Am NY Acad Sci.* 1952;55:330–344.
3. Cohen S. Purification of a nerve-growth promoting protein from the mouse salivary gland and its neurocytotoxic antiserum. *Proc Natl Acad Sci USA.* 1960;46:302–316.

Contents

     DAVID HAMERMAN and STEPHEN TAYLOR

     Structure–Function Relationships of Joint Components............. 267
     Interplay of Humoral Factors in the Pathogenesis of Osteoarthritis ... 276

13. Endocrine Control of Tooth and Periodontal Tissue
     Growth and Repair 286
     JOHN F. HELFRICK

     Pituitary Gland... 286
     Thyroid 290
     Parathyroid Hormone 293
     Adrenal Glands and Gonads 295
     Pancreas 297
     Epidermal Growth Factor 299

     Index 305

Contributors

BANKSOTA, NIRMAL K., M.D.
Department of Diabetes, Endocrinology and Metabolism, City of Hope
National Medical Center, Duarte, CA 91010, USA

BONILLA, MARY ANN, M.D.
Department of Pediatrics and Medicine, Memorial Sloan Kettering
Cancer Center, New York, NY 10021, USA

EPSTEIN, SOL, M.D.
Department of Medicine, Division of Endocrinology, Albert Einstein
Medical Center, Temple University, Philadelphia, PA 19141, USA

FERRINI, PIERLUIGI ROSSI
Department of Hematology, University of Florence and Unità Sanitaria
Locale, Florence, Italy

FRANK, ROBERT N., M.D.
The Kresge Eye Institute of Wayne State University School of Medicine,
Detroit, MI 48201, USA

GOLDSTEIN, DAVID, M.D.
Centre for Immunology, Australian Cancer Foundation for Medical
Research, St. Vincent's Hospital, Sydney, NSW 2010, Australia

GROSSI, ALBERTO
Department of Hematology, University of Florence and Unità Sanitaria
Locale, Florence, Italy

HAMERMAN, DAVID, M.D.
Departments of Medicine and Orthopaedic Surgery, Montefiore Medical
Center, Albert Einstein College of Medicine, Bronx, NY 10467, USA

HELFRICK, JOHN F., D.D.S., M.S.
Department of Oral and Maxillofacial Surgery, University of Texas
Health Science Center at Houston, Houston, TX, 77030 USA

HECHT, KARL, M.D.
Bereich Medizin (Charité), Humboldt University, Berlin, Germany

JAKUBOWSKI, ANN, M.D.
Department of Pediatrics and Medicine, Memorial Sloan Kettering
Cancer Center, New York, NY 10021, USA

KAPLAN, JOSEPH, M.D.
Departments of Pediatrics, Medicine, and Immunology-Microbiology,
Wayne State University School of Medicine, Detroit, MI 48201, USA

KATZ, IAN A., M.D.
The Royal North Shore Hospital, St. Leonards 2065, Sydney NSW,
Australia

KING, GEORGE L., M.D.
Research Division, Joslin Diabetes Center and Department of Medicine,
Brigham & Women's Hospital, Harvard Medical School, Boston, MA
02215, USA

KLEBER, BERND-MICHAEL, M.D.
Poliklinik für konservierend Stomatologie, Institut für Pathophysiologie,
Humboldt Universität, Berlin, Germany

KORKOR, ADEL B., M.D.
Department of Nephrology and Bone and Mineral Metabolism Center,
Waukesha Memorial Hospital, Waukesha, WI 53188, USA

LUMPKIN, MICHAEL D., PH.D.
Departments of Physiology and Biophysics, Georgetown University
School of Medicine, Washington, DC 20007, USA

RAFANELLI, DANIELA
Department of Hematology, University of Florence and Unità Sanitaria
Locale, Florence, Italy

SOTOLONGO, LAURA B., M.D.
The Kresge Eye Institute of Wayne State University School of Medicine,
Detroit, MI 48201, USA

TAYLOR, STEPHEN, M.D.
Departments of Medicine and Orthopaedic Surgery, Montefiore Medical
Center, Albert Einstein College of Medicine, Bronx, NY 10467, USA

VANNUCCHI, ALESSANDRO, M.
Department of Hematology, University of Florence and Unità Sanitaria
Locale, Florence, Italy

WILDING, GEORGE, M.D.
Department of Human Oncology, University of Wisconsin Clinical
Cancer Center, Madison, Wisconsin 53792, and William S. Middleton
Memorial Veterans Hospital, Madison, WI 53792, USA

ZERWEKH, JOSEPH E., PH.D.
Department of Internal Medicine, University of Texas Southwestern
Medical Center, Dallas, TX 75235-8885, USA

1
Angiogenesis: Its Regulation in Health and Disease

NIRMAL K. BANSKOTA and GEORGE L. KING

Angiogenesis is the process of new blood vessel formation. It occurs physiologically in the developing embryo where the process leads to establishment of both the microvasculature and macrovasculature. In the postnatal state, angiogenesis occurs in only a few selected tissues and in many pathologic conditions. It is only in the last two decades that attempts at identifying factors that regulate this process have yielded some information. This chapter summarizes some of the salient features of angiogenesis in health and disease and looks at the various factors that are involved in its regulation.

Angiogenesis in the Embryo

In the early stages of organogenesis, capillaries originate from cells of the yolk sac and invade the surrounding tissues.[1] These cells, called angioblasts, arrange themselves in a flat, epithelial layer surrounding a small lumen. Once this initial network is established, all further development and growth of new vessels always occurs from preexisting vessels.[2] This initial phase of angiogenesis occurs around the fourth week of gestation and is complete in about another 4 weeks. During this time, an extensive network of capillaries is established in each organ system (the microcirculation) as are the major trunk systems (the arterial, systemic venous and the portal venous systems). The factors that regulate embryonic angiogenesis are poorly understood. Flow is not essential to initiate this process because angiogenesis occurs before the establishment of a circulation[2] and, in experimental animals, removal of the embryonic heart does not deter the development of the vessels until the animal dies.[3,4]

Angiogenesis in the Postnatal State

Because postnatal angiogenesis occurs in only a few selected tissues and capillary endothelial cells represent a stable, nondividing population of cells, a variety of techniques have been developed to study capillary growth.[5]

Morphologic Studies with Light and Electron Microscopy

More than 150 years ago Platner described the morphologic features of growing capillaries in tadpole tails. Since then, similar observations have been extended to inflammation and wound healing, granulation tissue, and tumors.[6] The process begins with the formation of a sprout consisting of a cord of cells, usually at right angles to a parent capillary.[7] As the sprout extends, lumen formation is initiated at the base of the sprout, and extends to the tip of the sprout. Such sprouts often coalesce to form a small network.[8] If flow is not established, these new capillaries regress.[2] Detailed ultrastructural studies of growing capillaries are well documented for a variety of tissues such as rat myocardium[9] and brain.[10] These indicate that the growing capillary usually has an incomplete basement membrane (BM) and the luminal surface of the endothelial cell is characterized by large numbers of microvilli, loose junctions, and gaps between endothelial cells—characteristics of poorly differentiated cells. These cells also show signs of high metabolic activity such as few pinocytic vesicles, prominent Golgi bodies, large numbers of mitochondria, and extensive rough endoplasmic reticulum.[6] Extensive vacuolization of the central parts of endothelial cells in growing capillary sprouts has been observed and is thought to lead to or contribute to lumen formation.[11]

Tissue Cultures

In the adult animal the endothelial cells are very stable, having an estimated turnover time of more than 1000 days. Because of this, a variety of endothelial cells have been grown in tissue culture.[12] Although ultrastructural studies have shown these cultured cells to be quite similar to the cells of the same origin in vivo,[13] cells from different tissues demonstrate different growth potential in tissue culture.[14,15] Generally, cells derived from younger animals tend to grow better. Some endothelial cells in culture also tend to form capillaries.[7] These methods have provided invaluable means to study an otherwise stable population of cells and have allowed characterization of a variety of factors that regulate their growth in vitro.

Diffusion Chambers and Avascular Tissues

Prolonged observation of capillary growth in living animals under different stimuli has been possible using transparent chambers implanted into animals.[16] The rabbit cornea, which is normally avascular, has also been used extensively to study the effects of a variety of growth stimuli,[17] as has the rabbit ear.[18] The chick chorioallantoic membrane (CAM) provides a rapid, efficient, and economic method for the observation of vessel growth and has been used extensively to study tumor angiogenic factor.[19]

Thus, a variety of techniques are now available for the morphologic characterization of growing capillaries and for the study of factors that may regulate growth in vivo and in vitro.

Growth and Inhibitory Factors in Capillary Growth

Ever since capillary growth has been observed and the morphologic characteristics outlined, the search for the putative growth stimulators has been a serious subject of research.

The initial observations of capillary growth in the developing embryo lead to the consideration of mechanical factors, such as blood flow, as prime inducers of angiogenesis. Although flow clearly plays an important role in vascular growth, as evidenced by regression of bloodless capillaries, it is clear that it is not the prime stimulator of growth. This is clearly illustrated in the embryo, where angiogenesis appears before the circulation develops, and blood vessel growth continues after removal of the heart until the embryo dies.[3,4]

Early observations of capillary growth in hypoxic tissues lead to the hypothesis that hypoxia is an important stimulator of angiogenesis. In the normal adult animal, capillary growth is observed in the heart, brain, and skeletal muscle during exposure to hypoxia.[6] Capillary growth is also seen in chronically bradycardically paced hearts[20] and after long-term use of vasodilator drugs,[21] and in skeletal muscles[22] and brain[23] during exposure to high altitudes. Increased flow or hypoxia is thought to stimulate angiogenesis under these situations. Failure to understand the mechanism of the hypoxic stimulus and the observation of capillary growth in a variety of pathologic states have spurred the search for other stimuli. The following factors have been better characterized.

Angiogenin

The pioneering observation by Folkman that the growth of solid tumors was angiogenesis-dependent culminated in the purification and biochemical and genetic characterization of angiogenin by Fett et al.[24] This angiogenic factor was isolated from large quantities of conditioned media from the HT-29 colon adenocarcinoma. The CAM and rabbit cornea were used for the bioassays during the process of purification. Angiogenin derived from the HT-29 cells is a single-chain protein of 123 amino acids and an approximate molecular weight of 14,400. It has three disulfide bridges, is highly basic, and bears a remarkable overall sequence homology of 68% to the pancreatic ribonuclease family of enzymes.[25] However, despite strong conservation of residues known to be involved in ribonuclease activity [His 12, His 119, Lys 41], angiogenin does not produce small acid-soluble fragments from poly C or poly V as pancreatic ribonucleases do, although it does cleave 28S and 18S ribosomal RNA to large fragments of 1 to 500 nucleotides, suggesting only a partial cleavage.[26] It is a stronger inhibitor of protein synthesis than RNAase and seems to demonstrate specificity for ribosomal RNA only in the intact ribosome. This partial and weaker ribonucleolytic activity of angiogenin may be due to its specificty toward unidentified nucleolar substrates or to the fact that the precise characteristics and optimal experimental condition remain to be worked out.[27]

Angiogenin has also been identified in normal plasma and serum.[28] The first isolation of angiogenic cDNA from fetal liver[29] and its purification from nontumor sources clearly demonstrates that this molecule is not a tumor-specific product and that probably it has important, although yet unidentified, functions in normal blood vessel growth.

Heparin-Binding Growth Factors

The observation that mast cells aggregate around tumors led to the study of mast cell lysates in tumor angiogenesis.[30] Heparin, among a host of cell products, was identified as a potent angiogenesis regulator.[31] It does not initiate angiogenesis, but potentiates angiogenic activities of other molecules.[32] Shing et al. hypothesized that growth factors for endothelial cells may bind to the heparin present on endothelial surfaces and thus potentiate the angiogenesis induced by these growth factors. Guided by this hypothesis and using heparin-sepharose columns, they purified an endothelial growth factor from a rat chondrosarcoma.[33] Since these observations were made, many other endothelial mitogens have been identified using heparin-sepharose chromatography. These include brain-derived basic fibroblast growth factor, endothelial cell growth factor, retinal-derived growth factor, cartilage-derived growth factor, and basic and acidic fibroblast growth factors. These factors are all endothelial cell mitogens, induce angiogenesis, and demonstrate specific binding activity to heparin and not to other anionic glycosaminoglycans, such as hyaluronic acid. Although complete biochemical characterization of most of these molecules is not complete, heparin-binding endothelial cell growth factors fall into two main groups (Table 1.1). Most of their characteristics are

TABLE 1.1. Heparin-binding growth factors.

Heparin-binding growth factor 1 (acidic fibroblast growth factor)	Heparin-binding growth factor 2 (basic fibroblast growth factor)
Astroglial growth factor-I	Astroglial growth factor-II
Endothelial cell growth factor	Chrondrosarcoma growth factor
Endothelial growth factor	Cartilage-derived growth factor-I
Eye-derived growth factor-II	Endothelial growth factor
Glial growth factor	Eye-derived growth factor-I
Retina-derived growth factor	Hepatoma growth factor
	Macrophage growth factor
	Myogenic growth factor
	Prostate growth factor
	Tumor angiogenesis factor

Some names under which the two prototypes of heparin-binding growth factor have been described.

represented by those of acidic and basic fibroblast growth factor (FGF), respectively.[34] Both basic FGF and acidic FGF are single chain polypeptides of 146 and 140 amino acids, respectively,[35,36] sharing a 55% sequence homology at the amino acid and nucleotide level. They also have a common genomic organization, each gene consisting of three exons and two introns. The genes, however, are located in different chromosomes: basic FGB in chromosome 4 and acidic FGF in chromosome 5.[37,38] The primary translation product is a 155–amino acid peptide and lacks a signal peptide, a rare feature among secreted polypeptides.[39] Basic FGF has been isolated from a large number of cell types and from malignant tissues. Acidic FGF has been identified so far only in brain, retina, bone matrix, and osteosarcoma.[40]

Other Endothelial Cell Mitogens

Transforming growth factors (TGF), so called because of their ability to change phenotypic features of normal cells to those of transformed cells,[41] are also angiogenic in vivo. Two TGFs have been identified: α and β.[42,43] TGF-α, a 50–amino acid polypeptide with a 35% sequence homology with epidermal growth factor (EGF), causes endothelial growth but at higher concentrations than EGF.[44] TGF-β exhibits paradoxic effects on vascular endothelial cells, depending on experimental conditions: TGF-β stimulates capillary formation and matrix production in newborn mice,[45] but inhibits endothelial growth in tissue culture.[46] This paradox may be partially explained by the potent effect of TGF-β on macrophage migration, which could then account for the angiogenic activity seen in vivo.[47] As mentioned earlier, EGF is an endothelial mitogen, and as a platelet product that is released on platelet aggregation, may indeed play an important role in physiologic wound healing.[48] Platelet-derived growth factor (PDGF) is not an endothelial mitogen, but may play a role in blood vessel maturation as it is a mitogen for smooth muscles and fibroblasts, and increases intercellular matrix.[49] Insulin and insulinlike growth factors are also well characterized mitogens for capillary endothelial cells in vitro.[50] They are also potent stimulators of vascular smooth muscles in vitro.[51,52] However, their role in physiologic or pathologic angiogenesis needs further clarification.

Another macrophage product, tumor necrosis factor (TNF), has also been shown to induce angiogenesis in the rabbit corneal assays. Like TGF-β, it exhibits paradoxic growth inhibition of microvascular endothelial cells in culture.[53]

A variety of small molecular weight substances with angiogenic activity in vitro have been identified. Copper has been reported to have angiogenic activity.[54] The mechanism of action is not known, but it does modulate the synthesis of fibronectin by endothelial cells (EC), which in turn may regulate EC functions such as migration.[55] Prostaglandins of the E series are reported to have angiogenic activity; prostaglandin E_1 produces new vessel growth in the rabbit cornea, and prostaglandin E_2 stimulates angiogenesis in CAM assays. Other prostaglandins (A and F series) do not have angiogenic effects.[56] Nicotinamide has angiogenic properties and may be responsible for some of the angiogenic activity of certain tumors.[57]

The Role of the Extracellular Matrix on Angiogenesis

The extracellular matrix of endothelial cells is known to modulate endothelial cell growth and differentiation.[58,59] At the onset of angiogenesis, the basement membrane is degraded by the action of collagenase and other proteases, allowing endothelial migration.[60] Basement membrane components and interstitial collagens (I and III) allow high rates of EC migration whereas fibronectin lowers the rate of migration.[61] Endothelial cells when cultured on dishes coated with type IV–V collagen organize into capillarylike structures. In contrast, when grown in I–II collagen– coated dishes, they proliferate and form monolayers.[62] Thus, different types of matrix component may have a differential effect on growth and differentiation. In the amnion, endothelial cells exposed to the basement membrane form tubular structures, whereas on the stromal side they proliferate and migrate to the interstitial tissue.[62] This and similar observations led to the suggestion that basement membrane components direct the expression of a mature and stable phenotype of EC, whereas the interstitial collagens stimulate EC growth and migration. However, laminin, the most abundant glycoprotein of basement membrane, induces EC proliferation more than the interstitial collagens.[63] A fragment of laminin named P1, which is generated by partial proteolysis of laminin, has potent effects on EC growth. This fragment possesses cell binding sites, but lacks heparin or type IV collagen binding sites, suggesting that laminin does not require interaction with heparin or type IV collagen in promoting EC growth (at least in vitro). Endothelial cells from different sources, however, respond differently to the stimuli provided by basement components and interstitial collagens.[63] The preponderance of each class of molecules in the basement membrane and interstitium may thus modity the effects of various stimuli on EC growth and migration, thereby affecting angiogenesis. Of interest and possibly of great significance is that in embryogenesis and wound healing, matrix proteins appear to be synthesized in a temporal sequence, with laminin preceding the production of type IV collagen.[64,65] Such a sequence of events can easily be construed to have a major regulatory role in angiogenesis.

Heparin, a highly sulfated glycosaminoglycan, is also a component of BM and, as discussed earlier, has angiogenic properties. Its ability to promote angiogenesis without being a mitogen itself led to the characterization of a whole family of heparin-binding growth factors (HBGFs). The two classes of HBGFs have as their prototypes acidic FGF (HBGF-1) and basic FGF (HBGF-2), and have already been reviewed. Heparin is currently believed to have a role in angiogenesis by concentrating and stabilizing the HBGFs and potentiating their action. Tertiary structural changes of HBGF-1 after binding to heparin have been demonstrated.[66] Heparin also increases the resistance of HBGF-1 to the action of cellular proteases and protects it from heat and chemical denaturation in vitro.[67] Interaction with heparin enhances the mitogenic and chemotactic effects of HBGF-1 on ECs.[68] Interaction of HBGF-2 with heparin also occurs, resulting in stabilization and protection from proteolysis,[69] but not in an enhancement of the effect of HBGF-2 on EC growth or migration.[70] Heparin also acts as a copper chelator and has angiogenic properties when complexed to copper.[71] Components of the extracellular matrix clearly play an important role in angiogenesis by direct effects and by modulating the effects of various mitogens on EC. The mechanisms of their action, however, remain to be elucidated.

Inhibitors of Angiogenesis

In the postnatal state, the vascular endothelial cells in most tissues are stable and in a nondividing state[5] and must be in a perpetual state of inhibition. Physiologically, a few tissues in the adult body are avascular (e.g., cartilage and the cornea) and inhibitory factors from these tissues have been sought. Cartilage inhibits the neovascularization induced by tumors in experimental situations.[72] Cartilage contains collagenase inhibitors, which may be responsible for the inhibition of neovascularization.[73] Protamine, an arginine-rich basic protein, is a heparin antagonist and reverses the effect of heparin on angiogenesis. It also inhibits the action of various stimulants of angiogenesis in the corneal assays.[74] A class of steroids that inhibits angiogenesis in the presence of heparin has been described. Hydrocortisone and dexamethasone exhibit antiangiogenic activity in a limited dose range. Because of this unusual dose response curve, it was suggested that their antiangiogenic effects were not dependent on their glucocorticoid activity. Indeed, epicortisol, which is the biologically inactive stereoisomer of hydrocortisone, was found to be an inhibitor of angiogenesis. Further screening of a variety of steroids led to the identification of tetrahydrocortisol as the most potent inhibitor of angiogenesis.[75] It is unclear at this time whether these angiostatic steroids have a physiologic role.

Gamma-interferon is recognized as an endothelial cell growth inhibitor.[76] Phorbol esters induce capillarylike tube formations in vitro, but this effect on endothelial cell differentiation occurs without growth.[77] On the contrary, phorbol esters inhibit endothelial cell growth induced by other mitogens.[78] Both gamma-interferon and the phorbol ester phorbol myristate acetate (PMA) reduce growth factor receptor numbers on the endothelial cell surface.[78,79] Whether this accounts for the growth inhibition is not certain. Lymphotoxin (TNF-B) inhibits EC growth in vitro.[80] A placental inhibitor of pancreatic ribonuclease has potent inhibitory effects on both the angiogenic and ribonucleolytic properties of angiogenin.[81] It may play a role in embryonic and placental angiogenesis.

Angiogenesis in Health and Disease

The enormous interest in angiogenesis has stemmed not only from the study of developmental biology, but also from a variety of disease states in which neovascularization is an important feature. As mentioned previously, only a few organs in the postnatal stage demonstrate angiogenesis on a regular basis.

The development of the follicle in the maturing ovum during each menstrual cycle is associated with neovascularization. As soon as the antrum appears, the theca, especially the theca interna, of the developing follicle becomes vascularized. Angiogenic activity obtained from the granulosa and theca cells of the developing follicle seem to be regulated by gonadotrophins.[82] The nature of these angiogenic stimuli is not known, but a heparinlike material (which could potentiate the angiogenic factors) has long been identified in follicular fluid.[83] Within several days following ovulation, the vessels of the theca penetrate the granulosa membrane and vascularize the corpus luteum.[82] Conditioned media from cultured luteal cells induce endothelial cell growth in vitro, but the active molecules have not been characterized.[84] Basic FGF has been isolated and purified from luteal extracts and may be one of the mitogens involved in vascularization of the follicle and corpus luteum.[85]

The morphologic characteristics of the blood vessel growth of the endometrium during the menstrual cycle is well documented. The spiral arteries undergo significant changes in size and shape during the cycle and are influenced in a major way by estradiol and progesterone.[86] Estradiol receptors are present in endothelial cells and may regulate the vascular changes associated with menstruation.[87] Estradiol stimulates the synthesis of basic FGF in endometrial adenocarcinoma cell lines.[88] Heparinlike activity is present in endometrial fluid[89] and may potentiate other angiogenic factors.

Implantation of the embryo and the formation of placental vessels are other physiologic phenomena associated with angiogenesis. Although the morphologic characteristics are well documented, factors that trigger and maintain vascularization are not known. FGF has been identified and isolated from human placenta.[90]

Wound healing is associated with the accumulation of inflammatory cells, among them monocytes and macrophages. The secretory products of these cells, as well as of platelets, such as acidic and basic FGFs, TGF-α, and EGF may be involved in the angiogenesis of wound repair and granulation tissue. A nonmitogenic angogenic factor presumed to induce angiogenesis by inducing endothelial cell migration has been described but awaits further characterization.[91]

Neovascularization is implicated in the pathogenesis of several diseases. Osteoarthritis[92] and rheumatoid arthritis[93] demonstrate angiogenesis as components of the joint inflammation. The angiogenic responses seen in the inflammation of rheumatoid arthritis may result from the release of macrophage products,[94] whereas the therapeutic action of gold salts[95] and D-penicillamine[96] may result from their established antiangiogenic properties.

Diabetic retinopathy is the leading cause of blindness in industrialized societies. It is characterized by a constellation of pathologic changes in the retina, including loss of pericytes, endothelial proliferation, loss of permeability of blood vessels, thickening of the basement membrane, and neovascularization.[97] The causative factors resulting from the diabetic state have not been fully elucidated, and may include the direct toxic effects of hyperglycemia per se, altered hemodynamics of retinal vessels, or endocrine factors.[98] Growth hormone has been linked to diabetic retinopathy following observations of regression of the retinopathy after pituitary infarction[99] or hypophysectomy.[100] Insulin-like growth factor-I (IGF-I) has been implicated because it mediates most of the anabolic effects of growth hormone, and increased levels of IGFs have been detected in the vitreous of patients with severe retinopathy.[101] IGF-I is mitogentic for retinal capillary endothelial cella in vitro.[102] FGF is also present in human vitreous fluid.[103] These observations do not satisfactorily explain the mechanisms that may be involved in the pathogenesis of diabetic retinopathy, or how they may interact with the effects of hyperglycemia and altered hemodynamics of retinal vessels in diabetes.

Atherosclerosis is associated with abnormal growth of blood vessel wall smooth muscle cell.[104] This abnormal subintimal growth of smooth muscle cells in the diseased vessels is affected by a variety of growth factors,[105] some of which have been shown to induce proto-oncogenes in smooth muscle cell cultures.[106] Capillary growth within atherosclerotic plaques are often observed, and their rupture may lead to sudden occlusion of the artery.[107] Of interest is that myocardial infarct tissues release EC mitogens.[108] These mitogens may play a role in the development of collateral circulation in the ischemic heart.

The observation that sustained tumor growth requires angiogenesis has generated much interest in the regulation of blood vessel growth. Following studies of implanted tumors in animals, Folkman first postulated the existence of angiogenic factors,[32,109] now characterized as the heparin-binding growth factors. Angiogenin was also first isolated from the HT29 human adenocarcinoma cell line.[24] The dependence of tumors on angiogenesis is well established, and attempts at angiogenic inhibition have resulted in partial regression of tumors.[110] Clearly, the ability to modulate angiogenesis in tumors could have enormous practical implications in clinical medicine.

Summary

Angiogenesis is the result of a complex interaction among endothelial cells, pericytes, and cellular matrix. The stimulus for the vasculature to enter the proliferative phase appears to be a local need for nutrients or oxygen. The biochemical signals for the vascular cells to proliferate are not clear, but they are needed to activate or release angiogenic factors. Recent progress has provided a tremendous amount of information on a variety of factors that could mediate both proliferative or inhibitory actions, depending on tissue location and cell type.

The factors that promote these events are now being characterized in detail and include the family of heparin-binding growth factors, angiogenin, basement membrane, and interstitial components such as collagen, laminin, and heparin. Factors that inhibit angiogenesis are also increasingly recognized and include protamine, angiostatic steroids, and TNF, among others.

Even more complex are the factors that determine the role of these growth modulators in the disease state. It is likely that the significance of these growth modulators are not the same in all pathologic states marked by neovascularization. Innovative methods for assays for growth modulation are needed to evaluate the function of growth factor, since they generally act in a paracrine or autocrine manner. Therefore, correlation between systemic levels and local neovascularization is seldom helpful. However, the information obtained from cultured cells needs to be substantiated by in vivo studies. The understanding of the regulation of angiogenesis has important implications on such basic events as growth and differentiation of tissues, as well as essential clinical and therapeutic potentials in disease such as diabetic retinopathy, arthritis, and tumor growth.

Acknowledgments. The authors wish to express their appreciation to Leslie Balmat for her secretarial assistance. This work has been supported by NIH grants EY05110, DK36836, the Massachusetts Lions, and institutional funds from the Joslin Diabetes Center.

References

1. Tuchmann-Duplessis H, Haegel P. Organogenesis. In: *Illustrated Human Embryology*. Vol II. New York: Springer-Verlag; 1974.
2. Balinsky BI, Fabian BC. Organogenesis. In: *An Introduction to Embryology*. New York: Saunders; 1981.
3. Knower HME. Effects of early removal of the heart and arrest of the circulation on the development of frog embryos. *Anat Rec*. 1907;7:161–165.
4. Chapman WB. The effect of the heart beat upon the development of the vascular system in the chick. *Am J Anat*. 1918;23:175–203.
5. Wagner RC. Endothelial cell embryology and growth. *Adv Microcirc*. 1980;9:45–75.
6. Hudlicka O. Development of microcirculation: capillary growth and adaptation. In: Renkin EM, Michel CC. eds. *Handbook of Physiology. The Cardiovascular System. Vol IV. Microcirculation, Part 1.* pp165–216, Baltimore: Williams & Wilkins; 1984.
7. Clark ER, Clark EL. Microscopic observations on the growth blood capillaries in the living mammal. *Am J Anat*. 1939;64:251–299.
8. Clark ER, Clark EL. Microscopic observations on the extraendothelial cells of living mammalian blood vessels. *Am J Anat*. 1940;66:1–49.
9. Manasek FJ. The ultrastructure of embryonic myocardial blood vessels. *Dev Biol*. 1971;26:42–54.
10. Donahue S, Pappas GD. The fine structure of capillaries in the cerebral cortex of the rat at various stages of development. *Am J Anat*. 1961;108:331–347.
11. Aloisi M, Schiaffino S. Growth of elementary blood vessels in diffusion chambers. II. Electron microscopy of capillary morphogenesis. *Virchows Arch*. 1971;8:328–341.
12. McDonald RI, Shepro D, Rosenthal M, Boooyse FM. Properties of cultured endothelial cells. *Semin Haematol*. 1973;6:469–478.
13. Gimbrone MA Jr. Culture of vascular endothelium. *Prog Hemostasis Thromb*. 1976;3:1–28.

14. Nees S, Willershausen-Zonnchen B, Gerbes AL. Gerlach E. Studies on cultured coronary endothelial cells. *Folia Angiol.* 1980;28:64–68.
15. Frank RN, Kinsey VE, Frank KW, Mikus K, Randolph A. In vitro proliferation of endothelial cells from kitten retinal capillaries. *Invest Ophthalmol Vis Sci.* 1979;18:1195–1200.
16. Nims JC, Irwin JW. Technical report: chamber techniques to study the microvasculature. *Microvasc Res.* 1973;5:105–118.
17. Ausprink DH, Falterman K, Folkman J. The sequence of events in the regression of corneal capillaries. *Lab Invest.* 1978;38:284–294.
18. Sandison JC. Observations on the growth of blood vessels as seen in the transparent chamber implanted in the rabbit's ear. *Am J Anat.* 1928;41:475–496.
19. Folkman J, Cotran R. Relation of vascular proliferation to tumor growth. *Int Rev Exp Pathol.* 1976;16:207–248.
20. Wright AJA, Hudlicka O. Capillary growth and changes in heart performance induced by chronic bradicardial pacing in the rabbit. *Circ Res.* 1981;49:469–478.
21. Tornling G. Capillary neoformation in the heart of dipyridamole-treated rats. *Acta Pathol Microbiol Scand Sec A.* 1982;90:269–271.
22. Banchero N. Capillary density of skeletal muscle in dogs exposed to simulated altitude. *Proc Soc Exp Biol Med.* 1975;148:435–439.
23. Opitz E. Increased vascularization of the tissue due to acclimatization to high altitude and its significance for oxygen transport. *Exp Med Surg.* 1951;9:389–403.
24. Fett JW, Strydom DJ, Lobb RR, Alderman EM, Bethune, JL, Riordan JF, Vallee BL. Isolation and characterization of angiogenin, an angiogenic protein from human carcinoma cells. *Biochemistry.* 1985;24:5480–5486.
25. Strydom DJ, Fett JW, Lobb RR, Alderman EM, Bethune JL, Riordan JF, Vallee BL. Amino acid sequence of human tumor derived angiogenin. *Biochemistry.* 1985;24:5486–5494.
26. Shapiro R, Riordan JF, Vallee BL. Characteristic ribonuleolytic activity of human angiogenin. *Biochemistry.* 1986;25:3527–3532.
27. Vallee BL, Riordan JF. Chemical and biochemical properties of human angiogenin. *Adv Exp Med Biol.* 1988;234:41–53.
28. Shapiro R, Strydom DJ, Olson KA, Vallee BL. Isolation of angiogenin from normal human plasma. *Biochemistry.* 1987;26:5141–5146.
29. Kurachi K, Davie EW, Strydom DJ, Riordan JF, Vallee BL. Sequence of the cDNA and gene for angiogenin, a human angiogenesis factor. *Biochemistry.* 1985;24:5494–5499.
30. Kessler DA, Langer RS, Pless NA, Folkman J. Mast cells and tumor angiogenesis. *Int J Cancer.* 1976;18:703–709.
31. Azizkhan RG, Azizkhan JC, Zetter BR, Folkman J. Mast cell heparin stimulates migration of capillary endothelial cells in vitro. *J Exp Med.* 1980;152:931–944.
32. Folkman J. Tumor angigenesis. *Adv Cancer Res.* 1985;43:175–203.
33. Shing Y, Folkman J, Sullivan R, Butterfield C, Curray J, Klagsbrun M. Heparin affinity: purification of a tumor-derived capillary endothelial cell growth factor. *Science.* 1984;223:1296–1298.

34. Burgess WH, Maciag T. The heparin-binding (fibroblast) growth factor family of proteins. *Annu Rev Biochem.* 1989;58:575–606.
35. Esch F, Baird A, Ling N, Ueno N, Hill F, Deneroy L, Klepper R, Gospodarowicz D, Bohlen P, Guillemin R. Primary structure of bovine pituitary basic fibroblast growth factor (FGF) and comparison with the amino terminal sequence of bovine brain acidic FGF. *Proc Natl Acad Sci USA.* 1985;82:6507–6511.
36. Gimenez-Galeo G, Rodkey J, Bennett C, Rios Candelore M, Disalvo J, Thomas K. Brain-derived fibroblast growth factor: complete amino acid sequence and homologies. *Science.* 1985;230:1385–1388.
37. Abraham JA, Whang JL, Tuomolo A, Mergia A, Friedman J, Gospodarowicz D, Fiddes JC. Human basic fibroblast growth factor: nucleotide sequence and genomic organization. *EMBO J.* 1986;5:2523–2528.
38. Jaye M, Howk R, Burgess W, Ricca GA, Chiu IM, Ravera MW, O'Brien SJ, Modi WS, Maciag T, Drohan WN. Human endothelial cell growth factor: cloning, nucleotide sequence, and chromosomal localization. *Science.* 1986;233:541–544.
39. Gospodarowicz D, Ferrara N, Schweigerer L, Neufeld G. Structural characterization and biological functions of fibroblast growth factor. *Endocr Rev.* 1987;8:1–20.
40. Gospodarowicz D. Molecular and developmental biology aspects of fibroblast growth factor. *Adv Exp Med Biol.* 1988;234:23–39.
41. DeLarco JE, Todaro GJ. Sarcoma growth factor (SGF): specific binding to epidermal growth factor (EGF) membrane receptors. *J Cell Physiol.* 1980;102:267–277.
42. Marquardt H, Hunkapiller MW, Hood LE, Todaro GJ. Rat transforming growth factor Type 1: structure and relation to epidermal growth factor. *Science.* 1984;223:1079–1082.
43. Derynck R, Jarrett JA, Chen EY, Eaton DH, Bell JR, Assoian RK, Roberts AB, Sporn MB, Goeddel DV. Human transforming growth factor-β cDNA sequence and expression in tumor cell lines. *Nature.* 1985;316:701–705.
44. Schreiber AB, Winkler ME, Derynck R. Transforming growth factor: a more potent angiogenic mediator than epidermal growth factor. *Science.* 1986;232:1250–1253.
45. Roberts AB, Sporn MB, Assoian RK, Smith JM, Roche NS, Wakefied LM, Heine UI, Liotta LA, Falanga V, Kehrl JH, Fauci AS. Transforming growth factor type-β: rapid induction of fibrosis and angiogenesis in vivo and stimulation of collagen formation in vitro. *Proc Natl Acad Sci USA.* 1986;83:4167–4171.
46. Muller G, Behrens J, Nussbaumer U, Bohlen P, Birchmeier W. Inhibitory action of transforming growth factor on endothelial cells. *Proc Natl Acad Sci USA.* 1987;84:5600–5604.
47. Roberts AB, Thompson NL, Heine U, Flanders C, Sporn M. Transforming growth factor β: possible roles in carciogenesis. *Br J Cancer.* 1988;57:594–600.
48. Carpenter G. Receptors for epidermal growth factors and other polypeptide mitogens. *Annu Rev Biochem.* 1987;56:881–914.

49. Deuel TF, Huang JS. Platelet-derived growth factor. Structure, function and roles in normal and transformed cells. *J Clin Invest.* 1984;74:669–676.
50. King GL, Buzney SM, Kahn CR, Hetu N, Buchwald S, MacDonald SG. Differential responsiveness to insulin of endothelial and support cells from micro- and macrovessels. *J Clin Invest.* 1983;71:974–979.
51. Pfeifle B, Ditschuneit H. Effect of insulin on the growth of cultured smooth cells. *Diabetologia.* 1981;20:155–158.
52. Banskota NK, Taub R, Zellner K, Olsen P, King GL. Characterization of induction of protooncogene c-myc and cellular growth in human vascular smooth muscle cells by insulin and IGF-I. *Diabetes.* 1989;38:123–129.
53. Frater-Schroder M, Risau W, Hallmann R, Gautschi P, and Bohlen P. Tumor necrosis factor α, a potent inhibitor of endothelial cell growth in vitro, is angiogenic in vivo. *Proc Natl Acad Sci USA.* 1987;84:5277–5281.
54. Wissler JH, Logemann E, Meyer HE, Krutzfeldt B, Hockel M, Heilmeyer Jr LMG. Bioactive copper-ribonucleo-polypeptide complexes: angio-morphogens of porcinemonocytes. *Fed Proc.* 1987;46(Abstract No. 1557).
55. McAuslan BR, Reilly WG, Hanman GN, Gole GA. Angiogenic factors and their assay: activity of formyl methionyl leucyl phenylalanine, adenosine diphosphate, heparin, copper, and bovine endothelium stimulating factor. *Microvasc Res.* 1983;26:323–338.
56. Ziche M, Jones J, Gullino PM. Role of prostaglandin E, and copper in angiogenesis. *J Natl Cancer Inst.* 1982;69:475–482.
57. Kull FC, Brent DA, Parikh I, Cuatrecasas P. Chemical identification of a tumor-derived angiogenic factor. *Science.* 1987;236:843–845.
58. Kleinman HK, Klebe RJ, Martin GR. Role of collagenous matrices in the adhesion and growth of cells. *J Cell Biol.* 1981;88:473–485.
59. Madri JA, Pratt BM. Endothelial cell-matrix interactions: in vitro models of angiogenesis. *J Histochem Cytochem.* 1986;34:85–91.
60. Gross JL, Moscatelli D, Rifkin DB. Increased capillary endothelial cell protease activity in response to angiogenic stimuli in vitro. *Proc Natl Acad Sci USA.* 1983;80:2623–2627.
61. Pratt BM, Harris AS, Morrow JS, Madri JA. Mechanism of cytoskeletal regulation: modulation of aortic endothelial cell spectrin by the extracellular matrix. *Am J Pathol.* 1984;117:349–354.
62. Madri JA, Williams SK. Capillary endothelial cell cultures: phenotypic modulation by matrix components. *J Cell Biol.* 1983;97:153–165.
63. Form DM, Pratt BM, Madri JA. Endothelial proliferation during angiogenesis. *Lab Invest.* 1986;55:521–530.
64. Ekbloom K, Alitalo K, Vaheri A, Timpl R, Saxen L. Induction of a basement glycoprotein in embryonic kidney: possible role of laminin in morphogenesis. *Proc Natl Acad Sci USA.* 1980;77:485–489.
65. Wu T-C, Wan Y-J, Chung AE, Damjanov I. Immunohistochemical localization of enactin and laminin in mouse embryos and fetuses. *Dev Biol.* 1983;100:496–505.
66. Schreiber AB, Kenney J, Kowalski J, Friesel R, Mehlman T, Maciag, T. Interaction of endothelial cell growth factor with heparin: characterization by receptor and antibody recognition. *Proc Natl Acad Sci USA.* 1985;82:6138–6142.

67. Gospodarowicz D, Cheng J. Heparin protects basic and acidic FGF from inactivation. *J Cell Physiol*. 1986;128:475–484.
68. Thornton SC, Mueller SN, Levine EM. Human endothelial cells: use of heparin in cloning and long term serial cultivation. *Science*. 1983;222: 623–625.
69. Saksela O, Moscatelli D, Sommer A, Rifkin DB. Endothelial cell-derived heparan sulfate binds basic fibroblast growth factor and protects it from proteolytic degradation. *J Cell Biol*. 1988;107:743–751.
70. Maciag T. Molecular and cellular mechanisms of angiogenesis. Molecular and cellular mechanisms of angiogenesis. In: *Important advances in oncology*. 1990;85–98.
71. Raju KS, Allesandri G, Ziche M, Gullino PM. Ceruloplasmin, copper ions, and angiogenesis. *J Natl Cancer Inst*. 1982;69:1183–1188.
72. Brem H, Folkman J. Inhibition of tumor angiogenesis mediated by cartilage. *J Exp Med*. 1975;141:427–439.
73. Moses MA, Sudhalter J, Langer R. Identification of an inhibitor of neovascularization from cartilage. *Science*. 1990;248:1408–1410.
74. Taylor S, Folkman J. Protamine is an inhibitor of angiogenesis. *Nature*. 1982;297:307–312.
75. Folkman J, Ingber DE. Angiostatic steroids. *Ann Surg*. 1987;203:374–383.
76. Friesel R, Komoriya A, Maciag T. Inhibition of endothelial cell proliferation by gamma-interferon. *J Cell Biol*. 1987;104:689–696.
77. Montesano R, Orci L. Tumor-promoting phorbol esters induce angiogenesis in vitro. *Cell*. 1985;42:469–477.
78. Doctrow SR, Folkman J. Protein kinase C activators suppress stimulation of capillary endothelial cell growth by angiogenic endothelial mitogens. *J Cell Biol*. 1987;104:679–687.
79. Hoshi H, Kan M, Mioh H, Chen J-K, McKeehan WL. Phorbol ester reduces the number of heparin-binding growth factor receptors in human adult endothelial cells. *FASEB J*. 1988;2:2797–2800.
80. Tsuruoka N, Sugiyama M, Tawaragi Y, Tsujimoto M, Nishihara T, Goto T, Sato N. Inhibition of in vitro angiogenesis by lymphotoxin and interferon-Y. *Biochem Biophys Res Commun*. 1988;155:429–435.
81. Shapiro R, Vallee B. Human placental ribonuclease inhibitor abolishes both angiogenic and ribonucleolytic activities of angiogenin. *Proc Natl Acad Sci USA*. 1987;84:2238–2241.
82. Findlay JK. Angiogenesis in reproductive tissues. *J Endocrinol*. 1986;111:357–366.
83. Stangroom JE, de G Weevers R. Anticoagulant activity of equine follicular fluid. *J Reprod Fertil*. 1962;3:269–282.
84. Goodman AL, Rone JD. Detection of angiotropic (chemoattractant) released by rabbit luteal cell cultured in serum-free or serum-enriched media. *Biol Reprod*. 1985;32:Suppl 1, Abstract 296.
85. Gospodarowicz D, Cheng J, Lui G, Baird A, Esch F, Bohlen P. Corpus luteum angiogenic factor is related to fibroblast growth factor. *Endocrinology*. 1985;117:2383–2391.
86. Christiaens GCML, Sixma JJ, Haspels AA. Hemostasis in menstrual endometrium: a review. *Obstet Gynecol Surv*. 1982;37:281–303.

87. Colburn P, Buonassisi V. Estrogen-binding sites in endothelial cell cultures. *Science.* 1978;201:817–819.
88. Presta M. Sex hormones modulate the synthesis of basic fibroblast growth factor in human endometrial adenocarcinoma cells: implications for the neovascularization of normal and neoplastic endometrium. *J Cell Physiol.* 1988;137:593–597.
89. Foley ME, Griffin BD, Zuzel M, Aparicio SR, Bradbury K, Bird CC, Clayton JK, Jenkins DM, Scott JS, Rajah CM, McNicol GP. Heparin-like activity in uterine fluid. *Br J Med.* 1978;ii:322–324.
90. Gospodarowicz D, Cheng J, Lui GM, Fujii DK, Baird A, Bohlen P. Fibroblast growth factor in human placenta. *Biochem Biophys Res Commun.* 1985;128:554–562.
91. Hockel M, Sasse J, Wissler JH. Purified monocyte-derived angiogenic substance (angiotropin) stimulates migration, phenotypic changes, and "tube formation" but not proliferation of capillary endothelial cells in vitro. *J Cell Physiol.* 1987;133:1–13.
92. Brown RA, Weiss JR. Neovascularization and its role in the osteoarthritic process. *Ann Rheum Dis.* 1988;47:881–885.
93. Harris ED Jr. Mechanisms of disease: rheumatoid arthritis—pathophysiology and implications for therapy. *N Engl J Med.* 1990; 322:1277–1289.
94. Koch AE, Polverini PJ, Leibovich SJ. Stimulation of neovascularization by human rheumatoid synovial tissue macrophages. *Arthritis Rheum.* 1986;29:471–479.
95. Matsubara T, Ziff M. Inhibition of human endothelial cell proliferation by gold compounds. *J Clin Invest.* 1987;79:1440–1446.
96. Matsubara T, Saura R, Hirohata K, Ziff M. Inhibition of human endothelial cell proliferation in vitro and neovascularization in vivo by D-penicillamine. *J Clin Invest.* 1989;83:158–167.
97. Green WR. Systemic diseases with retinal involvement. In: Spenser WH, ed. *Ophthalmic pathology: an atlas and textbook.* 3rd ed. Philadelpia: WB Saunders; 1986;1034–1210.
98. Merimee TJ. Mechanisms of disease: diabetic retinopathy—a synthesis of perspectives. *N Engl J Med.* 1990;322:978–983.
99. Poulsen JE. The Houssay phenomenon in man: recovery from retinopathy in a case of diabetes with Simmond's disease. *Diabetes.* 1953;2:7–12.
100. Lundback K, Malmros R, Anderson HC. Hypophysectomy for diabetic angiopathy: a controlled clinical trial. In: Goldberg MF, Fine SL, eds. *Symposium on the treatment of diabetic retinopathy.* Washington, DC: Public Health Service; PHS publication no. 1890. 1989.
101. Grant M, Russel B, Fitzerald C, Merimee TJ. Insulin-like growth factors in vitreous: studies in controll and diabetic subjects with neovascularization. *Diabetes.* 1986;35:416–420.
102. King GL, Goodman AD, Buzney SM, Moses A, Kahn CR. Receptors and growth-promoting effects of insulin and insulin-like growth factors on cells from bovine retinal capillaries and aorta. *J Clin Invest.* 1985;75:1028–1036.
103. Baird A, Culler F, Jones KL, Guillemin R. Angiogenic factor in human ocular fluid [letter]. *Lancet.* 1985;2(8454):563.

104. Ross R. The pathogenesis of atherosclerosis: an update. *N Engl J Med.* 1986;314:488–497.
105. Schwartz SM, Campbell GR, Campbell JH. Replication of smooth muscle cell in vascular disease. *Circ Res.* 1986;58:427–444.
106. Banskota NK, Taub R, Zellner K, King GL. Insulin, insulin-like growth factor I and platelet-derived growth factor interact additively in the induction of the protooncogene c-myc and cellular proliferation in cultured bovine aortic smooth muscle cells. *Mol Endocrinol.* 1989;3:1183–1190.
107. Barger AC, Beeuwkes III R, Lainey LL, Silverman KJ. Hypothesis: vasa vasorum and neovascularization of human coronary arteries. *N Engl J Med.* 1984;310:175–177.
108. D'Amore PA, Thompson RW. Mechanisms of angiogenesis. *Annu Rev Physiol.* 1987;49:453–464.
109. Folkman J. How is blood vessel growth regulated in normal and neoplastic tissue? G.H.A. Clowes Memorial Award Lecture. *Cancer Res.* 1986;46: 467–473.
110. Langer R, Conn H, Vacanti J, Haudenschild C, Folkman J. Control of tumor growth in animals by infusion of an angiogenesis inhibitor. *Proc Natl Acad Sci USA.* 1980;77:4331–4335.

2
Growth Factors and the Retina: Normal Vascularization and Pathologic Neovascularization

Robert N. Frank and Laura B. Sotolongo

Normal and Pathologic Anatomy of the Retinal and Choroidal Blood Vessels

Growth of blood vessels within the human retina is normally completed at birth, and in other mammalian species that have been studied, including dogs, cats, mice, and rats, retinal vascularization stops within a few weeks after birth.[*][1] Studies of the retinas of normal adult mice and rats, using [³H]-thymidine autoradiography, have shown minimal labeling (0.01– 0.1%) of vascular cell nuclei.[5,6] This indicates that retinal vascular cells normally turn over scarcely at all during adult life. Thus, whenever vascular cell proliferation occurs in the adult retina, it is pathologic. The new vessels that are observed are always structurally and functionally abnormal. Although new vessels derived from the retinal circulation by definition begin their growth within the retina,[7] eventually they always grow inwardly, break through the inner limiting membrane of the retina, and continue their growth on the vitreal surface of the inner limiting membrane, or actually within the vitreous (Fig. 2.1). Unlike normal vessels, retinal new vessels have thinned and often fenestrated endothelial

---

[*]This statement refers, of course, only to those mammals that have a fully vascularized (euangiotic, also called holangiotic) retina. Humans and other primates, rats, mice, dogs, cats, cattle, and pigs are included in this group. There are three other classes of mammalian retinas, based on their degree of vascularization. These are: (a) angiotic, also called merangiotic, in which only a portion of the retina is vascularized; the rabbit is an example of this type; (b) pseudoangiotic, also called paurangiotic, in which minute vessels are found only around the optic nervehead, as in the horse, guinea pig, or elephant; and (c) anangiotic, in which there are no retinal vessels at all, as in the rhinocerus, hippopotamus, or in marsupials or certain rodents.[2] Nonmammalian species generally have avascular retinas, although birds possess a pecten, a vascularized membrane that extends into the vitreous, and some species, including various fish, amphibians, and reptiles, have a network of vessels on the posterior surface of the vitreous adjacent to the innermost layers of the retina. All vertebrates

cell cytoplasm, may lack "tight" intracellular junctions, and are often surrounded by a loose extracellular matrix lacking well defined basement membrane structures. Functionally, they show a breakdown of the blood–retinal barrier, as demonstrated by their ready permeability to the dye sodium fluorescein in the commonly used clinical test, intravenous fluorescein angiography (Figs. 2.2 and 2.3). Retinovitreal new vessels are also fragile and can rupture easily, producing often extensive hemorrhages that may severely impair visual acuity.

There have been no studies of choroidal vascular cell regeneration equivalent to those conducted in the retina. Although the choroidal vascular network in all species is much more dense than that of the retina and carries a much larger volume of blood, there is no reason a priori to assume that the cells of its vasculature normally turn over more frequently than do those of the neural retina. Whenever choroidal neovascularization has been demonstrated in the human eye, it has been pathological. Like new vessels arising from the retinal circulation, those that derive from the circulation of the choroid—primarily from its capillary layer, the choriocapillaris, which lies immediately adjacent to the retinal pigment epithelium—are extremely "leaky," permitting extravascular accumulation of noncellular elements of the blood underneath the retinal pigment epithelium and the neurosensory retina, and in addition, they are highly fragile, capable of sudden ruptures with severe and vision-threatening hemorrhages underneath the retina (Figs. 2.4 and 2.5).

Neovascularization of the retinal and of the choroidal circulations occurs in many diseases in humans. Retinal neovascularization (including new vessels arising from the optic nerve head) is a prominent feature of diabetic retinopathy, as well as of retinal vein occlusions, sickle cell retinopathy, retinopathy of prematurity, and of chronic uveal and retinal inflammatory disease. Choroidal neovascularization is a major reason for visual loss in age-related (senile) macular degeneration, the most common cause of legal blindness in the over-60 age group in the United

have a richly vascularized choroid that abuts the retinal pigment epithelium. It is important to recognize that these differences in retinal vascularization among species exist, because some investigators have occasionally used species such as the rabbit, in which only a small portion of the retina is vascularized—and that by vessels lying in a nerve fiber layer on the inner surface of the retina, rather than within the nuclear and plexiform layers—to study processes of retinal neovascularization. This would appear to be inappropriate. In the only comparative biochemical studies conducted on retinas of species with different degrees of vascularization, Lowry et al. showed that the euangiotic retina of the monkey is largely dependent on aerobic glucose metabolism for energy production, including the tricarboxylic acid cycle and the pentose shunt,[3] whereas the merangiotic retina of the rabbit depends primarily on anaerobic glycolysis.[4] One would suppose that the processes that induce normal and pathologic retinal vascularization might be very different in species whose retinal vascular anatomy and retinal metabolism differ so greatly.

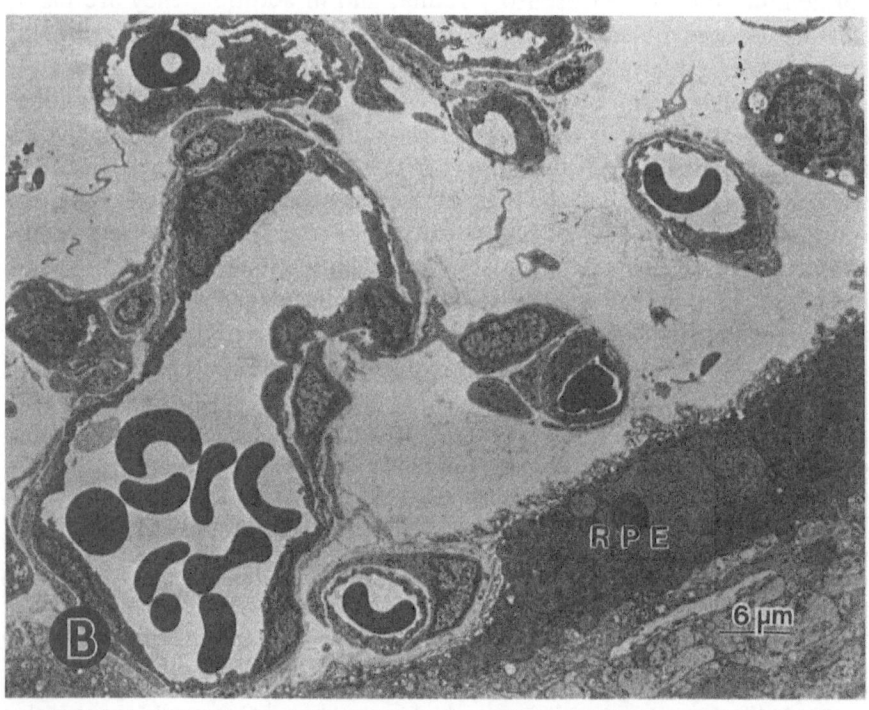

States, and it is also a prominent feature of the macular degeneration that occurs in the presumed ocular histoplasmosis syndrome, angioid streaks, and traumatic rupture of the choroid, among other disorders. In recent years several randomized, controlled clinical trials have demonstrated the efficacy of laser photocoagulation for reducing vision loss in several of these diseases, including proliferative diabetic retinopathy,[8] retinal branch vein occlusion,[9] age-related macular degeneration,[10] the presumed ocular histoplasmosis syndrome,[11,12] and idiopathic choroidal neovascularization.[13] Results of a treatment trial for proliferative sickle cell retinopathy have been more equivocal.[14] Yet even in the most successful of these trials, photocoagulation has not prevented vision loss in all cases, but only in most of those treated, and in some of the diseases studied, most notably age-related macular degeneration, the beneficial effect of photo-coagulation has been modest and has not been sustained with longer follow-up. It has therefore become clear both to clinicians and to basic investigators that fundamental studies of the pathogenesis of retinal and choroidal neovascularization must be continued, with the ultimate goals being prevention in those at risk when the disorder has not yet developed, and cure when it has.

The Role of Growth Factors in Normal and Pathologic Retinal Vascularization

The idea that a diffusible factor might stimulate retinal neovascularization is not new, having first been suggested by Isaac Michaelson as long ago as 1948.[15] Subsequently, Wise developed this concept more fully in an

---

FIGURE 2.1. **A:** Light microscopic histopathological preparation showing a tuft of new vessels extending through the inner limiting membrane of the retina into the vitreous cavity in a human patient with proliferative sickle cell retinopathy secondary to hemoglobin SC disease. Hematoxylin-eosin stain. From Romayananda N, Goldberg MF, Green WR. Histopathology of sickle cell retinopathy. *Trans Am Acad Ophthalmol Otolaryngol.* 1973;77:OP-652–OP-676, by permission of the authors, the editor, and the American Academy of Ophthalmology.

B: Electron micrograph of new blood vessels extending through the inner limiting membrane of the retina into the vitreous of a spontaneously hypertensive rat with retinal dystrophy and secondary retinovitreal neovascularization. The appearance is quite similar to that of the human case in shown in A, save that ectopic retinal pigment epithelial (*RPE*) cells accompany the new vessels as far as the inner limiting membrane. From Frank RN, Mancini MA. Presumed retinovitreal neovascularization in dystrophic retinas of spontaneously hypertensive rats. *Invest Ophthalmol Vis Sci.* 1986;27:346–355, by permission of the editor.

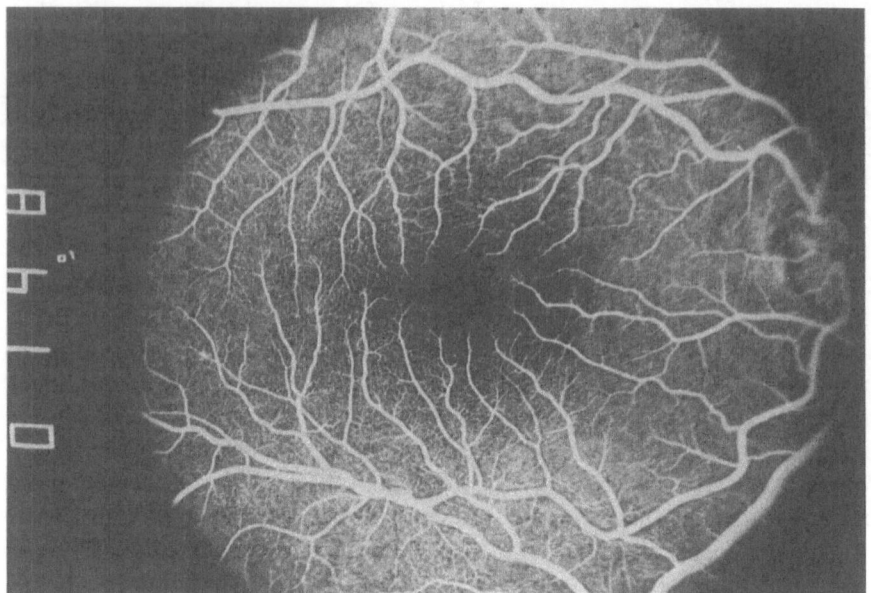

FIGURE 2.2. Early arteriovenous phase of a normal intravenous fluorescein angiogram, showing the macular region of the right eye of a human patient. The numbers at the left show the time, in seconds, from the beginning of the injection into an antecubital vein. Note the round, black zone in the center of the photograph, showing the avascular zone at the fovea, the region of the retina with the highest acuity. Note also that this technique is capable of clearly resolving vessels of capillary caliber, and that the borders of the vessels are sharp, since normally the dye is retained within the vascular lumina because of the barrier function of the retinal vasculature.

extensive thesis presented in 1956.[16] During the early 1970s a number of growth factors were discovered: all were relatively low molecular weight (<20,000 D) peptides with relatively limited target cell specificity, such as the "nerve growth factor,"[17] and the "epidermal growth factor."[18] In 1974, Gospodarowicz and associates reported the isolation of a new polypeptide growth factor from bovine pituitary, with mitogenic properties toward cells of neuroectodermal and mesodermal origin.[19] In particular, because this substance stimulated the mitosis of 3T3 fibroblasts, it was designated "fibroblast growth factor," or FGF. Subsequent work demonstrated that there were two fibroblast growth factors, one with an acidic and one with a basic isoelectric point. They were therefore designated "acidic fibroblast growth factor," or aFGF, and "basic fibroblast growth factor," or "bFGF," respectively.[20,21] These polypeptides have now been purified and sequenced, and their respective cDNAs have been cloned and sequenced. Acidic FGF is a 140–amino acid polypeptide, whereas

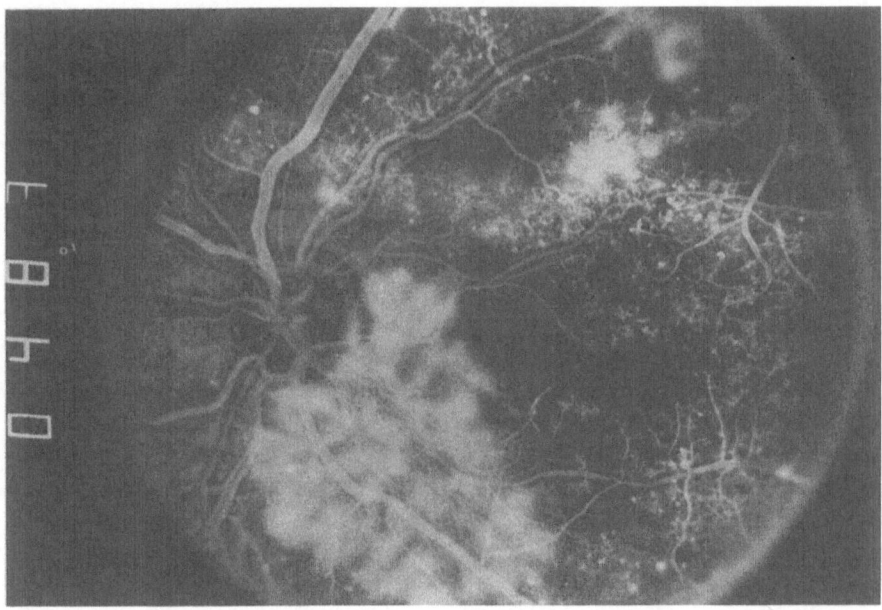

FIGURE 2.3. Late arteriovenous phase of a fluorescein angiogram of the posterior retina of a human patient with proliferative diabetic retinopathy. Two tufts of new vessels are seen; in particular, note the one extending inferiorly from the optic nerve head. Its contours are blurred because of the leakage of the dye from the vascular lumina, indicating that a breakdown of the blood–retinal barrier occurs in retinal neovascularization. Note also the punctate, hyperfluorescent spots representing microaneurysms, and the enlargement of the foveal avascular zone due to nonfunctioning capillaries.

basic FGF contains 146 amino acids. There is 53% homology in the polypeptide sequences of the two molecules, and there is wide conservation of these sequences among species.

In 1980 D'Amore and associates reported that aqueous extracts of bovine retinas contained a mitogen with high activity for cultured aortic endothelial cells.[22] Subsequent work by others demonstrated that these extracts contained both aFGF and bFGF,[23,24] and still later, Schweigerer and associates reported that bFGF was produced in vitro by cultured retinal vascular endothelial cells[25] and retinal pigment epithelium.[26] In addition, we found that aqueous extracts of bovine retinas contain a glial modulating factor that promotes the development of long, astrocytelike processes in cultured retinal or cerebral glial cells.[27] Some years previously, Lim and Mitsunobu had described and partially purified a similar factor from extracts of bovine hypothalamus.[28] It is entirely possible that still other factors involved in cell proliferation and differentiation will be found in retinal extracts.

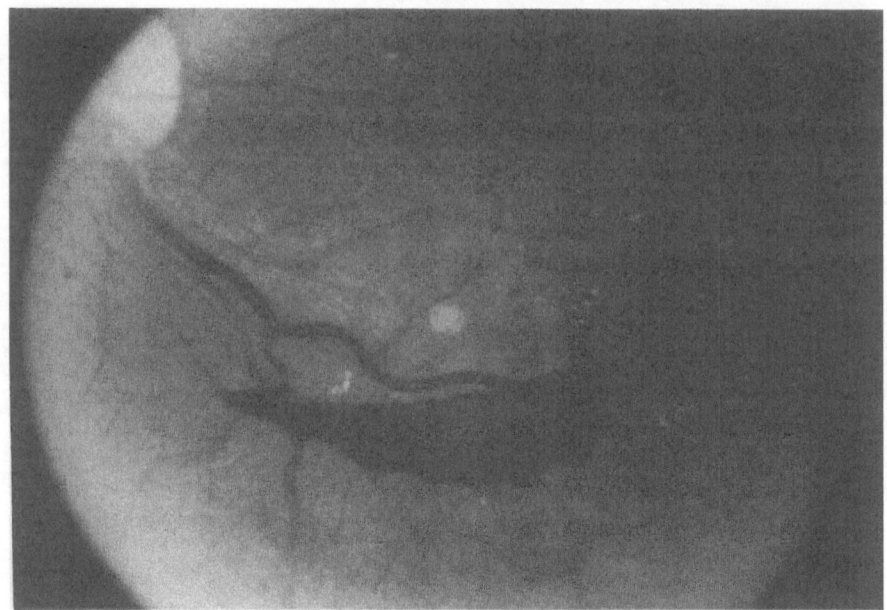

FIGURE 2.4. Photograph of a preretinal hemorrhage in the left eye of a patient with diabetic retinopathy. The presence of the hemorrhage is strong, indication that new vessels are present. The hemorrhage is preretinal because it clearly lies in front of the retinal vessels, which it partly covers, and because it has sunk by gravity to the bottom of a fluid space between the posterior face of the vitreous gel and the inner limiting membrane of the retina, to which the vitreous remains largely attached, limiting the migration of the preretinal blood and causing it to assume a "boat-shaped" configuration with a flat top, due to gravity.

FIGURE 2.5. **A**: Intravenous fluorescein angiogram of the left eye of a human patient with subretinal neovascularization, arising from the choroidal circulation underneath the retina, and subretinal hemorrhage. These lesions are caused by age-related ("senile") macular degeneration. The neovascularization is recogniz-able by its lace doily–like appearance, but with fluorescence much more intense than that of the surrounding, normal vessels due to pooling of the dye, and blurred margins, indicating leakage of dye from the new vessels. The dark region with sharp margins, located to the right of the neovascular tuft, is a subretinal hemorrhage arising from the fragile new vessels. It is evident that both the new vessels and the hemorrhage are beneath the retina, because the retinal vessels pass over them.

B: Histopathological preparation of subretinal new vessels (*NV*) in a pigmented rat several months after retinal laser photocoagulation. The new vessels are located beneath the largely atrophic neural retina (photoreceptors are absent), and within a multilayered and abnormal retinal pigment epithelium, which has now largely lost its pigment. Another, smaller new vessel is indicated by the open

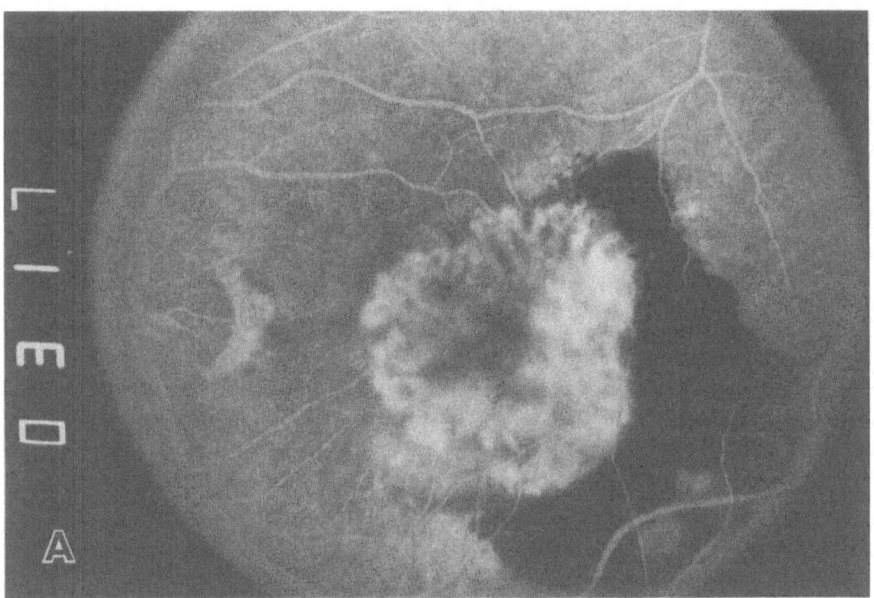

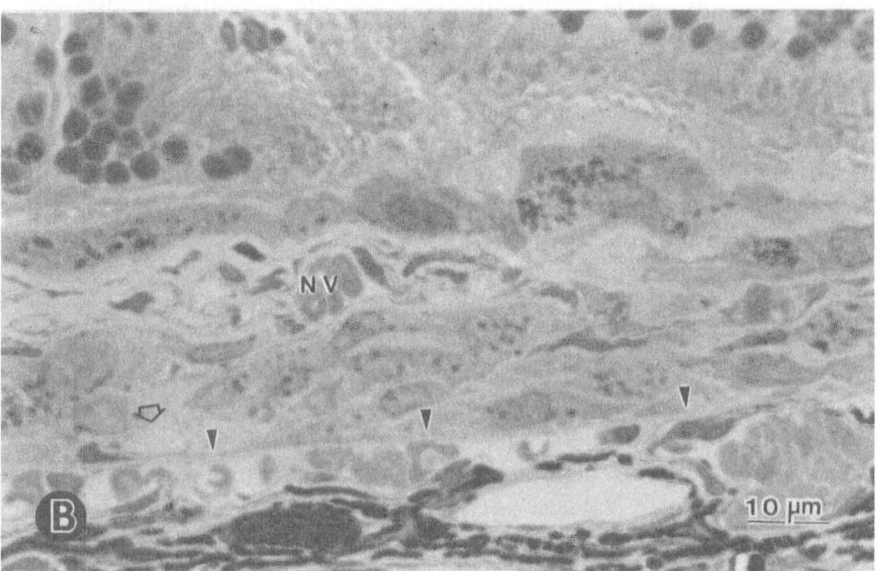

arrow. The filled arrowheads denote Bruch's membrane, a modified, multilaminar structure composed of basement membranes, elastic tissue, and fibrillar collagen that separates the retinal pigment epithelium from the choroidal capillaries that are visible in the lower part of the micrograph. This plastic-embedded section has been stained with toluidine blue. From Frank RN, Das A, Weber ML. A model of subretinal neovascularization in the pigmented rat. *Curr Eye Res.* 1989;8:239–247, by permission of the editors.

As noted above, retinal and, presumably, choroidal vascular cell proliferation is negligible during adult life. Yet an abundance of potent mitogens can be extracted from normal retinas. Since the vascular volume of these retinas is small and the enucleated globes are mostly devoid of blood, the growth factors must be present in the tissues and not in the minimal amount of entrapped blood. Thus, an immediate question arises: what can the function of these factors be, and what inhibitory processes must be present to prevent them from producing constant vasoproliferation?

Several answers suggest themselves. First, studies of cDNA sequences for aFGF and bFGF show that these DNA chains lack a signal sequence, which is always found in DNAs that encode for proteins that are secreted from cells.[29,30] Indeed, bFGF produced by cultured vascular endothelial cells is not secreted from the cells where it is synthesized, whereas another growth factor, platelet-derived growth factor, is so secreted.[31] Presumably, intracellular FGF molecules are physiologically inactive, and it is only those molecules that are released from their cellular sites of synthesis by pathologic stimuli, such as hypoxia, acidosis, or some other metabolic derangement, or by outright cell death, that are free to exert their mitogenic effects by binding to receptors on the cell membranes of appropriate cells nearby.[32-34] In this way, vascular endothelial cells might retain a reservoir of inactive FGF molecules, to be released only when there is a pathologic event, signaling a need for reparative neovascularization. Such an explanation might be satisfactory for endothelium of large vessels, such as the aorta, where localized damage may be followed by effective, localized endothelial repair. However, we are aware of no circumstances in which new vessel formation produces a physiologically useful result in the adult retina.

A second apparent brake on vascular proliferation in the retina appears to be the microvascular basement membrane. It is now widely appreciated that aFGF and bFGF have a strong affinity for the glycosaminoglycan heparin, and indeed, heparin-Sepharose chromatography is a commonly used preparative technique for these molecules.[32,35] Several recent studies have demonstrated that FGF molecules bind to basement membranes, apparently to heparan sulfate residues, which are chemically similar to heparin.[36-38] Thus, even if these molecules are released from their intracellular storage sites, they may be sequestered within basement membranes and thereby prevented from reaching cell membrane receptors, where they can initiate mitogenesis. It is not clear what the pathophysiologic stimulus might be for the release of FGF molecules from basement membranes. However, when microvascular endothelial cells are cultured on a basement membrane matrix, cellular proliferation is drastically slowed and the cells begin to organize into capillarylike tubes.[39,40] A similar result occurs when retinal pigment epithelial or lens epithelial cells are cultured on a matrix composed predominantly (>90%)

of basement membrane macromolecules, but fibroblast or tumor cell proliferation is not similarly retarded.[40] In diabetic retinopathy, a disease in which retinal neovascularization is an important lesion, thickening and other anatomic abnormalities of capillary basement membranes in the retina and other tissues occur early. A biochemical study of basement membrane macromolecules in a basement membrane–producing tumor that had been transplanted into mice that were made diabetic showed reduced synthesis of the heparan sulfate proteoglycan, by comparison with other macromolecular constituents.[41] We conducted a quantitative, electron microscopic immunocytochemical study of basement membrane components of nondiabetic but galactosemic rats, since basement membrane abnormalities virtually identical to those of diabetes occur in these animals, and they appear to represent an excellent model for at least some features of diabetic microangiopathy.[42] We found that the thickened basement membranes contained an increased complement of laminin and of type IV collagen, with no quantitative change in heparan sulfate proteoglycan core protein. The first result suggests that the basement membrane abnormalities of diabetes may decrease the inhibitory effects of these structures on neovascularization by reducing the heparan sulfate binding sites for FGF molecules. The second result, which is qualitatively similar to the first but quantitatively different, is more difficult to interpret in accord with this hypothesis, but perhaps the increase in molecules other than heparan sulfate proteoglycan in diabetic or galactosemic basement membranes may render the heparan sulfate binding sites less approachable by molecules of FGF. Of course, neither of these studies has dealt with another important and possibly relevant feature of the heparan sulfate residues, namely, their degree of sulfation in these various conditions.

Another powerful inhibitory influence on retinal neovascularization appears to be exerted by retinal capillary pericytes and by the closely related smooth muscle cells of the larger vessels. Nearly 30 years ago it was recognized that the pericytes are specifically lost from the retinal capillaries of individuals relatively early in the course of diabetic retinopathy.[43] The function of these cells has never been entirely clear, but their anatomic location, straddling the endothelial tube and with multiple, tentaclelike processes encircling the tube, together with their generous component of action fibrils, has led to the obvious conclusion that they are analogous to the smooth muscle cells of larger vessels, and that their constant tonus and/or periodic contraction regulates regional microcirculatory blood flow. Recently, Orlidge and D'Amore,[44] in a series of elegant cell culture experiments, showed that when retinal microvascular pericytes or large vessel smooth muscle cells are grown together with microvascular endothelial cells, the pericytes or smooth muscle cells inhibit endothelial cell proliferation. The effect is seen even when the pericyte: endothelial cell ratio is as low as 1:10, provided that pericyte processes are in contact or at least in close proximity with each endothelial cell. In more recent work, these authors showed that the molecular mechanism of this growth inhibition was based on the action of transforming growth factor β (TGFβ).[45] This molecule is produced in an inactive form by cultures of pericytes growing alone, and can be activated in medium removed from these cultures by acid treatment. However, when pericytes and endothelial cells are cultured together and allowed to make apparent physical contact, TGFβ is produced in active form.

These results suggest the physiologic importance of direct contact between pericyte processes and endothelial cells. Both Carlson[46] and Robison et al.[47] have published electron microscopic studies of retinal capillaries, demonstrating fenestrations in the basement membranes through which pericyte processes can make contact with endothelial cells. Robison et al.[47] reported that, in experimentally galactosemic rats, which, as noted above, develop capillary basement membrane thickening seemingly identical to that which occurs in diabetes, the number of pericyte–endothelial cell contacts in retinal capillaries is reduced. Recently, we measured the percent coverage of the circumference of the endothelial tube by pericyte processes in the capillaries of retina and several regions of the cerebral cortex from different species. In both the rat[48] and in the monkey,[49] there is substantially (and statistically highly significantly) greater pericyte coverage in the retinal capillaries than in capillaries from all regions of the cerebral cortex. Coverage of retinal capillaries by pericyte processes in human eyes is comparable to that in the monkey, but we were unable to get specimens of human brain cortex tissue suitable for electron microscopic examination to make comparable measurements.[49] In addition, those portions of the capillary basement membrane that lie between pericyte processes and endothelial cells are much thinner than the portions of the basement membrane that cover only pericytes, or only endothelial cells. These anatomic features would appear to facilitate contact between pericyte processes and endothelial cells in the retina, by contrast with the cerebral cortex. In 1966, De Oliveira reported histopathologic studies of human retinal and brain tissue in individuals dying with diabetes mellitus, indicating that loss of pericytes and other changes were present in the retinal capillaries of many of these individuals, but not in their cerebral capillaries.[50] Recently, a study by Kador et al. on retinal vessels from dogs with long-term galactosemia, either with or without treatment with aldose reductase inhibitors, provided evidence that endothelial cell proliferation and microaneurysm formation occurred in regions of the microcirculation where there was pericyte loss.[51] All of these findings strongly suggest that pericytes do inhibit endothelial cell proliferation in capillaries, and that the more extensive pericyte coverage in the retina than in the brain may help provide greater inhibition in a tissue where there may be greater mitogenic stimuli for microvascular endothelial cells.

What could be the physiologic function of potent vascular endothelial cell mitogens such as aFGF and bFGF in the retina? There are at least two plausible answers to this question. First, the growth factors may be necessary to facilitate retinal vascular growth in the developing retina either in utero (as in humans, whose retinal vasculature is fully developed in infants delivered at term) or in early postnatal life in the species whose retinal vessels are not completely formed at birth. For example, Hanneken et al.[52] found that the capillaries of the outer edge of the inner nuclear layer of the 5- to 6-months' gestation fetal bovine retina stained positively with bFGF antibodies (Fig. 2.6). Since these capillaries are embryologically the last to mature, these investigators interpreted their result to mean that these capillaries were still increasing the length of their endothelial tubes, thereby requiring a growth factor to facilitate endothelial cell proliferation, at the period of fetal life when the study was conducted. The persistence of these growth factors into adult life may be an atavistic remnant, with unfortunate consequences, at least in the retina, when certain disease processes supervene.

A second possible function of the FGFs in the retina may be to sustain morphologic and physiologic differentiation in neural and, perhaps, glial, cells. For example, Wagner and D'Amore[53] found that retinal-derived growth factor (which was later found to be bFGF) enhanced neurite process formation in certain neural cell cultures. Depending on the specific chemical mechanisms by which such physiologic functions are exerted in cells, it is entirely reasonable to suppose that FGF molecules could mediate these functions in retinal neurons, even while the mitogenic action of FGF is inhibited in vascular cells through the mechanisms described above.

Insulinlike Growth Factors

Insulinlike growth factors, or IGFs, are polypeptides with a close homology to the amino acid sequence of the A chain of insulin. They appear to be produced largely in the liver through the action of growth hormone (somatotropin), and they are effective in mediating the cellular action of some of the growth-promoting actions of somatotropin. Hence, they were previously called "somatomedins." In certain cell lines, IGFs have a powerful mitogenic action, at least in vitro. Insulin itself may be mitogenic to many types of cells, but usually only at unphysiologically high concentrations. In a comparative study, King et al.[54] found that insulin at nanomolar concentrations could stimulate DNA synthesis in cultured retinal vascular pericytes, retinal endothelial cells, and aortic smooth muscle cells, but not in aortic endothelial cells. All four of these types of cells have high-affinity receptors for insulinlike growth factors 1 and 2 (IGF-1 and IFG-2) on their cell membranes, with different specificities and ability to cross-react with each other and with insulin.

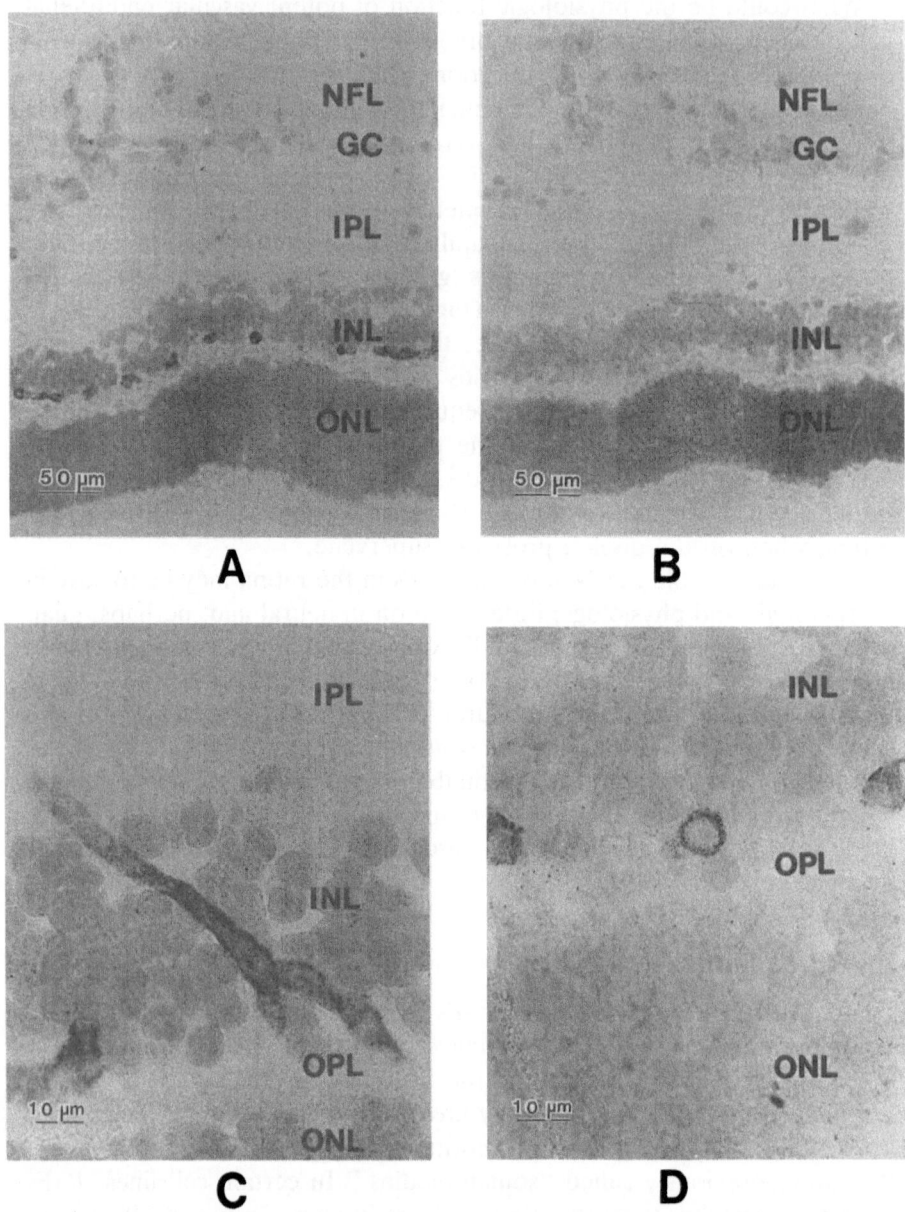

FIGURE 2.6. Immunohistochemistry of basic fibroblast growth factor (*bFGF*) in the retina of a fetal calf of 5–6 months estimated gestational age. The retinal layers are designated as follows: *NFL*, nerve fiber layer; *GC*, ganglion cell layer; *IPL*, inner plexiform layer; *INL*, inner nuclear layer; *ONL*, outer nuclear layer. Positive staining is seen as dark granules. **A**: Capillary-sized vessels at the outer margin of the inner nuclear layer are the only structures that stain positive. **B**: There is no positive staining in this preparation, treated with antibody that had

These retinal microvascular endothelial cells responded much more readily to IGF-1 and IGF-2 with increased DNA synthesis than did aortic endothelial cells, whereas insulin was similarly effective in stimulating DNA synthesis in retinal microvascular endothelial cells. Retinal pericytes and aortic smooth muscle cells also responded with markedly increased DNA synthesis to stimulation by IGF-1, but less readily to IGF-2.[55]

Whether the IGFs or insulin stimulate mitogenesis of retinal vascular cells in vivo is unclear at this time. Merimee et al.[56] found higher circulating levels of IGF-1 (somatomedin C) in patients with proliferative diabetic retinopathy than in control subjects with diabetes, but without proliferative retinopathy. Recently, Merimee has argued that the IGFs are critically important in the pathogenesis of proliferative diabetic retinopathy.[57] Some clinical observations give at least partial support to this hypothesis. First, diabetic retinopathy of any sort is rare before puberty, regardless of the antecedent duration of diabetes.[58-62] Second, a number of investigators in the past, before the introduction of retinal laser photocoagulation, had observed a probable beneficial effect of hypophysectomy on proliferative diabetic retinopathy,[63] and Kohner and colleagues have suggested that there are certain cases of "florid" proliferative disease that do not respond to photocoagulation, but whose retinopathy regresses following hypophysectomy.[64]

The Role of Ancillary Factors in Retinal Vascular Growth

A number of investigators, notably Patz[65] and Henkind,[66] have suggested that hypoxia is a major stimulus for retinal new vessel growth. It has been known since the work of Warburg in the 1920s that the retina has a very high rate of metabolism, as measured by glucose and oxygen consumption and CO_2 production.[67] It is reasonable to suppose that any disease process that reduces the retinal vascular supply, thereby decreasing oxygenation of the inner retina, might call forth a stimulus to produce new blood vessels to make up the deficit. (This may not apply to the

been preabsorbed with antigen to demonstrate the specificity of the staining. C: A capillary-sized vessel cut in longitudinal section within the inner nuclear layer stains in its entirety. D: Another vessel, cut in cross-section, also shows clear positive staining. The specific cells that stain cannot be determined for certain in these preparations, but they appear likely to be endothelial cells. All of the photomicrographs in this figure are taken from Hanneken A, Lutty GA, McLeod GS, Robey F, Harvey AK, Hjelmeland LM. Localization of basic fibroblast growth factor to the developing capillaries of the bovine retina. *J Cell Physiol.* 1989;138:115–120, by permission of the authors and the Editor.

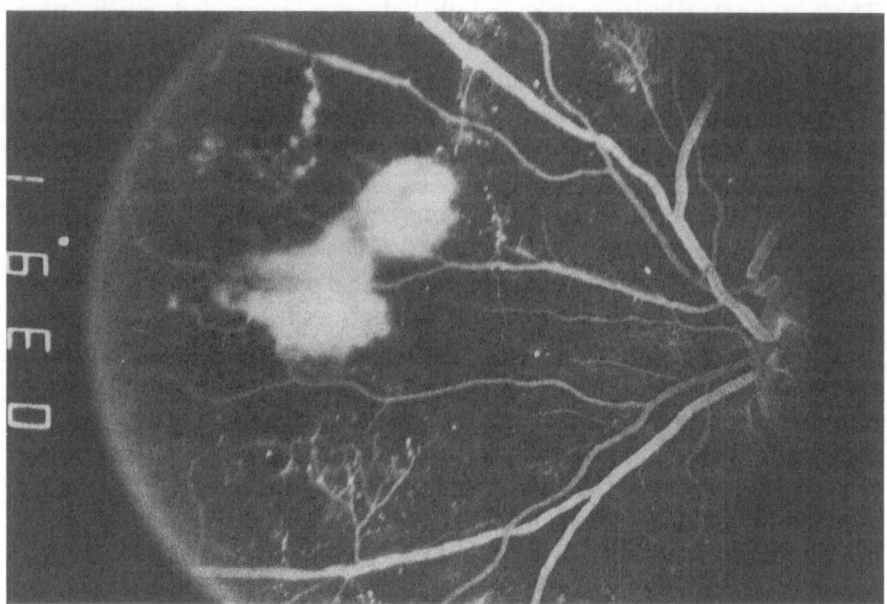

FIGURE 2.7. An intravenous fluorescein angiogram showing an intensely fluorescent, leaking tuft of new vessels located near a black zone, indicating capillary nonperfusion, in a human patient with proliferative diabetic retinopathy. The frequent juxtaposition of retinal neovascularization and zones of capillary nonperfusion is considered strongly suggestive that retinal hypoxia stimulates the development of retinal neovascularization.

outer layers of the retina, including the photoreceptors and the retinal pigment epithelium, which have the highest metabolic rates as judged morphologically by their very high concentrations of mitochondria and receive their blood supply from the choriocapillaris.) Clinical studies using intravenous fluorescein angiography have repeatedly demonstrated that retinal neovascularization forms adjacent to regions of vascular nonperfusion (Fig. 2.7), where localized hypoxia would seem to be greatest. However, this does not explain the frequent observation of optic nerve head neovascularization (the most frequent and, prognostically, the worst form of retinal neovascularization). Similarly, choroidal neovascularization does not appear to develop in response to a hypoxic stimulus.

Certain anatomic findings in the normal retina suggest that oxygen has a role in retinal vascularization, at least during the course of retinal development. For example, trypsin digest preparations of the retinal vasculature demonstrate a vessel-free zone alongside retinal arterioles, where the local pO_2 would appear to be greatest.[1] Retinal vessels normally do not grow in the retinal pigment epithelium, the photoreceptor layer, or the outer nuclear or outer plexiform layers of the retina, where the pO_2 is relatively high due to proximity to the oxygen-rich blood in the choriocapillaris. The fovea is normally avascular (Fig. 2.1), because only photoreceptor cells are present in this region of the retina and not the inner layers of the retina, which lie farther from the choriocapillaris.

In the mid-1960s Imre reported that intravitreal injections of lactic acid caused retinal neovascularization in cats,[68] a finding possibly related to the hypoxia theory. However, until now, this finding has not been confirmed by other investigators.

We have noted above that the vascular basement membrane may inhibit vascular proliferation by sequestering growth factors. Another, more obvious function of the basement membrane is simply to act as a barrier to proliferating cells. Ausprunk and Folkman,[69] in an electron microscopic study of corneal neovascularization in rabbits, found that the endothelial cells in the advancing tips of the new vessels appeared to digest the basement membrane in their path. Subsequently, Kalebic et al.[70] found that vascular endothelial cells proliferating in culture elaborated a collagenase. Plasminogen activators, which are known to be secreted by cultured endothelial cells, are thought to play an important role in neovascularization.[71] Plasminogen activators convert plasminogen to plasmin, a protease that can dissolve extracellular matrix,[72] allowing for local vascular sprouting.[73] Plasmin also prevents clot formation in the newly formed vessels, which is important because these vessels have highly permeable, "leaky" endothelial cells that allow exposure of the luminal contents to thrombogenic substances in the extravascular environment. Plasmin formation has been localized to the advancing tips of new blood vessels in the developing retina.[74]

The effectiveness of panretinal laser photocoagulation in eliminating retinal neovascularization in proliferative diabetic retinopathy is now well known. The most widely accepted, although unproven, hypothesis explaining this result, which has been most succinctly enunciated by Weiter and Zuckerman,[75] is that the destruction of large portions of the metabolically highly active outer retinal layers by photocoagulation substantially reduces outer retinal oxygen utilization. The oxygen from the choriocapillaris, therefore, can diffuse more readily to the oxygen-starved inner retina, thereby reducing its hypoxic stimulus to new vessel formation and causing the regression of newly formed vessels. More recently, Glaser et al.[76] proposed another hypothesis, namely, that the retinal pigment epithelial cells that proliferate following such extensive laser treatment elaborate a factor that causes such regression. This factor is claimed to be a protease inhibitor, the so-called pigment epithelial protease inhibitor, or PEPI, which inhibits plasminogen activation. Other investigators have failed to confirm Glaser's findings. As we have noted above, Orlidge and D'Amore[44] found that microvascular endothelial cell proliferation was inhibited during coculture with retinal microvascular pericytes or with smooth muscle cells from larger vessels. However, when microvascular endothelial cells were cocultured with several other types of cells, in particular with retinal pigment epithelial cells, endothelial cell proliferation was stimulated. Wong and colleagues also found that retinal pigment epithelial cells stimulated the proliferation of retinal microvascular endothelial cells.[77] In an animal model of retinovitreal neovascularization to be discussed further below, my coinvestigator and I have found strong anatomical evidence that proliferating retinal pigment epithelial cells stimulate the proliferation of adjacent vascular endothelial cells.[78] Similarly, we have the strong impression that in a model of subretinal neovascularization, choriocapillary endothelial cell proliferation is promoted by the proliferation of adjacent retinal pigment epithelial cells, which occurs to replace cells that have been damaged or destroyed by photocoagulation.[79] Henkind and Gartner[80] have pointed out that retinal pigment epithelial cells and the adjacent choriocapillaris exist in a kind of symbiotic relationship, in which each cellular layer supports the other. If either layer is destroyed by disease or by experimental manipulation, the other layer degenerates and eventually disappears. Thus, normally the retinal pigment epithelium appears to support rather than suppress adjacent vascular endothelium and, in certain pathological situations, even to stimulate its proliferation. In response to such arguments, Glaser[34] has proposed that the retinal pigment epithelium may elaborate several kinds of molecules, depending on circumstances, some of which may support proliferation of vascular endothelial cells, whereas others may inhibit such proliferation. This hypothesis remains unproven.

Models of Retinal and Subretinal Neovascularization

The role of growth factors and other determinants of retinal and choroidal neovascularization can best be tested in animal models. Such models have been difficult to develop, but a few do exist. The first model of retinal neovascularization was developed by Ashton and colleagues[81] and by Patz and colleagues[65,82] in order to study the pathologic changes of retinopathy of prematurity (formerly called retrolental fibroplasia). This model is based on the fact that the retinal vessels of the newborn kitten or puppy are not fully developed at birth, much like the premature human infant (but not the human infant born at full term) and like the retinal vessels of the premature human infant, are highly sensitive to high concentrations of ambient oxygen. Indeed, when the animals are placed in a hyperoxic atmosphere, the vessels of the peripheral retina contract, producing peripheral retinal ischemia. Presumably because of the localized hypoxia, there is a secondary stimulus to neovascularization. The newly formed vessels are relatively short lived, however, and these animals do not develop progressive retinal lesions with eventual traction retinal detachment, such as occurs in some human infants with retinopathy of prematurity. Flower and Blake[83] have suggested that progressive neovascularization and traction retinal detachments may develop in these animals if they are also dosed with aspirin. It is not clear why this should occur, and this observation has not yet been confirmed by others.

Several years ago Shabo and Maxwell[84] reported that, when monkeys were sensitized to bovine insulin or serum albumin injected in the footpad with Freund's adjuvant, a subsequent intravitreal injection of the same antigen resulted in a severe intraocular inflammation, followed by neovascularization of the optic nerve head. Unfortunately, this observation was interpreted to suggest that retinal or optic nerve neovascularization in diabetes might be a hypersensitivity reaction to insulin. There is, however, no evidence that intraocular inflammation occurs during the course of proliferative diabetic retinopathy in humans, and Shabo and Maxwell's observations probably represent only an example of retinal or (in their case) optic nerve head neovascularization following intraocular inflammation, a well known phenomenon in humans, rather than a model of proliferative diabetic retinopathy. This is not to deny the potential value of this model, however, which has received too little attention since it was first described, most likely because of the unfortunate emphasis on the incorrect interpretation.

We have recently observed the development of retinovitreal new vessels in the dystrophic retinas of older (>1 yr) rats of the spontaneously hypertensive (SHR) and of the Royal College of Surgeons (RCS) strains.[78,85,86] As the photoreceptors degenerate early in the course of these retinal dystrophies, vessels from the inner retinal layers appear to grow outward to the retinal pigment epithelium, where they then tunnel through the layer of pigment epithelial cells, but never break through Bruch's membrane at the base of these cells to anastomose with the choriocapillary vessels. Frequently, these intrapigment epithelial vessels lose the anatomic characteristics of normal retinal vessels and develop a very thin endothelium with frequent fenestrations, similar to those seen in the choriocapillaris, or in newly forming capillaries elsewhere. Similar observations have been made in rats with retinal photoreceptor degenerations induced by prolonged exposure to bright light, or to the anesthetic agent urethane.[87-89] However, we have observed that if the dystrophic animals are followed for a longer period, the intrapigment epithelial vessels in some regions develop extensions that proliferate back through the retina toward the inner layers. These appear to be surrounded nearly always by tubes of proliferating retinal pigment epithelium.[78] When these extensions reach the inner limiting membrane of the retina, the vessels break through and proliferate into the vitreous, while the pigment epithelial cells form a monolayer on the retinal surface of the inner limiting membrane. Both the anatomic features of the intravitreal vessels and the fact that their nuclei frequently can be labeled with [³H]-thymidine strongly suggest that they truly represent retinovitreal neovascularization. A puzzling finding, however, is that in our initial observations we only found these vessels in about 20% of the eyes of dystrophic RCS or SHR rats >1 year of age.[78,85] More recently, we have observed that if the dystrophic animals receive a prolonged (5-day) exposure to bright, white light, there is a substantial increase in the number of eyes with retinovitreal neovascularization.[86] We interpret these findings to mean that factors elaborated by proliferating retinal pigment epithelial cells initiate the proliferation of the nearby retinal vascular endothelial cells. When the pigment epithelial cells are relatively quiescent, vascular endothelial cells also do not multiply. Any stimulus to pigment epithelial cell proliferation, however, will result in a secondary proliferation of vascular endothelial cells. Prolonged white light exposure may represent such a stimulus, even in the absence of a retinal photoreceptor layer, although the light-absorbing structure(s) are not immediately apparent, since the animals we have studied thus far have all been albinos, in whom the only plausible candidates for such structures might be mitochondria. It is also not clear why the choriocapillaris does not proliferate in this model. Perhaps the presence of an intact Bruch's membrane serves as a barrier, and/or the release of the angiogenic molecules occurs asymmetrically in the polarized pigment epithelial cells.

Subretinal neovascularization (i.e., new vessels arising from the choroidal circulation) has been produced experimentally in monkeys and rats following photocoagulation. Ryan[90] demonstrated this lesion in monkeys following intense, small (50 μm diameter) argon green laser burns to the macular region. To produce new vessels, the laser burns had to rupture Bruch's membrane, the multilaminar extracellular matrix between the base of the RPE cells and the choroidal capillaries (or choriocapillaris). Thirty-nine percent of laser burns in the macula produced subretinal neovascularization, whereas only 3% of peripheral burns resulted in such lesions. However, unlike the majority of subretinal neovascular lesions that occur clinically in humans, the choroidal new vessels occurring after laser treatment in monkeys are transient, developing within 3 weeks after treatment, increasing in size for approximately 13 weeks (but with a wide range), and then involuting. Similar subretinal neovascular lesions were reported by Archer and Gardiner following intense, large size, white burns to the midperiphery of the monkey retina, using the xenon arc photocoagulator.[91] Further details of these new vessels, including their time course of occurrence, progression, and regression, have not been reported.

Another model of subretinal neovascularization was produced in the posterior retinas of pigmented rats (rats do not have maculas) using a similar focal burn with the krypton red laser of sufficient intensity to rupture Bruch's membrane.[92] The natural history of this lesion has also not yet been reported.

Pollack et al.[93] first showed that milder photocoagulation with the krypton red laser could produce subretinal new vessels in pigmented rats, even when there was no rupture of Bruch's membrane. We confirmed this finding[79] but found that new vessels were produced only following krypton red, but not argon blue-green plus green, laser treatment. We are in the process of studying these lesions in detail, and as yet have no information about their natural history.

There are a number of questions arising from these animal models of subretinal neovascularization. What are the cellular mechanisms that produce the new vessels? Why are the lesions predominantly formed in the macula of monkeys? Why are they of such short duration, whereas subretinal new vessels in humans persist indefinitely, expand, and bleed? What explains the different incidence of subretinal new vessels in rats following photocoagulation with red or green lasers? Further study of these animal models should have considerable value in understanding this vascular lesion, which has such clinical importance in humans.

Localization of Growth Factors in the Eye

Several techniques have been employed to determine the cellular localization of aFGF and bFGF in the eye. We have already discussed the work of Hanneken et al.,[52] who used antibodies against bFGF to localize this molecule to retinal capillaries of the inner nuclear layer in the developing bovine eye. More recently, Baudouin et al.[94] used an anti-aFGF antibody to localize this growth factor by light microscopic immunocytochemical methods in tissues of human eyes obtained post mortem. Strongly positive cells included the corneal, conjunctival, and lens epithelium, the pigmented and nonpigmented epithelium of the ciliary body, and all layers of the retina, in particular the photoreceptor cell layer and the inner and outer plexiform layer. Neither the retinal pigment epithelium nor the retinal or choroidal blood vessels reacted with this antibody. Fayein et al.[95] performed autoradiography on fetal and postnatal mouse ocular tissues following incubation with radiolabeled bFGF. Heavy labeling was observed in all basement membranes and could be removed by pretreatment with heparitinase, leading to the conclusion that the binding site is closely associated with the heparan sulfate proteoglycan of these basement membranes. Beginning at day 16 of embryonic life, there was heavy labeling of mast cells, identified by their metachromatic staining with toluidine blue, together with a generalized, diffuse labeling of all tissues but in particular of ectodermal and neuroectodermal tissues. Still later, at embryonic days 17 and 18, labeling within the retina appeared concentrated in the retinal pigment epithelium, in the inner and outer plexiform layers, and in the photoreceptor inner segments. This labeling was resistant to treatment with heparitinase but was sensitive to treatment with N-glycanase. Recently Hageman et al.[96] reported bFGF antigenicity associated with the inter-photoreceptor matrix, an extracellular matrix rich in glycosaminoglycans and in particular chondroitin sulfates, which lies between the retinal photoreceptor cell outer segments and the retinal pigment epithelium, in cynomolgus monkey retinas. The role of this growth factor in this location is not clear, but in another recent, remarkable study, Faktorovich et al.[97] reported that subretinal or intravitreal injection of bFGF greatly delayed or prevented the inherited retinal degeneration in Royal College of Surgeons rats. This finding gives strong support to the hypothesis that FGF molecules function to maintain retinal cell differentiation. Finally, Noji et al.[98] performed in situ hybridization using radiolabeled mRNA probes for the genes for aFGF and bFGF in adult rat eyes. They found different labeling patterns for the mRNAs for these two molecules. The aFGF gene was produced by the photoreceptor cells and by cells in the inner nuclear and ganglion cell layers, as well as in the pigment epithelial cells of the retina, iris, and ciliary body, the epithelial cells of the cornea, conjunctiva, and lens. By contrast, the bFGF gene was localized solely in the photoreceptor cells. In a clinical study, Sivalingam et al.[99] have recently reported finding elevated levels of bFGF in the vitreous of patients undergoing vitrectomy surgery for "active" proliferative diabetic retinopathy, by comparison with levels in diabetic patients who did not have neovascular lesions defined as active, or in nondiabetic patients undergoing vitrectomy for other reasons. This finding would suggest a role for bFGF in the neovascular process in the retinas of diabetic individuals. However, results described by Hanneken et al.[100] are more difficult to reconcile with this hypothesis. These authors performed light microscopic immunocytochemistry for bFGF and for heparan sulfate proteoglycan in eyes of normal humans and of humans with diabetes and varying degrees of diabetic retinopathy that had been removed post-mortem. Similar studies were also performed on proliferative vascular fronds that had been removed at vitrectomy surgery from diabetic patients. In normal subjects and in diabetic individuals without prolifer-ative retinopathy, the authors found that bFGF antigenicity co-localized with heparan sulfate proteoglycan antigenicity in the vascular base-ment membranes, and there was no intracellular bFGF staining. With increased thickening of the retinal vascular basement membranes in diabetic patients, there was increased staining for both heparan sulfate proteoglycan and for bFGF. However, in neovascular fronds, there was little or no bFGF staining despite increased heparan sulfate proteoglycan immunoreactivity. Rather, some vascular endothelial cells in these fronds now stained intracellularly for bFGF. The loss of bFGF immunoreactivity of the basement membranes cannot be attributed to loss of growth factor binding sites, since Hanneken et al. could demonstrate that bFGF staining could easily be restored by incubating the tissues with exogenous bFGF. The low levels of immunoreactive bFGF demonstrable in these neovascular fronds suggest that, contrary to the results of Sivalingam, et al.,[99] diffusible bFGF is not a likely angiogenic stimulus in these eyes. However, these low levels demonstrable immunochemically may simply be due to rapid utilization of the factor by the rapidly proliferating vessels, or to its post-translational modification.

In another study from the same laboratory, TGFβ was localized immunocytochemically to photoreceptor cells in human retinas, including eyes from subjects both with and without diabetes.[101] There was no apparent difference in staining between retinas from diabetic and non-diabetic subjects. No other retinal structures, specifically not retinal vessels, stained for this growth factor, and the role of TGFβ in this location in the retina is unclear. The interpretation of these various results is unclear at this time. Localization of a molecule by immunocyto-chemistry does not, of course, demonstrate that the molecule is produced at that location, but only that it is present at that site physiologically and may have a function there. Specific localization by incubation with radiolabeled molecules only indicates that receptors for those molecules are present, but proves nothing about physiologic function. Localization of the gene by in situ hybridization only demonstrates that the labeled cells have the genetic capacity to synthesize the molecule, but does not prove that the molecule is actually being produced at a given time or, if it is produced, whether it is biologically active. The presence of elevated levels of a growth factor in the vitreous of patients with active retinal or iris neovascularization is compatible with a causal role for the factor in producing the neovascularization, but does not prove it, since the factor may have originated elsewhere and reached the vitreous by leakage through the defective blood–retinal barrier. Thus, these results, although interesting, leave many questions unanswered.

As yet, no studies of tissue sections from intact eyes have been reported to localize other growth factors, for example the IGFs, to ocular tissues.

Summary

Normal vascular growth in the retina is completed at birth in the full-term human infant, and in the days to weeks following birth in many non-human species. All retinal (or choroidal) vascular growth after this time is pathologic. Such "new" vessels are fragile and can rupture easily, causing potentially serious hemorrhages. They seem to serve no useful physiologic function. There is now good, although indirect, evidence that both normal and pathologic retinal vascular growth is mediated through small molecular weight polypeptide growth factors. The precise factors that may be involved are not known for certain, although a number of plausible candidates have been identified.

The best evidence suggests that among the known growth factors, acidic and basic fibroblast growth factors are normally present in several ocular tissues, and in particular in the neural retina and the retinal pigment epithelium. These factors can cause proliferation of vascular endothelial cells in vitro. The precise localization of these individual factors in specific cell types and cell layers is less clear, since studies of cultured cells, and histochemical, autoradiographic, and in situ hybridization studies have yielded results that are in partial, but not complete, agreement. In one reported study, TGFβ has been localized immunochemically to human photoreceptor cells, where its function is unclear. A series of striking cell culture experiments demonstrates that activated TGFβ, produced by the cooperative action of retinal microvascular pericytes or vascular smooth musle cells and microvascular endothelial cells, can dramatically inhibit endothelial cell proliferation. Similar studies with other growth factors, in particular the IGFs, have not been described, but retinal vascular cells do have receptors for IGFs. Whether other growth factors exist in the retina and have a role in normal retinal vascularization or in pathologic neovascularization is unclear.

A number of other factors have been identified as possible stimulators or inhibitors of normal and pathologic vascular proliferation. Among the putative inhibitors are oxygen and basement membranes (and perhaps especially, their heparan sulfate proteoglycan residues), while among the putative stimulators are hypoxia, (lactic) acidosis, and certain collagenases or other proteases, such as plasmin.

Although the study of normal and pathologic vascularization of the retina and other ocular tissues has made great advances in recent years, much remains to be learned, and because of its importance for the understanding of normal retinal development and several common and blinding diseases, this remains a highly active field of research.

Acknowledgments. This work was supported in part by research grants RO1 EY-01857 and RO1 EY-02566 to Dr. Frank, and by a training grant T32 EY 07093, all from the National Eye Institute, U.S. National Institutes of Health, and by a departmental unrestricted grant from Research to Prevent Blindness, Inc., New York.

References

1. Wise GN, Dollery CT, Henkind P. *The retinal circulation*. Hagerstown, MD: Harper & Row; 1971: Development of retinal vessels. 1–18.

2. Wise GN, Dollery CT, Henkind P. *The retinal circulation*. Hagerstown, MD: Harper & Row; 1971: Phylogeny and retinal vessels in selected animals. 68–82.
3. Lowry OH, Roberts NR, Lewis C. The quantitative histochemistry of the retina. *J Biol Chem*. 1956;220:879–892.
4. Lowry OH, Roberts NR, Schulz DW, Clow JE, Clark JR. Quantitative histochemistry of retina. II. Enzymes of glucose metabolism. *J Biol Chem*. 1961;236:2813–2820.
5. Engerman RL, Pfaffenbach D, Davis MD. Cell turnover of capillaries. *Lab Invest*. 1967;17:738–743.
6. Wise GN, Dollery CT, Henkind P. *The retinal circulation*. Hagerstown, MD: Harper & Row; 1971: Chapter 3, Structure of retinal vessels. 34–54.
7. DeVenecia G, Davis MD. Histology and fluorescein angiography of microaneurysms in diabetes mellitus. *Invest Ophthalmol*. 1967;6:555.
8. The Diabetic Retinopathy Study Research Group. Photocoagulation treatment of proliferative diabetic retinopathy: clinical application of Diabetic Retinopathy Study (DRS) findings. DRS Report Number 8. *Ophthalmology*. 1981;88:583–600.
9. Branch Vein Occlusion Study Group. Argon laser scatter photocoagulation for prevention of neovascularization and vitreous hemorrhage in branch vein occlusion: a randomized clinical trial. *Arch Ophthalmol*. 1986;104:34–41.
10. Macular Photocoagulation Study Group. Argon laser photocoagulation for senile macular degeneration: results of a randomized clinical trial. *Arch Ophthalmol*. 1982;100:912–918.
11. Macular Photocoagulation Study Group. Argon laser photocoagulation for ocular histoplasmosis: results of a randomized clinical trial. *Arch Ophthalmol*. 1983;101:1347–1357.
12. Macular Photocoagulation Study Group. Krypton laser photocoagulation for neovascular lesions of ocular histoplasmosis: results of a randomized clinical trial. *Arch Ophthalmol*. 1987;105:1499–1507.
13. Macular Photocoagulation Study Group. Argon laser photocoagulation for neovascular maculopathy: three-year results from randomized trials. *Arch Ophthalmol*. 1986;104:694–701.
14. Condon P, Jampol LM, Farber MD, Rabb M, Serjeant G. A randomized clinical trial of feeder vessel photocoagulation of proliferative sickle cell retinopathy. II. Update and analysis of risk factors. *Ophthalmology*. 1984;91:1496–1498.
15. Michaelson IC. The mode of development of the vascular system of the retina: with some observations on its significance in certain retinal diseases. *Trans Ophthalmol Soc*. 1948;68:137–180.
16. Wise GN. Retinal neovascularization. *Trans Am Ophthalmol Soc*. 1956;54:729–826.
17. Levi-Montalcini R. The nerve growth factor 35 years later. *Science*. 1987;237:1154–1162.
18. Cohen S, Taylor JM. Epidermal growth factor: chemical and biological characterization. *Rec Prog Hormone Res*. 1974;30:533–550.
19. Gospodarowicz D. Purification of a fibroblast growth factor from bovine pituitary. *J Biol Chem*. 1975;250:2515–2519.

20. Bohlen P, Esch F, Baird A, Gospodarowicz D. Acidic fibroblast growth factor from bovine brain. Amino terminal sequence and comparison to basic fibroblast growth factor. *EMBO J*. 1985;4:1951–1956.
21. Abraham JA, Whang JL, Tumolo A, Fiddes JC. Human basic fibroblast growth factor: nucleotide sequence, genomic organization, and expression in mammalian cells. *Cold Spring Harbor Symp Quant Biol*. 1986;51:657–668.
22. D'Amore PA, Glaser BM, Brunson SK, Fenselau AH. Angiogenic activity from bovine retina: partial purification and characterization. *Proc Nat Acad Sci USA*. 1981;78:3068–3072.
23. Schreiber AB, Kenney J, Kowalski WJ, Thomas KA, Giminez Gallego G, Rios Candelore M, DiSalvo J, Barritault D, Courty J, Courtois Y, Moenner M, Loret C, Burgess WH, Mehlman T, Friesel R, Johnson W, Maciag T. A unique family of endothelial cell polypeptide mitogens: the antigen and receptor cross reactivity of bovine endothelial cell growth factor, brain derived acidic fibroblast growth factor, and eye derived growth factor II. *J Cell Biol*. 1985;101:1623–1626.
24. Baird A, Esch F, Gospodarowicz D. Retina- and eye-derived endothelial cell growth factors: partial molecular characterization and identity with acidic and basic fibroblast growth factors. *Biochemistry*. 1985;24:7855–7860.
25. Schweigerer L, Neufeld G, Friedman J, Abraham JA, Fiddes JC, Gospodarowicz D. Capillary endothelial cells express basic fibroblast growth factor, a mitogen that promotes their own growth. *Nature*. 1987;325:257–259.
26. Schweigerer L, Malerstein B, Neufeld G, Gospodarowicz D. Basic fibroblast growth factor is synthesized in cultured retinal pigment epithelial cells. *Biochem Biophys Res Commun*. 1987;143:934–940.
27. Reidy CA, Frank RN, Kennedy A, Das A. Aqueous extracts of bovine retinas contain a glial modulating factor. *Invest Ophthalmol Vis Sci*. 1988;29(Suppl):243.
28. Lim R, Mitsunobu K. Brain cells in culture: morphological transformation by a protein. *Science*. 1974;185:63–66.
29. Jaye M, Howk R, Burgess W, Ricca GA, Chiu I-M, Ravera MW, O'Brien SJ, Modi WS, Maciag T, Drohan WN. Human endothelial cell growth factor: cloning, nucleotide sequence, and chromosome localization. *Science*. 1986;233:541–545.
30. Abraham JA, Mergia A, Whang JL, Tumolo A, Friedman J, Hjerrild KA, Gospodarowicz D, Fiddes JC. Nucleotide sequence of a bovine clone encoding the angiogenic protein, basic fibroblast growth factor. *Science*. 1986;233:545–548.
31. Vlodavsky I, Fridman R, Sullivan R, Sasse J, Klagsbrun M. Aortic endothelial cells synthesize basic fibroblast growth factor which remains cell associated and platelet-derived growth factor-like protein which is secreted. *J Cell Physiol*. 1987;13:402–408.
32. Folkman J, Klagsbrun M. Angiogenic factors. *Science*. 1987;235:442–447.
33. D'Amore PA. Antiangiogenesis as a strategy for antimetastasis. *Semin Thromb Hemost*. 1988;14:73–78.
34. Glaser BM. Extracellular modulating factors and the control of intraocular neovascularization. *Arch Ophthalmol*. 1988;106:603–607.

35. Burgess WH, Maciag T. The heparin-binding (fibroblast) growth factor family of proteins. *Annu Rev Biochem.* 1989;58:575–606.
36. Folkman J, Klagsbrun M, Sasse J, Wadzinski M, Ingber D, Vlodavsky I. A heparin-binding angiogenic protein—basic fibroblast growth factor—is stored within basement membrane. *Am J Pathol.* 1988;130:393–400.
37. Vigny M, Ollier-Hartmann MP, Lavigne M, Fayein N, Jeanny JC, Laurent M, Courtois Y. Specific binding of basic fibroblast growth factor to basement membrane-like structures and to purified heparan sulfate proteoglycan of the EHS tumor. *J Cell Physiol.* 1988;137:321–328.
38. Gonzalez A-M, Buscaglia M, Ong M, Baird A. Distribution of basic fibroblast growth factor in the 18-day rat fetus: localization in the basement membranes of diverse tissues. *J Cell Biol.* 1990;110:753–765.
39. Kubota Y, Kleinman HK, Martin GR, Lawley TJ. Role of laminin and basement membrane in the morphological differentiation of human endothelial cells into capillary-like structures. *J Cell Biol.* 1988;107:1589–1598.
40. Kennedy A, Frank RN, Sotolongo LB, Das A, Zhang NL. Proliferative response and macromolecular synthesis by ocular cells cultured on extracellular matrix materials. *Curr Eye Res.* 1990;9:307–322.
41. Rohrbach DH, Wagner CW, Star VL, Martin GR, Brown KS, Yoon JW. Reduced synthesis of basement membrane heparan sulfate proteoglycan in streptozotocin-induced diabetic mice. *J Biol Chem.* 1983;258:11672–11677.
42. Das A, Frank RN, Zhang NL, Samadani E. Increases in collagen type IV and laminin in galactose-induced retinal capillary basement membrane thickening—prevention by an aldose reductase inhibitor. *Exp Eye Res.* 1990;50:269–280.
43. Cogan DG, Toussaint D, Kuwabara T. Retinal vascular patterns. IV. Diabetic retinopathy. *Arch Ophthalmol.* 1961;66:366–378.
44. Orlidge A, D'Amore PA. Inhibition of capillary endothelial cell growth by pericytes and smooth muscle cells. *J Cell Biol.* 1987;105:1455–1462.
45. Antonelli-Orlidge A, Saunders KB, Smith SR, D'Amore PA. An activated form of transforming growth factor beta is produced by cocultures of endothelial cells and pericytes. *Proc Natl Acad Sci USA.* 1989;86:4544–4548.
46. Carlson E. Fenestrated subendothelial basement membranes in human retinal capillaries. *Invest Ophthalmol Vis Sci.* 1989;30:1923–1932.
47. Robison WG Jr, Nagata M, Tillis TN, Laver N, Kinoshita JH. Aldose reductase and pericyte-endothelial contacts in retina and optic nerve. *Invest Ophthalmol Vis Sci.* 1989;30:2293–2299.
48. Frank RN, Dutta S, Mancini MA. Pericyte coverage is greater in the retinal than in the cerebral capillaries of the rat. *Invest Ophthalmol Vis Sci.* 1987;28:1086–1091.
49. Frank RN, Turczyn TJ, Das A. Pericyte coverage of retinal and cerebral capillaries. *Invest Ophthalmol Vis Sci.* 1990;31:999–1007.
50. DeOliveira F. Pericytes in diabetic retinopathy. *Br J Ophthalmol.* 1966;50:134–143.
51. Kador PF, Akagi Y, Takahashi Y, et al. Prevention of retinal vessel changes associated with diabetic retinopathy in galactose-fed dogs by aldose reductase inhibitors. *Arch Ophthalmol.* 1990;108:1301–1309.

52. Hanneken A, Lutty GA, McLeod DS, Robey F, Harvey AK, Hjelmeland LM. Localization of basic fibroblast growth factor to the developing capillaries of the bovine retina. *J Cell Physiol.* 1989;138:115–120.
53. Wagner JA, D'Amore PA. Neurite outgrowth induced by an endothelial cell mitogen isolated from retina. *J Cell Biol.* 1986;103:1363–1367.
54. King GL, Buzney SM, Kahn CR, Hetu N, Buchwald S, Macdonald SG, Rand LI. Differential responsiveness to insulin of endothelial and support cells from micro- and macrovessels. *J Clin Invest.* 1983;71:974–979.
55. King GL, Goodman AD, Buzney S, Moses A, Kahn CR. Receptors and growthpromoting effects of insulin and insulin-like growth factors on cells from bovine retinal capillaries and aorta. *J Clin Invest.* 1985;75:1028–1036.
56. Merimee TJ, Zapf J, Froesch ER. Insulin-like growth factors: studies in diabetics with and without retinopathy. *N Engl J Med.* 1983;309:527–530.
57. Merimee TJ. Diabetic retinopathy. A synthesis of perspectives. *N Engl J Med.* 1990;322:978–983.
58. Frank RN, Hoffman WH, Podgor MJ, Joondeph HC, Lewis RA, Margherio RR, Nachazel DP Jr, Weiss H, Christopherson KW, Cronin MA. Retinopathy in juvenile-onset diabetes of short duration. *Ophthalmology.* 1980;87:1–9.
59. Palmberg P, Smith M, Waltman S, Krupin T, Singer P, Burgess D, Wendtlant T, Achtenberg J, Cryer P, Santiago J, White N, Kilo C, Daughaday W. The natural history of retinopathy in insulin-dependent juvenile-onset diabetes. *Ophthalmology.* 1981;88:613–618.
60. Klein R, Klein BEK, Moss SE, Davis MD, DeMets DL. Retinopathy in young-onset diabetic patients. *Diabetes Care.* 1985;8:311–315.
61. Murphy RP, Nanda M, Plotnick L, Enger C, Vitale S, Patz A. The relationship of puberty to diabetic retinopathy. *Arch Ophthalmol.* 1990;108:215–218.
62. Kostraba JN, Dorman JS, Orchard TJ, Becker DJ, Ohki Y, Ellis D, Doft BH, Lobes LA, LaPorte RE, Drash AL. Contribution of diabetes duration before puberty to development of microvascular complications in IDDM subjects. *Diabetes Care.* 1989;12:686–693.
63. Lundbaek K, Malmros R, Andersen HC, Rasmussen JH, Bruntse E, Madsen PH, Jensen VA. Hypophysectomy for diabetic retinopathy: a controlled clinical trial. In: Goldberg MF, Fine SL, eds. *Symposium on the Treatment of Diabetic Retinopathy*, Public Health Service Publication No. 1890. Washington, DC: U.S. Government Printing Office; 1969:291–311.
64. Kohner EM, Hamilton AM, Joplin GF, Fraser TR. Florid diabetic retinopathy and its response to treatment by photocoagulation or pituitary ablation. *Diabetes.* 1976;25:104–110.
65. Patz A. Studies on retinal neovascularization. The Friedenwald Lecture. *Invest Ophthalmol Vis Sci.* 1980;19:1133–1138.
66. Henkind P. Ocular neovascularization. *Am J Ophthalmol.* 1978;85:287–301.
67. Warburg OH. *The metabolism of tumors*, transl. by F. Dickens. New York: Richard R. Smith; 1931:237–238, 322–324.
68. Imre G. Studies on the mechanism of retinal neovascularization: role of lactic acid. *Br J Ophthalmol.* 1964;48:75–82.
69. Ausprunk DH, Folkman J. Migration and proliferation of endothelial cells in preformed and newly formed blood vessels during tumor angiogenesis. *Microvasc Res.* 1977;14:53–65.

70. Kalebic T, Garbisa S, Glaser B, Liotta LA. Basement membrane collagen: degradation by migrating endothelial cells. *Science.* 1983;221:281–283.
71. Gross JL, Moscatelli D, Jaffe EA, Rifkin DB. Plasminogen activator and collagenase production by cultured capillary endothelial cells. *J Cell Biol.* 1982;95:974–981.
72. Liotta LA, Goldfarb RH, Brundage R, Siegal GP, Terranova V, Garbisa S. Effect of plasminogen activator (urokinase), plasmin, and thrombin on glycoprotein and collagenous components of basement membrane *Cancer Res.* 1981;41:4629–4636.
73. Jerdan JA, Kristensen P, Maglione A, Glaser BM. New blood vessel formation is associated with urokinase-type plasminogen activator. *Invest Ophthalmol Vis Sci.* 1988;29(Suppl):109.
74. Pandolfi M. Localization of fibrinolytic activity in the developing rat eye. *Arch Ophthalmol.* 1967;78:512–517.
75. Weiter JJ, Zuckerman R. The influence of the photoreceptor-RPE complex on the inner retina: an explanation for the beneficial effects of photocoagulation. *Ophthalmology.* 1980;87:1133–1139.
76. Glaser BM, Campochiaro PA, Davis J, Sato M. Retinal pigment epithelial cells release an inhibitor of neovascularization. *Arch Ophthalmol.* 1985;103:1870–1875.
77. Wong HC, Boulton M, McLeod D, Bayly M, Clark P, Marshall J. Retinal pigment epithelial cells in culture produce retinal vascular mitogens. *Arch Ophthalmol.* 1988;106:1439–1443.
78. Frank RN, Mancini MA. Presumed retinovitreal neovascularization in dystrophic retinas of spontaneously hypertensive rats. *Invest Ophthalmol Vis Sci.* 1986;27:346–355.
79. Frank RN, Das A, Weber ML. A model of subretinal neovascularization in the pigmented rat. *Curr Eye Res.* 1989;8:239–247.
80. Henkind P, Gartner S. The relationship between retinal pigment epithelium and the choriocapillaris. *Trans Ophthalmol Soc.* 1983;103:444–447.
81. Ashton N. Oxygen and the growth and development of retinal vessels. In: Kimura SJ, Caygill WM, eds. *Vascular complications of diabetes mellitus.* St. Louis: CV Mosby; 1967:3–32.
82. Patz A. The role of oxygen in retrolental fibroplasia. *Trans Am Ophthalmol Soc.* 1968;66:940–985.
83. Flower RW, Blake DA. Retrolental fibroplasia: evidence for a role of the prostaglandin cascade in the pathogenesis of oxygen-induced retinopathy in the newborn beagle. *Pediatr Res.* 1981;15:1293–1302.
84. Shabo AL, Maxwell DS. Experimental immunogenic proliferative retinopathy in monkeys. *Am J Ophthalmol.* 1977;83:471–480.
85. Weber ML, Mancini MA, Frank RN. Retinovitreal neovascularization in the Royal College of Surgeons rat. *Curr Eye Res.* 1989;8:61–74.
86. Frank RN, Zhang NL, Das A, Miller T. Light exposure stimulates retinovitreal neovascularization in dystrophic RCS rats. *Invest Ophthalmol Vis Sci.* 1990;31(Suppl):195.
87. Bellhorn RW, Bellhorn M, Friedman AH, Henkind P. Urethane-induced retinopathy in pigmented rats. *Invest Ophthalmol Vis Sci.* 1973;12:65–75.
88. Bellhorn RW, Burns MS, Benjamin JV. Retinal vessel abnormalities of phototoxic retinopathy in rats. *Invest Ophthalmol Vis Sci.* 1980;19:584–595.

89. Shiraki K, Burns MS. Neovascularization in urethane rat retinopathy demonstrated by thymidine labelling. *Curr Eye Res*. 1986;5:683–695.
90. Ryan SJ. The development of an experimental model of subretinal neovascularization in disciform macular degeneration. *Trans Am Ophthalmol Soc*. 1979;77:707–745.
91. Archer DB, Gardiner TA. Experimental subretinal neovascularization. *Trans Ophthalmol Soc*. 1980;100:363–368.
92. Dobi ET, Puliafito CA, Destro M. A new model of experimental choroidal neovascularization in the rat. *Arch Ophthalmol*. 1989;107:264–269.
93. Pollack A, Heriot WJ, Henkind P. Cellular processes causing defects in Bruch's membrane following krypton laser photocoagulation. *Ophthalmology*. 1986;93:1113–1119.
94. Baudouin C, Fredj-Reygrobellet D, Caruelle J-P, Barritault D, Gastaud P, Lapalus P. Acidic fibroblast growth factor distribution in normal human eye and possible implications in ocular pathogenesis. *Ophthalmol Res*. 1990;22:73–81.
95. Fayein NA, Courtois Y, Jeanny JC. Ontogeny of basic fibroblast growth factor binding sites in mouse ocular tissues. *Exp Cell Res*. 1990;188:75–88.
96. Hageman GS, Kirchoff-Rempe MA, Lewis GP, Fisher SK, Anderson DH. Sequestration of basic fibroblast growth factor in the primate retinal interphotoreceptor matrix. *Proc Natl Acad Sci USA*. 1991;88:6706–6710.
97. Faktorovich EG, Steinberg RH, Yasumura D, Yamaai T, Nohno T, Matsuo N, Taniguchi S. Photoreceptor degeneration in inherited retinal dystrophy delayed by basic fibroblast growth factor. *Nature*. 1990;347:83–86.
98. Noji S, Matsuo T, Koyama E, Yamaai T, Nohno T, Matsuo N, Taniguchi S. Expression pattern of acidic and basic fibroblast growth factor genes in adult rat eyes. *Biochem Biophys Res Commun*. 1990;168:343–349.
99. Sivalingam A, Kenney J, Brown GC, Donoso L. Basic fibroblast growth factor levels in the vitreous of patients with proliferative diabetic retinopathy. *Arch Ophthalmol*. 1990;108:869–872.
100. Hanneken A, De Juan E Jr, Lutty GA, Fox GM, Schiffer S, Hjelmeland LM. Altered distribution of basic fibroblast growth factor in diabetic retinopathy. *Arch Ophthalmol*. 1991;109:1005–1011.
101. Lutty G, Ikeda K, Chandler C, McLeod DS. Immunohistochemical localization of transforming growth factor beta in human photoreceptors. *Curr Eye Res*. 1991;10:61–74.

3
The Humoral Regulation of Normal and Pathologic Erythropoiesis

ALBERTO GROSSI, ALESSANDRO M. VANNUCCHI,
DANIELA RAFANELLI, and PIERLUIGI ROSSI FERRINI

About 40 years ago it was recognized that a humoral factor, later identified as erythropoietin (Epo), was able to regulate the response of bone marrow to changes in red cell mass.[1,2] It was also found that Epo was produced by the kidney in response to renal tissue hypoxia,[3,4] and a quantitative assay based on the measurement of newly produced red cells labeled with ^{59}Fe was developed.[2,5] The introduction of assays that allow cells to proliferate and differentiate in culture has led to the conclusion that Epo primarily acts on cells morphologically unrecognizable as erytroblasts. Semisolid cultures using methylcellulose, plasma-clot, and agar[6-9] showed that two classes of erythroid progenitors can be observed at different times of incubation.[7,8,10] Human bone marrow mononuclear cells in culture give rise to erythroid colonies after 4 to 5 days of incubation,[11] while a longer (8–9 days) incubation time determines the appearance of larger colonies,[7] and even larger and more hemoglobinized ones appear after 14 days. The first type is currently defined as colony-forming unit–erythroid (CFU-E)[8]; 14 and 8- to 9-day colonies are referred to as burst-forming unit–erythroid (BFU-E)[12] and intermediate BFU-E,[7] respectively (Fig. 3.1). Murine counterparts of these cells grow after 2 (CFU-E), 3 to 4 (intermediate BFU-E), and 8 (BFU-E) days.[13] In addition, a murine progenitor with higher proliferative and differentiative capacity is seen after 9 to 12 days of incubation.[7] It is interesting to note that BFU-E and CFU-E require different levels of regulators and also different regulators to proliferate and differentiate in vitro. In fact, relatively small amounts of Epo are sufficient to induce the proliferation of CFU-E, and its dependence on Epo in vivo is demonstrated by the fact that plethora reduces the CFU-E compartment and expands bone marrow BFU-E.[14,15] On the other hand, in order to proliferate, BFU-E requires higher levels of Epo,[16] as well as the presence of another factor(s), initially termed burst-promoting activity (BPA).[17,18] Now it seems evident that interleukin-3 (IL-3), granulocyte macrophage–colony stimulating factor (GM-CSF), and interleukin-4 (IL-4) all have BPA-like activity[19-21] (Table 3.1). The different time of appearance in culture, different sen-

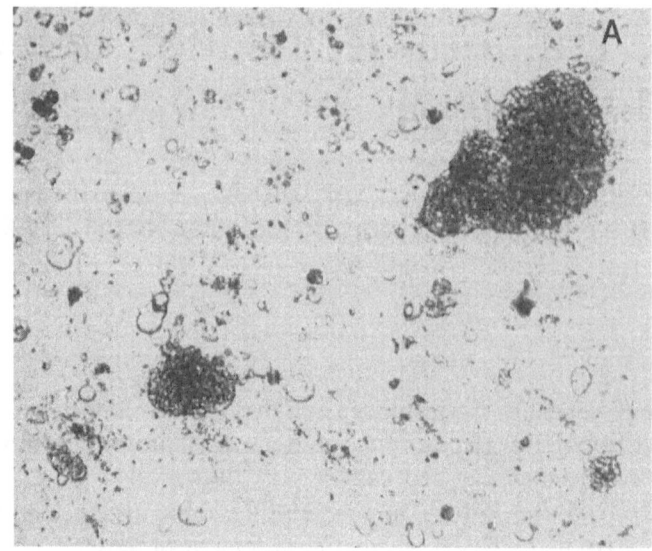

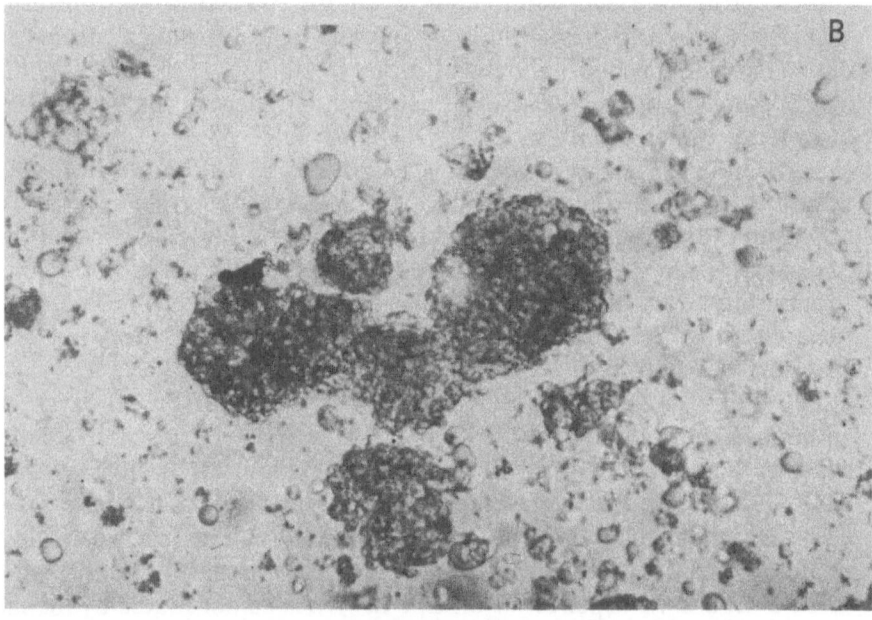

FIGURE 3.1. Morphologic appearance of erythroid colonies from human bone marrow mononuclear cells grown in methylcellulose. A: Two CFU-E–derived colonies (day 7). B: BFU-E (day 14).

TABLE 3.1. Characteristics of the hemopoietic growth factors active in erythropoiesis.

Growth factor	Molecular weight	Chromosomal location of gene	Cellular sources	Target erythroid progenitor cells
IL-3	14–28	5q23–31	T lymph	CFU-GEMM, BFU-E
GM-CSF	14–35	5q23–31	T lymph, Mo, fibr, endt	CFU-GEMM, BFU-E
Epo	34–39	7q11–22	Peritubular kidney cells, Kupffer cells, macrophages	CFU-E, late BFU-E
IL-4	20	5	T lymph Mast cells	CFU-GEMM, BFU-E

Abbreviations: T lymph, T lymphocytes; CFU-GEMM, colony forming unit—granulocyte, erythrocyte, macrophage, megakaryocyte; Mo, monocytes; fibr, fibroblasts; endt, endothelial cells.

sitivity to Epo, and progressively decreasing size of the colonies indicate that BFU-E to CFU-E represents two stages of the process that leads to red cell production. A schematic representation of erythropoiesis is presented in Figure 3.2.

Erythropoietin

The observation in 1863, by Denis Jourdanet, that the blood of individuals living in the highlands of Mexico contained an increased number of red cells[22] is generally considered to represent the first indication of a feedback mechanism regulating erythropoiesis. However, it was only in 1953 that Erslev demonstrated that erythropoiesis was regulated by a humoral factor, named erythropoietin, contained in the plasma of anemic rabbits[2]; this factor was later shown to be produced in the kidney by Jacobson et al.,[3] who thus provided a rationale for the association of anemia and end-stage renal diseases, which has been known from the beginning of this century.

Epo exists in such small quantities in normal blood that attempts to purify it from the plasma of experimentally induced anemic animals[23] were unsuccessful. However, the hormone was found in relatively high amounts in the urine of aplastic patients, and it was from this source that Epo was first purified to homogeneity by Miyake and coworkers in 1977.[24] Purified human urinary Epo is a glycoprotein with a molecular weight of about 34,000 daltons, of which two forms (α and β) can be separated in the last stages of purification. These two forms have different behaviors on hydroxyapatite chromatography and on gel electrophoresis, and differ in carbohydrate composition,[25] but have identical biologic activity and amino acid composition. Indeed, it is possible that the β form may derive from the breakdown of the intact Epo molecule during

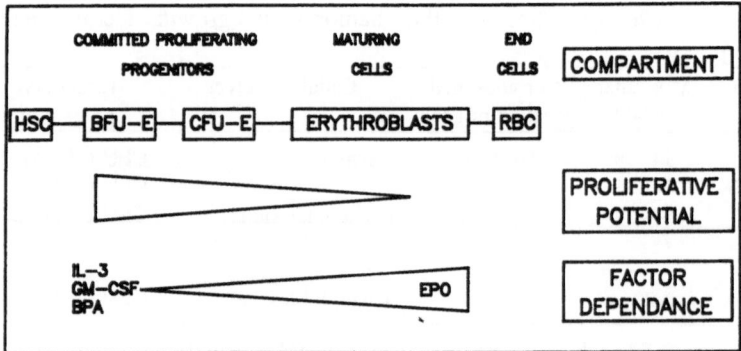

FIGURE 3.2. Cellular compartments and regulators in erythropoiesis. HSC, hematopoietic stem cells; BFU-E, burst-forming unit—erythroid; CFU-E, colony-forming unit—erythroid; RBC, red blood cells; IL-3, interleukin-3; GM-CSF, granulocyte macrophage—colony stimulating factor; BPA, burst-promoting activity (?others than IL-3 or GM-CSF; i.e., IL-4); Epo, erythropoietin.

purification. However, the amount of Epo recovered with this procedure was extremely low, sufficient only for in vitro experiments; moreover, it was available to only a few laboratories. The breakthrough was provided by the identification, cloning, and expression of the Epo gene. Starting from the data obtained with microsequentiation of the Epo derived from human urine, oligonucleotide probes were prepared that allowed the isolation of clones from a human genomic DNA library[26]; these clones were then used to screen for cDNA clones from a fetal liver cDNA library. The gene was spliced into a plasmid expression vector that was used to transfect the nuclei of a mammalian cell line, the Chinese hamster ovary (CHO). After transfection, Epo activity was detectable in the supernatant,[27] from which purified recombinant human erythropoietin (rhEpo) was prepared. The use of a mammalian expression system allows for the posttranslational processing of the protein, namely for the glycosylation of Epo,[28] which is essential for the in vivo activity of the hormone. Moreover, the cDNA for human Epo has also been expressed in *Escherichia coli*.[29]

The Epo gene has been located in the human chromosome 7q11–q22 (and in the mouse chromosome 5)[30] by analysis of somatic cell hybrids and by in situ hybridization.[31] The gene is organized in five exons and four intervening sequences, and exists as a single copy in the human genome. The Epo coding information is contained in 579 nucleotides, corresponding to a 193–amino acid polypeptide; the first 27 amino acids of this peptide form a hydrophobic leader sequence, which is cleaved upon secretion from the cell.[26,27] A further posttranslational processing consists in the removal of the C-terminal amino acid, giving rise to the mature functional protein of 165 amino acids, with a calculated molecular

TABLE 3.2. Amino acid sequence of the mature molecule of human Epo.

1

ALA	PRO	PRO	ARG	LEU	ILE	CYS■	ASP	SER	ARG	VAL	LEU▲
GLU	ARG	TYR	LEU	LEU■	GLU	ALA	LYS	GLU■	ALA	GLU	ASN
ILE	THR▲	THR	GLY	CYS	ALA	GLU	HIS	CYS	SER	LEU	ASN
GLU	ASN	ILE	THR	VAL	PRO	ASP	THR	LYS	VAL	ASN	PHE
TYR	ALA	TRP	LYS	ARG	MET	GLU	VAL	GLY	GLN	GLN	ALA
VAL	GLU	VAL	TRP	GLN	GLY	LEU	ALA	LEU	LEU	SER	GLU▲
ALA	VAL	LEU	ARG	GLY	GLN	ALA	LEU	LEU	VAL	ASN	SER
SER	GLN	PRO	TRP	GLU	PRO	LEU	GLN	LEU	HIS	VAL	ASP
LYS	ALA	VAL	SER	GLY	LEU	ARG	SER	LEU	THR	THR	LEU
LEU	ARG	ALA	LEU	GLY	ALA●	GLN	LYS	GLU	ALA	ILE	SER
PRO	PRO	ASP	ALA	ALA	SER	ALA	ALA	PRO	LEU	ARG	THR
ILE	THR	ALA	ASP	THR	PHE	ARG	LYS	LEU	PHE	ARG	VAL
TYR	SER	ASN	PHE	LEU	ARG	GLY	LYS	LEU	LYS	LEU	TYR
THR	GLY	GLU	ALA	CYS■	ARG	THR	GLY	ASP[165]			

▲, sites of N-linked glycosylation; ●, sites of O-linked glycosylation; ■, sites of internal disulfide bonds.
From Jacobs et al.[26], and Sasaki et al.[28], modified.

weight of 18,400 daltons[32] (Table 3.2). These are three sites of N-linked glycosylation at residues 24, 38, and 83, and a site of O-linked glycosylation is at the position 126[33]; the molecular weight of the fully glycosylated molecule is 36,000 daltons, as determined by sodium dodecyl sulfate polyacrylamide gel electrophoresis.[33] There are four cysteine residues, involved in two internal disulfide bonds (one at positions 7 and 161, the other at positions 29 and 33); these disulfide bonds are essential for the biologic activity, which is indeed lost upon reduction of the molecule. The main site of Epo production in the adult is the kidney, although extrarenal synthesis has been demonstrated in the liver,[34] which accounts for less than 15% of total Epo production[35]; on the other hand, the liver is the primary site of Epo production in the fetus.[36,37] Moreover, at least two human epatoma cell lines are able to produce Epo constitutively; the production of Epo by these cell lines is related to the proliferative activity of the cells in culture, and can be induced by exposure to hypoxia or to cobalt(II) chloride.[38,39] In addition, splenic and bone marrow macrophages have been implicated as producing Epo,[40] and more recently the expression of the Epo gene has been demonstrated in murine bone marrow macrophages by means of in situ hybridization techniques.[41]

The site of Epo production in the kidney has been located at the level of the endothelial cells of the peritubular capillaries in the cortex by in situ hybridization.[42,43] Studies performed in rats made acutely hypoxic by placing them in a hypobaric chamber have shown a significant accumulation of Epo mRNA in the kidney within 1 hr after the initiation of hypoxia, paralleling the increase in kidney and plasma Epo; mRNA and Epo levels quickly returned to baseline levels after the discontinuation of hypoxia.[44] There appears to be no baseline cellular accumulation of preformed Epo, and rodents pretreated with actinomycin D (an inhibitor of RNA synthesis) or puromycin (which inhibits protein synthesis) fail to show the expected increase in plasma Epo levels in response to hypoxia.[45] Finally, recent studies with in situ hybridization techniques have demonstrated that the rise of plasma and kidney Epo levels in mice made acutely anemic is due to an exponential increase in the number of cells expressing mRNA for Epo more than to an increase in the amount of Epo produced per single cell.[46] The mechanisms by which the de novo synthesis of Epo is stimulated in the intact animal in response to hypoxia or other stimuli are not known, but recent evidence in vitro[47] suggests that the oxygen sensor may be a heme protein, whose deoxy conformation, such as that prevailing in states of low oxygen tension, may induce the expression of the Epo gene.

Human urinary purified Epo, and the recombinant preparation derived from CHO cells, is a glycoprotein whose carboidrate component corresponds to about 40% of the molecular weight.[48] It is well known that the activity of Epo in vivo is completely dependent on the presence of sialic acid,[49] while the activity of asialoerythropoietin in vitro is indeed greater than that of the native molecule.[50] This apparent loss of activity of the asialoerythropoietin in vivo may be ascribed to the rapid clearance from the circulation by hepatic cells; in fact, the removal of the terminal moieties of sialic acid exposes galactose residues that can bind to specific galactose receptors on the hepatocytes.[51] This has been demonstrated with sialidase-treated preparations of rhEpo,[52] whose half-life in the circulation of rats decreased from 2 hr for the native molecule to less than 10 min; on the other hand, oxidation of the penultimate galactosyl residues of desialated rhEpo restores its half-life to normal.[53] The pattern of organ distribution of asialated rhEpo also differs significantly from that of the unmodified hormone, due to a greater accumulation in the kidney, bone marrow, and spleen.[53] These kinetic characteristics of rhEpo have important implications for the use of the hormone in therapy (see below): in a recent study in patients on continuous ambulatory peritoneal dialysis it has been calculated that the half-life of intravenously injected rhEpo is around 8.2 hr[54]; delayed peaks were obtained after subcutaneous administration (18 hr after injection), with biologically significant levels being maintained as long as 3 to 4 days. Therefore, it appears that the subcutaneous administration of fully glycosylated rhEpo may be the best form of treatment.

It is well known that Epo is required for the maturation of erythroid progenitors,[55] although the detailed mechanism(s) at the molecular level has not yet been fully elucidated. In 1971, Stephenson et al.[56] demonstrated the presence of CFU-E in livers of fetal mice. These are the most mature among erythroid colony-forming progenitors in culture, and are absolutely dependent on Epo for growth and maturation to erythroblasts. A more immature precursor is the BFU-E, which has a greater proliferative capacity than CFU-E and forms colonies containing several thousand cells (Figs. 3.1 and 3.2). Although a subset of BFU-E may respond to Epo alone, the growth of BFU-E requires BPA as an additional factor (see Fig. 3.2 and Table 3.1). Whereas CFU-Es are virtually confined to bone marrow, a substantial number of BFU-Es are found in the peripheral blood of normal subjects, and can be successfully grown in vitro using a BPA-like activity and Epo. On the other hand, in the myeloproliferative syndromes (including polycythemia vera, primary thrombocythemia, idiopathic myelofibrosis, and chronic granulocytic leukemia) the erythroid progenitors have an increased sensitivity to Epo and can grow in colonies in the absence of exogenous growth factors ("endogenous erythroid colonies").[57–60] The study of the in vitro growth pattern of erythroid progenitors obtained from the peripheral blood of patients suspected to have a myeloproliferative disorder may represent a useful diagnostic tool.[61]

The stimulation by Epo of responsive erythroid cells is mediated by specific cell surface receptors, which have been demonstrated on murine Friend virus-infected erythroid progenitors,[62] murine erythroleukemia cells,[63] on CFU-E derived from fetal mouse liver cells,[64] or from the spleen of adult anemic mice,[65] and more recently on human peripheral blood-derived erythroid cells grown in vitro.[66] The number of Epo receptors is high at the level of murine CFU-E and decreases in more mature erythroid cells,[65] which, however, still possess some Epo receptors.[65,67] It has been calculated that about 20% of human peripheral blood-derived BFU-E have some receptors for Epo[66] and that their number increases with the in vitro maturation and the loss of proliferative capacity of the cells. Both high- and low-affinity receptors may exist,[68] although only the high-affinity type has been identified in some studies[63,65] and is probably the physiologically important one.[69] The biochemical mechanisms of the Epo-derived signal transduction, which induces the erythroid maturation, are unknown, but receptor-mediated endocytosis may be an essential step.[70] Finally, specific high-affinity Epo receptors have been demonstrated on murine and rat megakaryocytes,[71] in accordance with in vivo[72,73] and in vitro[74] studies suggesting some effects of Epo on megakaryocyte maturation and on platelet production, although the physiologic importance of these effects remains questionable.[73] The murine Epo receptor has recently been expression cloned,[69] and a solid phase assay for its quantitation developed.[75] The cDNA sequence of the murine receptor encodes for a 507–amino acid polypeptide with a single

TABLE 3.3. Clinical uses of rhEPO.

With proven efficacy
Anemia of renal diseases
Anemia of chronic inflammatory disorders (e.g., rheumatoid arthritis)
Anemia of AIDS
Anemia of multiple myeloma
Autologous blood donation
With potential usefulness
Anemia of prematurity
Anemia of malignancy
Anemia of hemoglobinopathies (e.g., sickle cell disease)
Anemia of myelodysplastic syndromes
Anemia induced by radiochemotherapy
Anemia of patients undergoing bone marrow transplantation
Anemia of Gaucher's disease

membrane-spanning domain; the recombinant protein possesses both high- and low-affinity properties.[69]

After the stimulation with Epo, erythroid proerythroblasts isolated from the spleen of mice infected with the anemia-inducing strain of Friend virus undergo terminal erythroid differentiation which appears very close to the normal counterpart[76]; therefore, this is the most extensively studied model of erythroid differentiation in vitro. A number of biologic effects of Epo on target cells have been demonstrated: these include the induction and/or modulation of the synthesis of some erythroid-specific proteins,[77] the stimulation of the rate of RNA synthesis,[78] the stimulation of the synthesis of transferrin receptors,[79] alterations in Ca^{2+} metabolism, with an increase in both the rate of calcium efflux and the steady state intracellular calcium levels,[80] the stimulation of globin gene transcription,[81] and finally, the progression through morphologic differentiation stages that strictly corresponds to the ones observed in vivo in the spleen of mice.[82]

The availability of recombinant human Epo has permitted its use in some clinical conditions characterized by the presence of anemia (Table 3.3) most extensively in patients with anemia due to chronic renal failure. It has been calculated that only 3% of almost 100,000 patients on dialysis in the United States have normal hematocrit, most showing hematocrit levels between 20% and 30%.[83] The direct consequence of anemia (i.e., fatigue) is the main complaint of these patients, and represents one of the reasons for the failure of rehabilitation.[83] These patients usually require many red blood cell transfusions, which expose them to the risk of hepatitis and human immunodeficiency virus (HIV) transmission. Although more than one mechanism may be operative in the pathogenesis of the anemia of end-stage renal disease, relative Epo deficiency is the main one.[83] In fact, it has been demonstrated that patients with chronic renal failure have subnormal to normal levels of serum Epo, but these levels are 10 to 100 times lower than expected in individuals with similar degrees of anemia, but without renal disease.[84]

After phase I and II trials,[85,86] a number of clinical trials have clearly demonstrated the effectiveness of rhEpo in ameliorating anemia-related symptoms in patients with chronic renal failure.[87-89] The usual effective dose of rEpo is greater than 50 UI/kg IV 3 times a week after dialysis; the increase in hematocrit seems to be dose-dependent. The maintenance dose required to keep the hematocrit stable at 33% to 38% is around 25 to 100 U/kg IV 3 times a week, although at least 30% of patients may require greater doses. The stimulation of erythropoiesis may induce iron depletion requiring iron supplementation even in heavily (posttransfusion) iron-overloaded patients. The increase in the hematocrit is quickly followed by amelioration of many symptoms related to organ hypoxia, such as the correction of cardiomegaly, the improvement in central nervous symptoms such as anorexia, insomnia, depression, sexual dysfunctions, and the expected amelioration of fatigue, anginous symptoms, and shortness of breath. Moreover, the administration of rhEpo may correct the hemostatic defects of uremic patients.[90] Overall, it has been determined that, after 6 months of therapy with rhEpo, the quality of life of these patients greatly improves, and the benefits appear comparable with those obtained by patients successfully subjected to renal transplantation.[91]

Side effects are relatively uncommon and mild, and consist of flulike or other transient symptoms shortly after the injection; occasional thrombosis of arteriovenous shunts have been reported. However, most patients show an increase in diastolic blood pressure, most likely due to the increase in peripheral vascular resistance, and especially evident in the initial phase of treatment. The increase in blood pressure can be prevented, or at least reduced, by the judicious scheduling of the doses of rhEpo in order to achieve a gradual, as opposed to a rapid, increase in hematocrit. The development of antibodies to rhEpo has not been detected in more than 500 patients collected in a study.[92]

Although the anemic patient with relative Epo deficiency due to end-stage renal failure is the most obvious candidate for the treatment with rhEpo, it appears that rhEpo may be useful even in situations with normal to high serum Epo levels. These include the anemia accompanying chronic inflammatory conditions, such as rheumatoid arthritis,[93] and the zidovudine-induced anemia in patients with the acquired immunodeficiency syndrome (AIDS), especially when the pretreatment Epo levels are <500 mU/ml.[94] Recent evidence also suggests that rhEpo may ameliorate the anemia of patients with malignancy,[95] in whom serum Epo levels are inappropriately low for the degree of anemia,[96] and of patients with multiple myeloma.[97] A very interesting, although hitherto only experimental, indication of rhEpo may be sickle cell disorders, as an increase in the fetal hemoglobin levels has been demonstrated in baboons treated with the recombinant hormone.[98,99] A beneficial effect of rhEpo has also been demonstrated in an anemic patient with Gaucher's disease.[100] Finally, rhEpo has been used to increase the preoperative collection of autologous blood in patients undergoing elective surgery.[101]

Interleukin-3

This multipotent hemopoietin is produced by activated T lymphocytes, and in vitro it stimulates a broad range of hemopoietic precursors.[19] Both the human and murine form of interleukin-3 (IL-3), which show a strict species specificity, have been molecularly cloned.[102,103] IL-3 probably accounts for the most part of burst-promoting activity found in many conditioned media (c.m.), such as the murine pokeweed mitogen-stimulated spleen c.m., the c.m. of the murine WEHI-3 leukemia cell line, and the c.m. obtained by human leukocytes stimulated with phyto-hemagglutin or concanavalin A. In vitro IL-3 stimulates the proliferation of the most immature erythroid precursors (BFU-E), whereas Epo is required to obtain a complete erythroid maturation. Clinical trials with human rIL-3 are in progress, and no definitive data have been published about its effects on the erythroid compartment of patients treated with this multipotent hemopoietin.

Granulocyte Macrophage–Colony Stimulating Factor

The in vitro biologic activities of granulocyte macrophage–colony stimulating factor (GM-CSF) are broader than implied by its name, in that it can stimulate the proliferation of erythroid and megakaryocyte progenitors, in addition to the granulocyte–macrophage precursors.[20] The target cell in the erythroid pathway is probably the same as for IL-3, and Epo is required for the terminal maturation of the erythroid committed cells.[104,105] Animal studies in vivo have shown that rGM-CSF can also affect the erythroid and megakaryocytic precursors, although Epo and a thrombopoiesis-stimulating factor are necessary to obtain an increase of peripheral reticulocytes and platelets, respectively.[106,107] The GM-CSF is a strictly species-specific factor, and both the human[108,109] and murine[110] form have been molecularly cloned. The recombinant preparation of human GM-CSF has been introduced in the clinical setting for the treatment of patients with AIDS, myelodysplastic syndromes, or subjected to intensive chemotherapy for the treatment of solid neoplasia or for bone marrow transplantation.[111] In some of these patients, an effect on erythroid progenitors in addition to myeloid precursors was observed, thus suggesting that even in vivo GM-CSF may be active on the erythroid compartment.

Erythroblast-Enhancing Factor

Krystal[112,113] has described a factor enhancing the late stages of erythroblast maturation. The factor has a molecular weight of 100,000 daltons and has been called erythroblast-enhancing factor (EEF). It is interesting that EEF behaves similarly to Epo in response to anemia and erythrocytosis, indicating a true role in erythropoiesis. Similar active substances with approximately the same molecular weight, and whose concentration in the serum varies accordingly with hemoglobin levels have also been reported by others.[114-116] In addition, Mouchiroud and Blanchet[117] found that normal mouse serum contains another erythroid-enhancing activity with a molecular weight of 50,000 daltons, which is active in the presence of the c.m. of Mac-1 positive cells (granulocytes, monocytes–macrophages, and lymphoid cells). However, its serum levels are regulated differently, because anemia does not change the concentration of the factor. An erythroid-potentiating activity has been purified also from the c.m. of a human T-lymphoblast cell line (Mo), and the complementary cDNA has been obtained.[118] Recently, activin, a promoter of follicle-stimulating hormone secretion, has been shown to potentiate BFU-E and CFU-E colony formation. Its activity seems mediated through monocytes and T lymphocytes.[119]

Platelet-Derived Growth Factor

It has been shown that human bone marrow cells cultured in the presence of platelet-poor plasma-derived serum (PDS) give rise to fewer erythroid colonies than cells cultured in the presence of whole blood serum or fetal calf serum (FCS).[120] This indicates that a factor released from platelets may be involved in the biologic effect. Supernatant of platelets treated with thrombin applied to a Sepharose 4B column yields a peak with stimulatory activity coincident with that of PDGF, and capable of enhancing bone marrow CFU-E and BFU-E-derived, but not peripheral blood BFU-E-derived colonies. The same biologic activity is shown by purified platelet-derived-growth factor (PDGF) in the presence of Epo. However, PDGF cannot replace Epo, because it does not induce colony formation when used alone.[120] These results have been confirmed and extended by Delwiche et al.,[121] who observed that serum enhanced the growth of bone marrow CFU-E and BFU-E, as compared with plasma, and that colony growth of whole bone marrow cells in the presence of PDS and Epo was enhanced is a dose-dependent manner by increasing concentrations of PDGF. On the other hand, nonadherent human cells or peripheral blood BFU-E did not show any growth, suggesting that they are not the target for PDGF, and that an accessory cell population is required for the effect of this factor on hematopoietic progenitors. In fact, two-layer cultures containing fibroblasts or smooth muscle cells in the overlayer restored the ability of nonadherent bone marrow cells and of peripheral BFU-E to proliferate. This indicates that a diffusible factor more than a direct cell-to-cell interaction is responsible for this co-operation. Furthermore, a monospecific antihuman PDGF only partially reversed the effect of platelet-rich plasma-derived serum on erythroid progenitors, indicating that other products released from activated platelets, such as catecholamines and prostaglandins,[122,123] could play a role. However, we do not know if PDGF or PDGF-related molecules (deriving from endothelial and smooth muscle cells) or macrophages modulate erythropoiesis within the microenvironment.[121]

Prostaglandins

It has been shown that some prostaglandins (PG) stimulate erythropoiesis in vitro and in vivo,[124-126] and that the PGA_2 was the most effective, whereas $PGF_{2\alpha}$ and $PGF_{2\beta}$ showed no effect.[126] This is not a generalized effect on hematopoiesis, because granulopoiesis is inhibited in vitro.[127] Prostaglandin stimulation in vivo is dependent on the Epo levels, because it is not operative in the presence of large amounts of Epo,[126] and antierythropoietin immune serum inhibits the response of plethoric mice to prostaglandins.[125] The relationship between prostaglandin E class and Epo levels is also suggested by the in vivo models of renal artery constriction (RAC) and hypoxia in dogs. In both these models the released prostaglandins stimulate Epo production probably via an increase of cyclic adenosine monophosphate.[124-129] On this ground it would be reasonable to hypothesize that inhibitors of prostaglandin synthesis would block Epo response. In fact, plasma from dogs with RAC and from hypoxic dogs pretreated with the nonsteroidal antiinflammatory drug indomethacin reduces greatly the ^{59}Fe incorporation in polycythemic mice, as compared with plasma from non-pretreated animals.[128] Renovascular effects of prostaglandins of E class have also been considered, since a redistribution of blood flow with decreased O_2 tension brings about a stimulation of Epo production.[125] A and E classes of prostaglandins seem also capable of regulating Epo synthesis in extrarenal sites,[130] and finally an increased responsiveness of Epo target cells has been claimed as a possible mechanism of action of prostaglandins.[126]

Other Hormones

Clinical and experimental data support a role of steroid hormones on erythropoiesis. Addison's disease in man is associated with normochromic normocytic anemia, and glucocorticoids have been used in the treatment of congenital erythroid hypoplasia of children (Diamond-Blackfan syndrome)[131] and of aplastic anemia.[132] Besides the immunosuppressive activity, glucocorticoids and adrenocorticotropic hormone may act through a stimulation of erythroid progenitors.[133,134] In vitro experiments have shown that dexamethasone alone stimulates the growth of erythroid colonies from mouse fetal liver cells.[135] Since this response can be inhibited by an anti-Epo antibody, dexamethasone is likely to act via Epo endogenously produced by the liver cells, and not by a direct stimulatory effect. Adult mouse and human bone marrow cells also gave rise to erythroid colonies in the presence of dexamethasone with no added Epo,[135] but the effect of minimal amounts of the latter in FCS used for culture could not be excluded. Moreover, dexamethasone showed a potentiating activity when used together with Epo.

Androgenic steroids stimulate erythropoiesis,[136] and have long been used in the treatment of aplastic anemia with some success[132,137]; their use is based on experimental evidence that they enhance erythropoiesis in vivo in normal,[138] but not in anephric, animals,[139] and in vitro.[140] However, the various compounds do not seem to act by the same mechanism; in fact, some studies suggest that 5β configuration is linked to a stimulation of bone marrow progenitors,[141] whereas 5α is linked to a stimulation of Epo production.[142] Estrogens determine a moderate suppression of erytropoiesis,[143] but it is unclear whether this is due to an inhibition of Epo production[144] or of Epo activity on target cells.[145]

Hypothyroidism is often accompanied by a mild anemia and replacement therapy and experimental works indicate a role for thyroid hormones in the regulation of erythropoiesis. It seems likely that thyroxine (T_4) and 3,5,3'-triiodothyronine (T_3) potentiate the effect of Epo on erythroid progenitors,[146] the effect probably being mediated by receptors with adrenergic properties.[147] Moreover, thyroid hormone administration creates an oxygen deficit in the kidneys, which in turn stimulates Epo production.[148] Nevertheless, Malgor et al. have shown an increase in total nucleated erythroid bone marrow cells in nephrectomized rats whether treated with rabbit antiserum against Epo or not[149]; this suggests a direct effect of thyroid hormones on erythropoiesis.

The effect of growth hormone (GH) on erythropoiesis has been ascribed to a stimulation of Epo production,[150,151] and in one instance, human and bovine GH and the plasmin-cleaved hGH fragment Cys(Cam)$_{53}$-hGH (1–134)[152] were also capable of potentiating Epo activity on erythroid progenitors. However, an autonomous role of GH and of the cleaved peptide may also be envisaged, because both were able to stimulate the proliferation of Friend virus-infected erythroleukemia cells in a serum-free clonogenic assay.[153]

Anemia is sometimes present in diabetes, and Ritchey et al.[154] have shown that BFU-E proliferation is improved when metabolic control is achieved. In vitro studies indicate that purified and biosynthetic insulins stimulate BFU-E to proliferate,[155] and that the effect is independent of Epo in serum-free cultures.[156] It may be assumed that low affinity receptors for insulin, which also bind insulin growth factors (IGF) I and II, are present on red cell progenitors. An enhancement of CFU-E proliferation by IGF I has also been reported.[156]

Inhibitors

Macrophage-derived tumor necrosis factor α (TNFα) is a cytotoxic cytokine that has been shown to regulate erythropoiesis negatively. In fact, TNFα inhibits the in vitro growth of BFU-E and CFU-E,[157] although it is effective only on CFU-E[158] in vivo. In a clinical trial its use led to a significant decrease in hemoglobin levels.[159] In this respect, experiments in mice have not resolved the question of whether the simultaneous treatment with recombinant human Epo abrogates the suppressive effect of TNFα or not.[160,161] Other inhibitors of erythropoiesis include inhibin, which consists of the same A subunit of activin plus an α polypeptide, and interferon α.[162] Finally, human IL-1 has been shown to antagonize the stimulatory effects of Epo on erythroid precursors in vitro[163] and to suppress the CFU-E compartment in vivo in mice.[164] A role of this interleukin in the pathogenesis of some hypoplastic anemias has been suggested.[163]

Acknowledgments. This work was supported by the Associazione Italiana per le Leucemie (Firenze) and the Associazione Italiana per la Ricerca sul Cancro (AIRC).

References

1. Erslev AJ. Humoral regulation of red cell production. *Blood*. 1953;8:349–357.
2. Plzak LF, Fried W, Jacobson LO, Bethard WF. Demonstration of stimulation of erythropoiesis by plasma from anemic rats using ^{59}Fe. *J Lab Clin Med*. 1955;46:671–678.
3. Jacobson LO, Goldwasser E, Fried W, Plzak L. Role of the kidney in erythropoiesis. *Nature*. 1957;179:633–634.
4. Erslev AJ. The renal biogenesis of erythropoietin. *Am J Med*. 1975;58:25–30.
5. Filmanowicz E, Gurney C. Studies on erythropoiesis. XVI. The response to a single dose of erythropoietin in the polycythemia mouse. *J Lab Clin Med*. 1961;57:65–72.
6. Iscove NN, Sieber NF, Winterhalter KH. Erythroid colony formation in cultures of mouse and human bone marrow: analysis of the requirement for

erythropoietin by gel filtration and affinity chromatography on agarose-concanavalin-A. *J Cell Physiol*. 1974;83:309–320.

7. Gregory CJ. Erythropoietin sensitivity as a differentiation marker in the hemopoietic system studies of three erythropoietic colony responses in culture. *J Cell Physiol*. 1976;89:289–301.

8. Stephenson JR, Axelrad AA, McLeod DL, Shreeve MM. Induction of colonies of hemoglobin-synthesizing cells by erythropoietin in vitro. *Proc Natl Acad Sci USA*. 1971;68:1542–1546.

9. Johnson GR, Metcalf D. Pure and mixed erythroid colony formation in vitro stimulated by spleen conditioned medium with no detectable erythropoietin. *Proc Natl Acad Sci USA*. 1977;74:3879–3382.

10. McLeod DL, Shreeve MM, Axelrad AA. Improved plasma culture system for production of erythrocytic colonies in vitro; quantitative assay method for CFU-E. *Blood*. 1974;44:517–534.

11. Ogawa M, MacEachern MD, Avila L. Human marrow erythropoiesis in culture. II Heterogeneity in the morphology, time course of colony formation, and sedimentation velocity analysis of the colony-forming cells. *Am J Hematol*. 1977;3:29–36.

12. Axelrad AA, McLeod DL, Shreeve MM, Heat DS. Properties of cells that produce erythrocytic colonies in vitro. In: Robinson WA, ed. *Proceedings of the 2nd International Workshop on Hemopoiesis in culture*. Washington, DC: HEW Publication, NHI Government Printing Office; 1974:226–234.

13. Ogawa M, Leary AG. Erythroid progenitors. In: Golde DW, ed. *Hematopoiesis*. New York: Churchill Livingstone; 1984:123–132.

14. Johnson GR, Metcalf D. Nature of cells forming erythroid colonies in agar after stimulation by spleen conditioned medium. *J Cell Physiol*. 1978;94:243–252.

15. Aye MT. Erythroid colony formation in cultures of human marrow. Effect of leukocyte conditioned medium. *J Cell Physiol*. 1977;91:69–79.

16. Humphries RK, Eaves AC, Eaves CJ. Characterization of a primitive erythropoietic progenitor found in mouse marrow before and after several weeks in culture. *Blood*. 1979;53:746–763.

17. Iscove NN. Erythropoietin-indipendent stimulation of early erythropoiesis in adult marrow cultures by conditioned media from lectin-stimulated mouse spleen cells. In: Golde DW, Cline MJ, Metcalf D, Fox CF, eds. *Hematopoietic cell differentiation*. New York: Academic Press; 1978:37–52.

18. Tsang RW, Aye MT. Evidence for proliferation of erythroid progenitor cells in the absence of added erythropoietin. *Exp Hematol*. 1979;7:383–388.

19. Goodman JW, Hall EA, Miller KL, Shinpock SG. Interleukin-3 promotes erythroid burst formation in "serum-free" culture without detectable erythropoietin. *Proc Natl Acad Sci USA*. 1985;82:3291–3295.

20. Sieff CA, Emerson SF, Donahue RE, Nathan DG. Human recombinant granulocyte-macrophage colony stimulating factor: a multilineage hematopoietin. *Science*. 1985;230:1171–1173.

21. Peschel C, Paul WE, Ohara J, Green I. Effects of B cell stimulatory factor 1/interleukin-4 on hematopoietic progenitor cells. *Blood*. 1987;70:254–263.

22. Jourdanet D. *De l'anemie des altitudes et de l'anemie en general dans ses rapports avec la pression de l'athmosphere*. Paris: Bailliere; 1863.

23. Goldwasser E, Kung K-H. Purification of erythropoietin. *Proc Natl Acad Sci USA*. 1971;68:697–698.
24. Miyake T, Kung CKH, Goldwasser E. Purification of human erythropoietin. *J Biol Chem*. 1977;252:5558–5564.
25. Goldwasser E, Krantz SB, Wang FF. Erythropoietin and erythroid differentiation. In: Ford RJ, Maizel AL, eds. *Mediators in cell growth and differentiation*. New York: Raven Press; 1985:103–107.
26. Jacobs K, Shoemaker C, Rudersdorf R, Neill SD, Kaufman RJ, Mufson A, Seehra J, Jones SS, Hewick R, Fritsch EF, Kawakita M, Shimizu T, Miyake T. Isolation and characterization of genomic and cDNA clones of human erythropoietin. *Nature*. 1985;313:806–810.
27. Lin FK, Sugos S, Lin CH, Browne JK, Smalling R, Egrie JC, Chen KK, Fox GM, Martin F, Stabinsky Z, Badrawi SM, Lai P-H, Goldwasser E. Cloning and expression of the human erythropoietin gene. *Proc Natl Acad Sci USA*. 1985;82:7580–7585.
28. Sasaki H, Bothner B, Dell A, Fukuda M. Carbohydrate structure of erythropoietin expressed in Chinese hamster ovary cells by a human erythropoietin cDNA. *J Biol Chem*. 1987;262:12059–12076.
29. Lee-Huang S. Cloning and expression of human erythropoietin cDNA in *Escherichia coli*. *Proc Natl Acad Sci USA*. 1984;81:2708–2712.
30. Lacombe C, Tambourin P, Mattei MG, Simon D, Guenet JL. The murine erythropoietin gene is localized on chromosome 5. *Blood*. 1988;72:1440–1442.
31. Law ML, Cai GY, Lin FK, Wei Q, Huang SZ, Hartz JH, Morse H, Lin CH, Jones C, Kao FT. Chromosomal assignment of the human erythropoietin gene and its DNA polymorphism. *Proc Natl Acad Sci USA*. 1986;83:6920–6924.
32. Recny MA, Scoble HA, Kim Y. Structural characterization of natural human urinary and recombinant DNA-derived erythropoietin. Identification of des-arginine 166 erythropoietin. *J Biol Chem*. 1987;262:17156–17163.
33. Egrie JC, Strickland TW, Lane J, Aoki K, Cohen AM, Smalling R, Traie G, Lin FK, Browne JK, Hines DK. Characterization and biological effects of recombinant human erythropoietin. *Immunobiology*. 1986;172:213–224.
34. Fisher JW. Extrarenal erythropoietin production. *J Lab Clin Med*. 1979;93:695–703.
35. Bondurant MC, Koury MJ. Anemia induces accumulation of erythropoietin mRNA in the kidney and liver. *Mol Cell Biol*. 1986;6:2731–2733.
36. Zanjani ED, Poster J, Borlington H, Mann LI, Wassermann LR. Liver as the primary site of erythropoietin formation in the fetus. *J Lab Clin Med*. 1977;89:640–646.
37. Zucali JR, Mirand EA. Biosynthesis of erythropoietin by mouse fetal liver in culture. *Blood Cells*. 1975;1:485–493.
38. Nielsen OJ, Schuster SJ, Kaufman R, Erslev AJ, Caro J. Regulation of erythropoietin production in a human hepatoblastoma cell line. *Blood*. 1987;70:1904–1909.
39. Goldberg MA, Glass GA, Cunningham JM, Bunn HF. The regulated expression of erythropoietin by two human hepatoma cell lines. *Proc Natl Acad Sci USA*. 1987;84:7972–7976.

40. Rich IN, Heit W, Kubanek BK. Extrarenal erythropoietin production by macrophages. *Blood*. 1982;60:1007–1018.
41. Rich IN, Vogt C, Pentz S. Erythropoietin gene expression in vitro and in vivo detected by in situ hybridization. *Blood Cells*. 1988;14:505–520.
42. Lacombe C, Da Silva JL, Bruneval P, Fournier JC, Wendling F, Casadevall N, Camilleri JP, Bariety J, Varet B, Tambourin P. Peritubular cells are the site of erythropoietin synthesis in the murine hypoxic kidney. *J Clin Invest*. 1988;81:620–623.
43. Koury ST, Bondurant MC, Koury MJ. Localization of erythropoietin synthesizing cells in murine kidneys by in situ hybridization. *Blood*. 1988;71:524–527.
44. Schuster SJ, Wilson JH, Erslev AJ, Caro J. Physiologic regulation and tissue localization of renal erythropoietin messanger RNA. *Blood*. 1987;70:316–318.
45. Schooley JC, Mahlmann LJ. Evidence for the de novo synthesis of erythropoietin in hypoxic rats. *Blood*. 1972;40:662–670.
46. Koury ST, Koury MJ, Bondurant MC, Caro J, Graber SE. Quantitation of erythropoietin-producing cells in kidneys of mice by in situ hybridization: correlation with hematocrit, renal erythropoietin mRNA, and serum erythropoietin concentration. *Blood*. 1989;74:645–651.
47. Goldberg MA, Dunning SP, Bunn HF. Regulation of the erythropoietin gene: evidence that the oxygen sensor is a heme protein. *Science*. 1988;242:1412–1415.
48. Takeuchi M, Takasaki S, Miyazaki H, Kato T, Hoshi S, Kochibe N, Kobata A. Comparative study of the asparagine-linked sugar chains of human erythropoietins purified from urine and the culture medium of recombinant Chinese hamster ovary cells. *J Biol Chem*. 1988;203:3657–3664.
49. Lowry PH, Keighley G, Borsook H. Inactivation of erythropoietin by neuraminidase and mild substitution reactions. *Nature*. 1960;185:102–105.
50. Goldwasser E, Kung C KH, Eliason J. On the mechanism of erythropoietin-induced differentiation: XIII. The role of sialic acid in erythropoietin action. *J Biol Chem*. 1974;249:4202–4206.
51. Morell AG, Irvine KA, Sternlieb I, Sheimberg IH. Physical and chemical studies on ceruloplasmin: V. Metabolic studies on sialic acid-free ceruloplasmin in vivo. *J Biol Chem*. 1968;243:155–160.
52. Fukuda MN, Sasaki H, Lopez L, Fukuda M. Survival of recombinant erythropoietin in the circulation: the role of carbohydrates. *Blood*. 1989;73:84–89.
53. Spivak JL, Hogans BB. The in vivo metabolism of recombinant human erythropoietin in the rat. *Blood*. 1989;73:90–99.
54. MacDougall IC, Roberts DE, Neubert P, Dharmasena AD, Coles GA, Williams JD. Pharmacokinetics of recombinant human erythropoietin in patients on continuous ambulatory peritoneal dialysis. *Lancet*. 1989;i:425–427.
55. Spivak JL. The mechanism of action of erythropoietin. *Int J Cell Cloning*. 1986;4:139–154.
56. Stephenson JR, Axelrad AA, McLeod DL, Shreeve MM. Induction of colonies of hemoglobin-synthesizing cells by erythropoietin in vitro. *Proc Natl Acad Sci USA*. 1971;68:1542–1545.

57. Eaves AC, Henkelman DH, Eaves CJ. Abnormal erythropoiesis in the myeloproliferative disorders: an analysis of underlaying cellular and humoral mechanisms. *Exp Hematol.* 1980;8:235–247.
58. Lacombe C, Casadevall N, Varet B. Polycythaemia vera: in vitro studies of circulating erythroid progenitors. *Br J Haematol.* 1980;44:189–199.
59. Eridani S, Pearson TC, Sawyer B, Batten B, Wetherley-Mein E. Erythroid colony formation in primary proliferative polycythaemia, idiopathic erythrocytosis, and secondary polycythaemia: sensitivity to erythropoietic stimulating factors. *Clin Lab Haematol.* 1983;5:121–129.
60. Eridani S, Batten E, Sawyer B. Erythroid colony formation in primary thrombocythaemia: evidence of hypersensitivity to erythropoietin. *Br J Haematol.* 1983;55:157–161.
61. Reid CDL. The significance of endogenous erythroid colonies (EEC) in haematological disorders. *Blood Rev.* 1987;1:133–140.
62. Krantz SB, Goldwasser E. Specific binding of erythropoietin to spleen cells infected with the anemia strain of Friend virus. *Proc Natl Acad Sci USA.* 1984;81:7574–7578.
63. Mayeux P, Billat C, Jacquot R. Murine erythroleukemia cells (Friend virus) possess high-affinity binding sites for erythropoietin. *FEBS Lett.* 1987;211:229–233.
64. Fukamachi H, Saito T, Tojo A, Kitamura T, Urabe A, Takaku F. Binding of erythropoietin to CFU-E derived from fetal mouse liver cells. *Exp Hematol.* 1987;15:833–837.
65. Vannucchi AM, Grossi A, Rafanelli D, Vannucchi L, Rossi Ferrini P. Binding of recombinant human ^{125}I-erythropoietin to CFU-E from the spleen of anemic mice. *Haematologica.* 1990;75:21–26.
66. Sawada K, Krantz SB, Dai C-H, Koury ST, Horn ST, Glick AD, Civin CI. Purification of human blood burst-forming units-erythroid and demonstration of the evolution of erythropoietin receptors. *J Cell Physiol.* 1990; 142:219–230.
67. Fraser JK, Nicholls J, Coffey C, Lin FK, Berridge MV. Down-modulation of high-affinity receptors for erythropoietin on murine erythroblasts by interleukin-3. *Exp Hematol.* 1988;16:769–773.
68. Sawyer ST, Krants SB, Goldwasser E. Binding and receptor-mediated endocytosis of erythropoietin in Friend virus-infected erythroid cells. *J Biol Chem.* 1987;262:5554–5562.
69. D'Andrea AD, Lodish HF, Wong GG. Expression cloning of the murine erythropoietin receptor. *Cell.* 1989;57:277–285.
70. Mufson RA, Gesner TG. Binding and internalization of recombinant human erythropoietin in murine erythroid precursor cells. *Blood.* 1987;69:1485–1490.
71. Fraser JK, Tan AS, Lin F-K, Berridge MV. Expression of high-affinity binding sites for erythropoietin on rat and mouse megakaryocytes. *Exp Hematol.* 1989;17:10–16.
72. McDonald TP, Cottrel MB, Clift RE, Cullen WC, Lin FK. High doses of recombinant erythropoietin stimulate platelet production in mice. *Exp Hematol.* 1987;15:719–721.
73. Grossi A, Vannucchi AM, Rafanelli D, Rossi Ferrini P. Recombinant human erythropoietin has little influence on megakaryocytopoiesis in mice. *Br J Haematol.* 1989;71:463–468.

74. Ishibashi T, Koziol JA, Burstein SA. Human recombinant erythropoietin promotes differentiation of murine megakaryocytes in vitro. *J Clin Invest.* 1987;79:286–290.
75. Vannucchi AM, Grossi A, Rafanelli D, Vannucchi L, Rossi Ferrini P. A dot assay for the erythropoietin receptor using human recombinant ^{125}I-erythropoietin. *Anal Biochem.* 1989;182:182–186.
76. Koury MJ, Bondurant MC, Atkinson JB. Erythropoietin control of terminal erythroid differentiation: maintenance of cell viability, production of hemoglobin, and development of the erythrocyte membrane. *Blood Cells.* 1987;13:217–226.
77. Koury MJ, Bondurant MC, Mueller TJ. The role of erythropoietin in the production of principal erythrocyte proteins other than hemoglobin during terminal erythroid differentiation. *J Cell Physiol.* 1986;126:259–265.
78. Koury MJ, Bondurant MC. Maintenance by erythropoietin of viability and maturation of murine erythroid precursor cells. *J Cell Physiol.* 1988;137:65–74.
79. Sawyer ST, Krantz SB. Transferrin receptor number, synthesis, and endocytosis during erythropoietin-induced maturation of Friend-virus infected erythroid cells. *J Biol Chem.* 1986;261:9187–9195.
80. Sawyer ST, Krantz SB. Erythropoietin stimulates $^{45}Ca^{2+}$ uptake in Friend virus-infected erythroid cells. *J Biol Chem.* 1984;259:2769–2774.
81. Bondurant MC, Lind RN, Koury MJ, Ferguson ME. Control of globin gene transcription by erythropoietin in erythroblasts from Friend-virus infected mice. *Mol Cell Biol.* 1985;5:675–682.
82. Koury ST, Koury MJ, Bondurant MC. Morphological changes in erythroblasts during erythropoietin-induced terminal differentiation in vitro. *Exp Hematol.* 1988;16:758–763.
83. Eschbach JW. The anemia of chronic renal failure: pathophysiology and the effects of recombinant erythropoietin. *Kidney Int.* 1989;35:134–148.
84. McGorrigh RJS, Wallin JD, Shadduck RK, Fisher JW. Erythropoietin deficiency and inhibition of erythropoiesis in renal insufficiency. *Kidney Int.* 1984;25:437–444.
85. Eschbach JW, Egrie JC, Downing MR, Browne JK, Adamson JW. Correction of the anemia of end-stage renal disease with recombinant human erythropoietin: result of the Phase I and II clinical trial. *N Engl J Med.* 1987;316:73–78.
86. Winearls CG, Oliver DO, Pippard MJ, Reid C, Downing MR, Cotes PM. Effect of human erythropoietin derived from recombinant DNA on the anaemia of patients maintained by chronic haemodialysis. *Lancet.* 1986;ii:1175–1178.
87. Casati S, Passerini P, Campise MR, Graziani G, Gesana B, Perisic M, Ponticelli C. Benefits and risks of protracted treatment with human recombinant erythropoietin in patients having haemodialysis. *Br Med J.* 1987;295:1017–1020.
88. Bommer J, Alexiou C, Muller-Buhl U, Eifer J, Ritz E. Recombinant human erythropoietin therapy in hemodyalisis patients- dose determination and clinical experience. *Nephr Dialysis Trans.* 1987;2:238–242.
89. Eschbach JW, Abdulhadi MH, Browe JK, Delano BG, Downing MR, Egrie JC, Evans RW, Friedman EA, Graber SE, Haley R, Korbet S, Krantz SB, Lundin AP, Nissenson AR, Ogden DA, Paganini EP, Rader B, Rutsky EA,

Stivelman J, Stone WJ, Teschan P, Van Stone JC, Van Wyck DB, Zuckerman K, Adamson JW. Recombinant human erythropoietin in anemic patients with end-stage renal disease. Results of a Phase III multicenter clinical trial. *Ann Int Med*. 1989;111:992–1000.

90. Moia M, Vizzotto L, Cattaneo M, Mannucci PM, Casati S, Ponticelli C. Improvement in the haemostatic defect of uraemia after treatment with recombinant human erythropoietin. *Lancet*. 1987;ii:1227–1229.

91. Adamson JW. The promise of recombinant human erythropoietin. *Semin Hematol*. 1989;26:5–8.

92. Eschbach JW, Adamson JW. Recombinant human erythropoietin: implications for nephrology. *Am J Kidney Dis*. 1988;1:203–209.

93. Means RT, Olsen NJ, Krantz SB, Dessypris EN, Graber SE, Stone WJ, O'Neil VL, Pincus T. Treatment of the anemia of rheumatoid arthritis with recombinant human erythropoietin: clinical and in vitro studies. *Arthritis Rheum*. 1989;32:638–642.

94. Groopman JE. Retroviral infection and haemopoiesis. In: Bock G, Marsh J, eds. *Molecular control of haemopoiesis*. Ciba Foundation Symposium 148. London: J Wiley & Sons; 1990:173–185.

95. Henry DH, Rudnick SA, Bryant E, Abels RI, Denna RP, Staddon AP, Mason BA, Ortho Cancer Treatment Group. Preliminary report of two double-bind, placebo controlled studies using recombinant human erythropoietin in the anemia associated with cancer. *Blood*. 1989;74(Suppl):6a.

96. Miller CB, Jones RJ, Piantadosi S, Abeloff MD, Spivak JL. Decreased erythropoietin response in patients with the anemia of cancer. *N Engl J Med*. 1990;322:1689–1692.

97. Ludwig H, Fritz E, Kotzmann H, Hocker P, Gisslinger H, Baruas U. Erythropoietin treatment of anemia associated with multiple myeloma. *N Engl J Med*. 1990;322:1693–1699.

98. Al-Katti A, Veith RW, Papayannopoulou T, Fritsch EF, Goldwasser E, Stomatoyannopoulos G. Stimulation of fetal hemoglobin synthesis by erythropoietin in baboons. *N Engl J Med*. 1987;317:415–420.

99. Levine EA, Rosen AL, Sehgal LR, Gould SA, Moss GS. Fetal hemoglobin and treatment of sickle cell disease. *N Engl J Med*. 1988;319:118.

100. Rodgers GP, Lessin LS. Recombinant erythropoietin improves the anemia associated with Gaucher's disease. *Blood*. 1989;73:2228–2229.

101. Goodnough LT, Rudnick S, Price TH, Ballas SK, Collins ML, Crowley JP, Kosmin M, Kruskall MS, Lenes BE, Menitove JE, Silberstein LE, Smith KJ, Wallas CH, Abels R, Von Tress M. Increased preoperative collection of autologous blood with recombinant human erythropoietin therapy. *N Engl J Med*. 1989;321:1163–1168.

102. Yang YC, Ciarletta AB, Temple PA, Chung MP, Kovacic S, Witek-Giannotti JS, Leary AC, Kriz R, Donahue RE, Wong GG, Clark SC. Human interleukin-3 (multi-CSF): identification by expression cloning of a novel hematopoietic growth factor related to murine IL-3. *Cell*. 1986;47:3–10.

103. Fung MC, Hapel AJ, Ymer S, Cohen DR, Johnson RM, Campbell HD, Young IG. Molecular cloning of cDNA for mouse interleukin-3. *Nature*. 1984;307:233–238.

104. Donahue RE, Emerson SG, Wang EA, Wong GG, Clark SC, Nathan DG. Demonstration of burst-promoting activity of recombinant human GM-CSF on circulating erythroid progenitors using an assay involving the delayed addition of erythropoietin. *Blood*. 1985;66:1479–1481.
105. Migliaccio AR, Bruno M, Migliaccio G. Evidence of direct action of human biosynthetic (recombinant) GM-CSF on erythroid progenitors in serum-free cultures. *Blood*. 1987;70:1867–1871.
106. Donahue RE, Wang EA, Stone DK, Kamen R, Wong GG, Sehgal PK, Nathan DG, Clark SC. Stimulation of haematopoiesis in primates by continuous infusion of recombinant human GM-CSF. *Nature*. 1986;321:872–875.
107. Vannucchi AM, Grossi A, Rafanelli D, Rossi Ferrini P. In vivo stimulation of megakaryocytopoiesis by recombinant murine granulocyte-macrophage colony-stimulating factor. *Blood*. 1990;76(8):1473–1480.
108. Lee F, Yokota T, Otsuka T, Gemmel L, Larson N, Luh J, Arai K, Rennick D. Isolation of cDNA for a human granulocyte-macrophage colony-stimulating factor by functional expression in mammalian cells. *Proc Natl Acad Sci USA*. 1985;82:4360–4364.
109. Wong GG, Witek JAS, Temple PA, Wilkens KM, Luxenberg DP, Jones SS, Brown EC, Kay RM, Orr EC, Shoemaker C, Golde DW, Kaufman RJ, Hewik RM, Wang EA, Clark SC. Human GM-CSF: molecular cloning of the complementary DNA and purification of the natural and recombinant proteins. *Science*. 1985;228:810–815.
110. Gough NM, Gough J, Metcalf D, Kelso A, Grail D, Nicola NA, Burgess AW, Dunn AR. Molecular cloning of cDNA encoding a murine hematopoietic growth regulator, granulocyte-macrophage colony-stimulating factor. *Nature*. 1984;309–314.
111. Groopman JE, Molina J-M, Scadden DT. Hematopoietic growth factors. Biology and clinical applications. *N Engl J Med*. 1989;321:1449–1459.
112. Krystal G. Human serum levels of erythroblast enhancing facotr (EEF). *Exp Hematol*. 1982;10(Suppl 12):295.
113. Krystal G. Physical and biological characterization of erythroblast enhancing factor (EEF), a late acting erythropoietic stimulator in serum distinct from erythropoietin. *Exp Hematol*. 1983;11:16–19.
114. Udupa KB, Lipschitz DA. Proerythroblast stimulating activity: its purification from mouse serum and its effect on mouse erythroid cell proliferation in vitro. *Br J Haematol*. 1988;76:157–162.
115. Blanchet JP, Arnaud S, Samarut J, Bouabdelli M. A factor present in normal mouse stimulates late erythroid precursor proliferation. *Exp Hematol*. 1985;12:595–602.
116. Arnaud S, Mouchiroud G, Arnaud M, Niveleau A, Blanchet JP. Mouse serum contains two erythroid progenitor-stimulating activities, one being highly increased in serum from anemic mice. *Exp Hematol*. 1989;17:765–768.
117. Mouchiroud G, Blanchet JP. Normal mouse serum contains two erythroid stimulating activities differing by their mode of action. *Eur J Haematol*. 1990;44:51–55.
118. Niskanen E, Gasson J, Teates CD, Golde DW. In vivo effect of human erythroid-potentiating activity on hematopoiesis in mice. *Blood*. 1988;72:806–810.

119. Yu J, Shao L, Vaughan J, Vale W, Yu Al. Characterization of the potentiation effect of activin on human erythroid colony formation in vitro. *Blood.* 1989;73:952–960.

120. Dainiak N, Davies G, Kalmanti M, Lawler J, Kulkarni V. Platelet-derived growth factor promotes proliferation of erythropoietic progenitor cells in vitro. *J Clin Invest.* 1983;71:1206–1214.

121. Delwiche F, Raines E, Powell J, Ross R, Adamson J. Platelet-derived growth factor enhances in vitro erythropoiesis via stimulation of mesenchymal cells. *J Clin Invest.* 1985;76:137–142.

122. Rossi GB, Migliaccio AR, Lettieri F, Di Rosa M, Mastroberardino G, Peschle C. In vitro interactions of PGE and cAMP with murine and human erythroid precursors. *Blood.* 1980;56:74–79.

123. Brown J, Adamson JW. Modulation of in vitro erythropoiesis: the influence of beta-adrenergic agonist on erythroid colony formation. *J Clin Invest.* 1977;60:70–77.

124. Dukes PP. Potentiation of erythropoietin effects in marrow cultures by prostaglandin E1 or cyclic 3'5'-AMP. *Blood.* 1971;38:822a.

125. Schooley JC, Mahlmann LJ. Stimulation of erythropoiesis in plethoric mice by prostaglandins and its inhibition by antierythropoietin. *Proc Soc Exp Biol Med.* 1971;138:523–524.

126. Dukes PP, Shore NA, Hammond D, Ortega JA, Datta MC. Enhancement of erythropoiesis by prostaglandins. *J Lab Clin Med.* 1973;82:704–712.

127. Kurland JL, Bokman RS, Broxmeyer HE, Moore MAS. Limitation of excessive myelopoiesis by the intrinsic modulation of macrophage-derived prostaglandin E. *Science.* 1978;199:552–555.

128. Fisher JW, Gross DM. Renal prostaglandins and kidney production of erythropoietin. In: Fisher JW, ed. *Kidney hormones.* New York: Academic Press; 1977:357–385.

129. Chan HS, Saunders EF, Freedman MH. Modulation of human erythropoiesis by prostaglandins and lithium. *J Lab Clin Med.* 1980;95:125–132.

130. Naughton BA, Naughton GK, Liu P, Arce JM, Piliero SJ, Gordon AS. The effect of prostaglandins on extrarenal erythropoietin production. *Proc Soc Exp Biol Med.* 1982;170:231–236.

131. Erslev AJ. Pure red cell aplasia In: Williams WJ, Beutler E, Erslev AJ, Lichtman MA, eds. *Hematology.* New York: McGraw-Hill; 1990:430–438.

132. Adamson JW, Erslev AJ. Aplastic anemia. In: Williams WJ, Beutler E, Erslev AJ, Lichtman MA, eds. *Hematology.* New York: McGraw-Hill; 1990:158–174.

133. Peschle C, Sasso F, Mastroberardino G, Condorelli M. The mechanism of endocrine influences on erythropoiesis. *J Lab Clin Med.* 1971;78:20–29.

134. Malgor LA, Torales PR, Klainer E, Barrios L, Blanc CC. Effects of dexamethasone on bone marrow erythropoiesis. *Horm Res.* 1974;5:269–277.

135. Golde DW, Bersch N, Cline MJ. Potentiation of erythropoiesis in vitro by dexamethasone. *J Clin Invest.* 1976;57:57–62.

136. Shahidi NJ. Androgens and erythropoiesis. *N Engl J Med.* 1973;289:72–80.

137. French Cooperative Group for the study of Aplastic and Refractory Anemias. Androgen therapy in aplastic anemia: a comparative sudy of high and low doses and of 4 different androgens. *Scand J Haematol.* 1986;36:346–352.

138. Fried W, Gurney CW. Erythropoietic effect of plasma from mice receiving testosterone. *Nature*. 1965;206:1160–1161.
139. Fried W, Kilbridge T. Effect of testosterone and of cobalt on erythropoietin production by anephric rats. *J Lab Clin Med*. 1969;74:623–629.
140. Singer JW, Samuels AI, Adamson JW. The effect of steroids on in vitro erythroid colony growth: structure activity/relationship. *J Cell Physiol*. 1976; 88:127–134.
141. Gorshein D, Hait WN, Besa EC, Jepson JH, Gardner FH. Rapid stem cell differentiation induced by 19-nortestosterone decanoato. *Br J Haematol*. 1974;26:215–225.
142. Paulo LG, Fink GD, Roh BL, Fisher JW. Effects of several androgens and steroid metabolites on erythropoietin production in the isolated perfused dog kidney. *Blood*. 1974;43:39–47.
143. Dukes PP, Goldwasser E. Inhibition of erythropoiesis by estrogens. *Endocrinology*. 1961;69:21–29.
144. Piliero SJ, Medici PT, Haber C. The interrelationship of the endocrine and erythropoietic system in the rat with special reference to the mechanism of action of estradiol and testosterone. *Ann NY Acad Sci*. 1968;149:336–355.
145. Jepson JH, Lowenstein L. Inhibition of the stem-cell action of erythropoietin by estradiol. *Proc Soc Exp Biol Med*. 1966;123:457–460.
146. Golde DW, Bersch N, Chopra JJ, Cline MJ. Thyroid hormones stimulate erythropoiesis in vitro. *Br J Haematol*. 1977;37:173–177.
147. Popovich WJ, Brown JE, Adamson JW. The influence of thyroid hormones on in vitro erythropoiesis. Modulation by a receptor with beta adrenergic properties. *J Clin Invest*. 977;60:907–912.
148. Lucarelli G, Ferrari V, Rizzoli A, Porcellini A, Carnevali C, Monica C, Tanzi B, Butturini U. The effect of triiodothyronine on the erythropoiesis: Assay in the normal, starved, polycythemic and nephrectomized rat. *Biochim Biol Sper*. 1966;5:475–480.
149. Malgor LA, Blanc CC, Klainer E, Irizar SE, Torales PR, Barrios R. Direct effects of thyroid hormones on bone marrow erythroid cells of rats. *Blood*. 1975;45:671–679.
150. Jepson JH, McGarry EE. Hemopoiesis in pituitary dwarfs treated with human growth hormone and testosterone. *Blood*. 1972;39:238–248.
151. Peschle C, Rappaport IA, Sasso GF, Gordon AS, Condorelli M. Mechanism of growth hormone (GH) action on erythropoiesis. *Endocrinology*. 1972;91:511–517.
152. Golde DW, Bersch N, Li CH. Growth hormone: species-specific stimulation of erythropoiesis in vitro. *Science*. 1977;196:1112–1113.
153. Golde DW, Berch N, Li CH. Growth hormone modulation of murine erythropleukemia cell growth in vitro. *Proc Natl Acad Sci*. 1978;75:3437–3439.
154. Ritchey AK, Tamborlane WV, Gertner J. Improved diabetic control enhances erythroid stem cell proliferation in vitro. *J Clin Endocrinal Metab M*. 1985;60:1257–1260.
155. Bersch N, Groopman JE, Golde DW. Natural and biosynthetic insulin stimulates the growth of human erythroid progenitors in vitro. *JCE M*. 1982;55:1209–1211.

156. Kurtz A, Jelkmann W, Bauer C. Insulin stimulates erythroid colony formation indepoendently of erytropoietin. *Br J Haematol.* 1983;53:311–316.
157. Roodman GD, Bird A, Hutzler D, Montgomery W. Tumor necrosis factor-alpha and hemopoietic progenitors: effects of tuimor necrosis factor on the growth of erythroid progenitors CFU-E and BFU-E and the haematopoietic cell lines K562, HL60 and HEL cells. *Exp Hematol.* 1987;15:928–935.
158. Johnson CS, Chang MJ, Furmansky P. In vivo hematopoietic effects of tumor necrosis factor- in normal and erythroleukemic mice: characterization and therapeutic application. *Blood.* 1988;72:1875–1883.
159. Blick M, Sherwin SA, Rosenblum M, Gutterman J. Phase I study of recombinant tumor necrosis factor in cancer patients. *Cancer Res.* 1987;47:2986–2989.
160. Johnson CS, Cook CA, Furmansky. In vivo suppression of erythropoiesis by tumor necrosis factor-(TNF-α): reversal with exogenous erythropoietin (EPO). *Exp Hematol.* 1990;18:109–113.
161. Clibon U, Bonewald L, Caro J, Roodman JD. Erytrhropoietin fails to reverse the anemia in mice continuously exposed to tumor necrosis factor-alpha in vivo. *Exp Hematol.* 1990;18:438–441.
162. Orlic D, Gill R, Feldschuh R, Quaini F, Malice A, Sandoval C. Molecular mechanism for the inhibitory action of interferon on hematopoiesis. *Ann NY Acad Sci.* 1989;554:36–48.
163. Schooley JC, Kullgren B, Allison AC. Inibition by interleukin-1 of the action of erythropoietin on erythroid precursors and its possible role in the pathogenesis of hypoplastic anaemias. *Br J Haematol.* 1987;67:11–17.
164. Johnson CS, Keckler DJ, Topper MI. In vivo hematopoietic effects of recombinant interleukin-1a in mice: stimulation of granulocytic, monocytic, megakaryocytic, and early erythroid progenitors, suppression of late-stage erythropoiesis, and reversal of erythroid suppression with erythropoietin. *Blood.* 1989;73:678–683.

4
Colony-Stimulating Factors in Leukopoiesis

Mary Ann Bonilla and Ann Jakubowski

Normal hematopoiesis entails a constant equilibrium. It aims to maintain a given number of red cells, white cells, and platelets and adjusts to the body's requirement for these cells in response to different stimuli such as infection, trauma, or stress. The basic pluripotent stem cell is capable of giving rise to committed progenitor cells, which in turn differentiate into the mature blood cells. This pluripotent stem cell is capable of self-renewal, whereas the committed progenitor cell is not. In humans hematopoiesis occurs in the bone marrow, whereas in the mouse it occurs both in the bone marrow and in the spleen.

In the 1960s the techniques for culturing hematopoietic cells in semi-solid culture media were developed.[1,2] Early studies with these systems used murine spleen or bone marrow progenitor cells capable of differentiating into granulocytes and macrophages, giving rise to the so-called colony-forming unit in culture, or CFU-C. The processes of proliferation, differentiation, and modulation were found to depend on regulatory proteins known as colony-stimulating factors (CSF; Fig. 4.1),[3] produced by a variety of cell types including monocyte-macrophages, T and B lymphocytes, endothelial cells, fibroblasts, dendritic cells, keratinocytes, and certain malignant tissue cell lines. These CSFs may act synergistically or interact in positive and negative feedback systems (Fig. 4.2).[4] Indeed, cellular interactions between stromal and hematopoietic cells may be the primary means of regulating hematopoiesis.

The CSFs that are present in small quantities in the supernatants of some cloned cell lines have been purified to homogeneity and structurally analyzed to yield the amino acid sequence. The messenger RNAs have been used to identify the complementary DNA (cDNA) and subsequently genomic DNA clones were isolated. The genes for CSFs have been expressed in bacterial, eukaryotic, and mammalian systems, increasing the availability of the recombinant protein and allowing the direct study of the CSF's biological activities in semisolid culture. Currently available human CSFs involved in leukopoiesis include granulocyte macrophage–CSF (GM-CSF), granulocyte-CSF (G-CSF), macrophage-CSF (M-CSF),

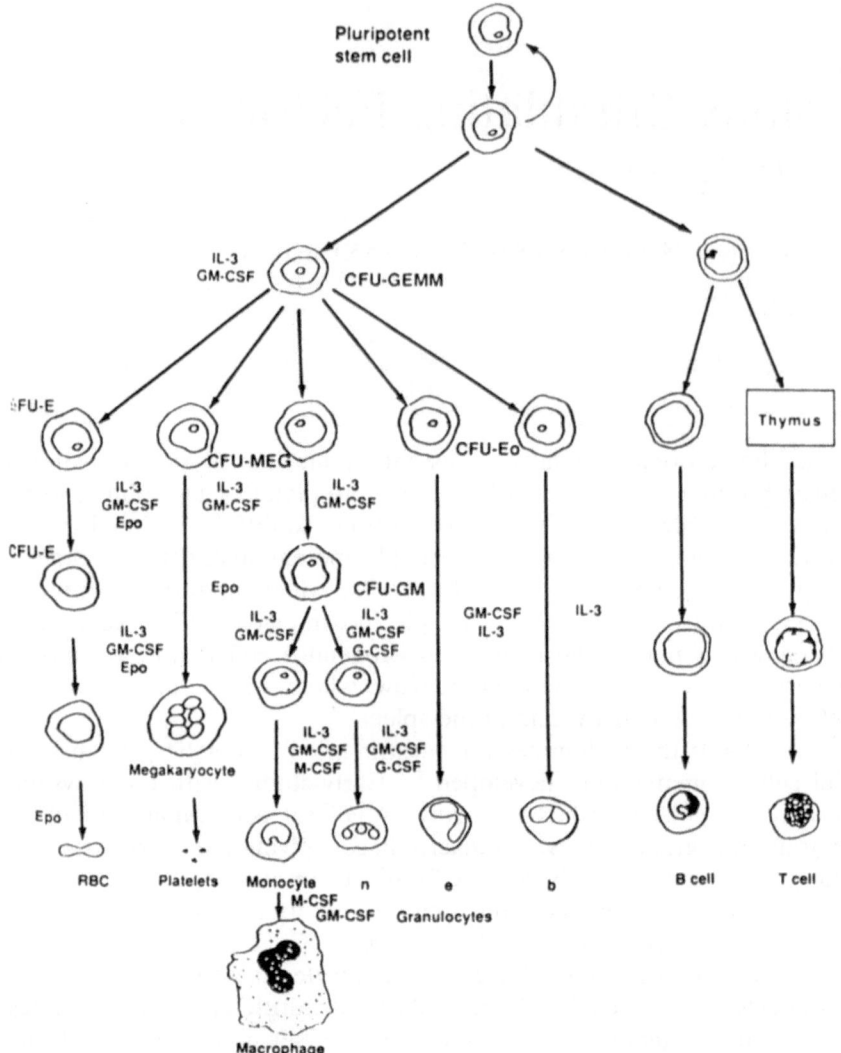

FIGURE 4.1. Interaction of the CSFs with hematopoietic cells. (Adapted with permission from ref. 3.)

interleukin-1 (IL-1), and interleukin-3 (IL-3, multi-CSF). The CSFs and their tissue sources are listed in Table 4.1.[5]

Each CSF has its own specific receptor. In general, only a few hundred to a few thousand receptors have been detected on any of the target cells studied thus far. Mature neutrophils express receptors for GM-CSF and G-CSF, eosinophils have receptors for GM-CSF, IL-3, and IL-5, and monocytes have receptors for G-CSF, GM-CSF, M-CSF, IL-2, and IL-4.

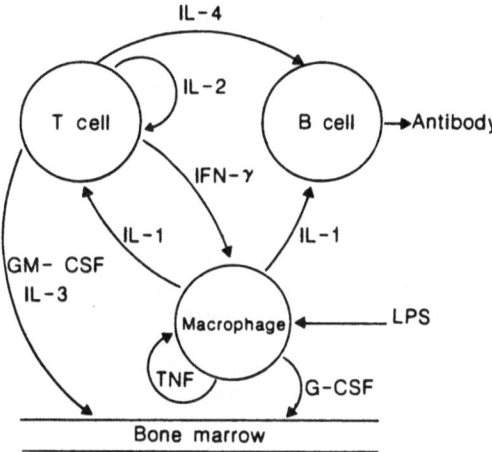

FIGURE 4.2. Network of cytokine interaction, positive and negative feedback. (Adapted with permission from ref. 4.)

M-CSF receptors are among the most abundant, as more than 10,000 have been found on a monocyte-macrophage. CSFs also have the capacity to down-regulate other CSF receptors on bone marrow cells in a rapid, dose-dependent, and hierarchical manner.[6]

Although CSFs have been named according to their ability to stimulate hematopoietic cells, they also promote cell survival and stimulate the function of mature cells. These CSFs have been studied in vitro, in vivo in preclinical trials, and are actively undergoing clinical trials. A review of these studies follows.

Granulocyte Macrophage–Colony Stimulating Factor

Granulocyte macrophage–colony stimulating factor (GM-CSF) was purified to homogeneity from the conditioned medium of the Mo hairy T leukemia cell line by Gasson and colleagues.[7] DNA clones have been isolated and transfected into monkey COS cells[8,9] and yeast cells.[10] The recombinant protein contains 127 amino acids and has a molecular weight of 14,000 to 35,000 daltons, depending on the degree of glycosylation. Natural GM-CSF has both N- and O-glycosylation, which modifies its in vitro activity. Thus, molecules lacking the N-glycosylation have greater biologic activity than the heavily glycosylated ones.[3] The gene encoding GM-CSF has been localized to the chromosome region 5q23-32.[11]

Recombinant human GM-CSF (rhGM-CSF) supports the growth of macrophage and granulocyte colonies in semisolid culture.[3] It acts synergistically with erythropoietin in stimulating the proliferation of

TABLE 4.1. Characteristics of the Human Colony Stimulating Factors. (Adapted from Groopman et al. 1989)[5]

CSF	Molecular weight kd	Chromosomal location of gene	Cellular source	Mature cell target
GM-CSF	14–35	5q23–31	T lymphocytes, monocytes, fibroblasts, endothelial cells	granulocytes, eosinophils monocytes
G-CSF	18–22	17q11.2–21	monocytes, fibroblasts, endothelial cells	granulocytes
M-CSF	70–90	5q33.1	monocytes, fibroblasts endothelial cells	monocytes
IL-1	12–19	2q23–34	monocytes	monocytes, granulocytes lymphocytes
IL-3	14–28	5q23–31	T lymphocytes	eosinophils, basophils monocytes

megakaryocyte and erythroid progenitors. It also influences the activity of mature cells including neutrophils, eosinophils, and monocyte-macrophages. Other reported effects include enhanced monocyte-macrophage tumoricidal and phagocytic activity,[12] intracellular killing, antibody-dependent cell-mediated cytotoxicity, chemoattractant-induced superoxide production,[13] complement-mediated phagocytosis (opsonization), and chemotaxis by neutrophils.[14] GM-CSF increases granulocyte aggregation in vitro and the induction of cell surface adhesion molecules.[15] In one clinical study, GM-CSF was shown to inhibit in vivo and in vitro neutrophil migration.[16]

In vivo administration of rhGM-CSF results in a dose-dependent increase in neutrophil counts. In patients with metastatic carcinoma, a profound leukopenia was noted within 30 min of a bolus infusion of rhGM-CSF. A leukocytosis, due to an increase in neutrophils, eosinophils, monocytes, and lymphocytes, was noted by 24 hr and was followed by a decline of the white cell count after discontinuation of therapy.[17] Recombinant human GM-CSF has also been shown to be effective when given by subcutaneous injection.[18] Therapy with rhGM-CSF causes an increase in circulating immature myeloid forms, including myelocytes, promyelocytes, and myeloblasts,[19,20] as well as hypersegmentation, and toxic granulation of peripheral blood neutrophils.[21]

The first clinical trial with rhGM-CSF was conducted in 16 patients with the acquired immunodeficiency syndrome (AIDS).[22] Leukopenia is a common feature in patients with AIDS and the AIDS-related complex. It contributes to the development of opportunistic infections, and is often the dose-limiting toxicity of drugs such as azidothymidine, currently used as an antiviral drug in this disease. Groopman et al. evaluated the effects of rhGM-CSF given as a single intravenous dose, followed 48 hr later by a 14-day continuous infusion with the same total daily dose. The doses were 1.3×10^3 U, 2.6×10^3 U, 5.2×10^3 U, and 1.0×10^4 U per kilogram per day. With doses of 1.3×10^3 U and 1.0×10^4 U, the mean peak leukocyte counts during the second week of therapy were 4575 ± 2397 cells/µl to $15{,}267 \pm 3885$ cells/µl, respectively. The peripheral blood revealed primarily an increase in neutrophils and esoinophils but an associated two- to fourfold increase in monocyte counts was also noted. No changes in reticulocyte or platelet counts were observed. Mild side effects included low-grade fever, myalgia, phlebitis, and flushing.

A phase I/II trial of rhGM-CSF, given by continuous intravenous infusion, was conducted in patients receiving myelosuppressive chemotherapy for advanced malignancy. Doses of rhGM-CSF between 4 and 32 µg/kg were well tolerated. At each dose level in the phase I trial, rhGM-CSF resulted in a rapid increase in peripheral blood neutrophils accompanied by bone marrow hyperplasia. A regimen of rhGM-CSF administered together with chemotherapy resulted in a shorter and less severe neutropenia than did treatment with an identical course of chemotherapy alone.[23] With doses of 64 µg/kg, two of four patients developed thromboses of their central venous catheters, fluid retention, and pericardial and pleural effusions.

The effect of rhGM-CSF in the setting of decreased marrow reserve was studied in patients with aplastic anemia and myelodysplastic syndrome (MDS).[24-26] In 10 aplastic anemia patients, rhGM-CSF, given in doses of 60 to 500 µg/m² by continuous infusion, increased the white cell count (1.6- to 10-fold), eosinophils (12- to >70-fold), and monocytes (2-to 32-fold).[27] There was no significant effect on erythroid or platelet counts. In the MDS patients, a similar increase in granulocytes and monocytes was studied and the appearance of immature myeloid cells (metamyelocytes to myeloblasts) in the peripheral blood was noted. Ganser et al. also noted an increase in the myeloblasts in the bone marrow. Toxic side effects included low-grade fever, phlebitis at the infusion site, and bone pain.

The use of rhGM-CSF in patients with lymphoid malignancies undergoing autologous marrow transplantation was investigated by Nemunaitis and colleagues.[28] The rhGM-CSF treatment, given as a 2-hr infusion, was well tolerated. Patients who received doses ≥60 µg/m²/day achieved an absolute neutrophil count (ANC) of 500 cells/µl by day 14 posttransplant. This resulted in fewer days of fever and an earlier hospital discharge. In addition, the platelet count recovery appeared to be accelerated by GM-CSF therapy. Brandt et al. studied the role of rhGM-CSF on hematopoietic recovery after myeloablative chemotherapy and autologous bone marrow transplantation in breast cancer and melanoma patients. In this phase I trial, rhGM-CSF was administered for 14 days by continuous intravenous infusion.[29] The mean leukocyte counts at the end of the rhGM-CSF infusion were 1511 ± 1003 cells/μl in patients treated with 2 to 8 μg/kg/day, 2575 ± 2304 cells/μl with 16 μg, 3120 ± 1744 cells/μl with 32 μg, and 863 ± 645 cells/μl in the untreated controls. There were no differences in the platelet counts. At higher doses, toxicity included generalized edema, pleural and pericardial effusions, myalgias, and erythroderma.

Granulocyte–Colony Stimulating Factor

G-CSF was first isolated from the conditioned medium of the human bladder carcinoma cell line 5637.[30] It has the ability to stimulate granulocyte colony formation, to potentiate erythroid formation in the presence of erythropoietin, and to induce the differentiation of HL60 and WEHI-3B colonies.[31] It has subsequently been detected in other epithelial cell lines, in placenta-conditioned medium, and in a hepatoma cell line.[30,32] Cellular sources of G-CSF include the monocyte-macrophage, endothelial cells, certain types of epithelial cells, bone marrow stromal cells, fibroblasts, and neutrophils. The gene for human G-CSF has been localized to chromosome region 17q11.2–q21.[33]

The cDNA has been expressed in *Escherichia coli* to yield a recombinant product with the same biologic activities as the highly purified natural form. The recombinant glycoprotein derived from *Escherichia coli* consists of 174 amino acids with a molecular weight of 18,800, and has a specific activity of 1×10^8 U/mg of protein.[34] A second human G-CSF has been cloned by Nagata et al. from a squamous carcinoma cell line.[35] Both the natural and recombinant forms of G-CSF support the growth of CFU-GM colonies, which contain predominantly neutrophils. The type of colony generated in semisolid culture depends on the concentration of G-CSF; high concentrations produce mixed neutrophil/macrophage, low concentrations yield pure neutrophil colonies.[36] In combination with IL-3, G-CSF supports the growth of megakaryocyte colonies.[37]

G-CSF has specific high affinity receptors in low numbers (50–500 per cell) on mature human and murine neutrophils. Receptors for G-CSF have been noted also on WEHI 3B leukemia cells and on myeloid cells.[34,38] Like GM-CSF, G-CSF enhances the chemotactic function and the antibody-dependent cellular cytotoxicity of mature neutrophils, and the chemotactic peptide binding of f-met-leu-phe.[39]

In a phase I/II clinical trial, an intravenous bolus of rhG-CSF given in the phase I portion before chemotherapy produced a dose-dependent rise in the peripheral white blood cell count attributable to an increase in the absolute neutrophil count. The mean increase (±SE) in the ANC at each dose level was 1.9 ± 0.1 times with a dose of 1 μg/kg, 4.3 ± 0.4 times with 10 μg, 6.2 ± 2.1 times with 30 μg, and 10.2 ± 2.5 times with 60 μg.

These results confirm those results previously obtained in the non human primate studies. Notable side effects included mild to moderate transient bone pain localized to the lower back, posterior iliac crests, and sternum in 72% of patients and an increase in leukocyte alkaline phosphatase.[40] Similar results were obtained in patients with advanced malignancy,[19] including small cell lung cancer.[41] In these clinical trials rhG-CSF was effective in shortening the period of neutropenia after chemotherapy.

The ability of rhG-CSF to enhance hematopoietic engraftment following high-dose chemotherapy and autologous bone marrow transplantation was investigated by Taylor et al.[42] Eighteen patients with Hodgkin's disease received 60 µg of rhG-CSF/kg/day as a 30-min infusion for a maximum of 28 days. An absolute granulocyte count of 500 cells/µl was noted after a median of 13 days (vs. 22 days in the control group). Platelet recovery, however, was not significantly different. Toxic effects included myalgias and bone pain, but no evidence of capillary leak syndrome was reported.

MDS is a group of diseases with refractory cytopenia and a characteristically dysplastic marrow involving at least two hematopoietic cell lineages. Patients are at an increased risk of infection, bleeding episodes requiring erythrocyte and platelet transfusions, and transformation to acute leukemia. In a phase I/II study 12 patients with MDS were treated with subcutaneous doses of rhG-CSF increasing from 0.1 to 3.0 µg/kg over an 8-week period. In five of seven patients the neutrophil counts increased from $<0.5 \times 10^9$/L to 1.2–16.3×10^9/L. An increased reticulocyte count was noted in five patients resulting in a decrease in red blood cell transfusion requirements in two patients. Three of the responding patients had abnormal cytogenetics that persisted after rhG-CSF therapy. After 6 months of rhG-CSF maintenance therapy on patient (RAEB-T) out of nine transformed to Acute Myeloid Leukemia (AML) and had an increase in vitro in proliferative response to G-CSF.[43] Eight patients continued to have a marked neutrophil response and four patients had a greater than 20% rise in their hematocrit.

Severe chronic neutropenias are a spectrum of diseases characterized by marked underproduction of mature neutrophils in the peripheral blood. This in turn results in frequent bacterial infections, particularly otitis media, pneumonia, gingivitis, urinary tract infections, and abscesses. Progression to septicemia, meningitis, peritonitis, or enteritis results in a high mortality. The ability of rhG-CSF to ameliorate the neutropenia and associated infections was investigated in patients with cyclic, congenital, and idiopathic neutropenia.

Hammond and colleagues treated six patients with cyclic neutropenia with rhG-CSF at doses of 3 to 10 µg/kg/day given intravenously or subcutaneously. In five out of six patients; the therapy decreased the length of cycling from 21 to 14 days but did not abolish it. The number of days of severe neutropenia (<200 neutrophils/mm^3) was reduced from 12.7 ±

1.7 days to 0.7 ± 0.4 days per month during treatment.[44] This resulted in a reduction in the frequency of oropharyngeal inflammation, infection, and fever.

In a pilot study at Memorial Sloan-Kettering Cancer Center, five patients with congenital agranulocytosis (Kostmann syndrome) were, treated with 3 to 60 μg/kg/day of rhG-CSF intravenously or subcutaneously. The absolute neutrophil counts rose from less than 100 to between 1300 and 9500 cells/mm³. Bone marrow revealed maturation to the mature neutrophil stage. There was a decrease in the requirement for intravenous antibiotics from 90.8 ± 28.0 days (in the year preceding treatment) to 9.6 days during 383.8 days of rhG-CSF therapy. All patients have enrolled in a subcutaneous maintenance regimen with sustained neutrophil responses for over 24 months.[45,46] A similar response to rhG-CSF has been reported by Welte and colleagues.[47]

One patient with idiopathic neutropenia experienced an increase in the number of neutrophils to ≥1500 cells/mm³ when treated with rhG-CSF at a dose of 1 to 3 μg/kg/day subcutaneously. Since the initiation of therapy, there have been no episodes of infection, and chronic mouth ulcerations healed.[48] In patients with chronic neutropenia, long-term theapy with rhG-CSF has been well tolerated. Side effects include medullary pain, splenomegaly, and a rise in leukocyte alkaline phosphatase.

Macrophage–Colony Stimulating Factor

M-CSF, also known as CSF-1, is another of the more lineage-specific growth factors. It stimulates proliferation and differentiation of monocyte progenitors and peripheral blood monocytes, and promotes the survival and functional activation of mature monocytes and macrophages. M-CSF is produced by a variety of cell types, including activated mononuclear phagocytic cells, fibroblasts, endothelial cells, and bone marrow stroma.[49,50] It can be produced constitutively by many mesenchymal cells. Expression of its gene is inducible by a number of agents, including lecithin and possibly ovarian hormones.[51,52]

This glycoprotein is cleaved proteolytically to a homodimer from a membrane-bound precursor.[53,54] The homodimers are linked by disulfide bridges that can be split by reducing agents, yielding biologically inactive monomeric subunits. The molecular weight of various preparations ranges from 47 to 76 kDa. The carbohydrate moieties are not required for biologic activity.[55] The human protein is encoded for by a single gene of approximately 21 kilobases on the long arm of chromosome 5 at band q33.1 and is not species restricted.[56,57] It has been detected in the serum of normal animals and in the serum of patients with a variety of hematopoietic disorders.[58–60]

The cell-surface receptor for M-CSF is a product of the proto-oncogene c-fms.[61] It is a transmembrane glycoprotein with tyrosine kinase activity that is stimulated by binding of the M-CSF ligand.[62] After ligand binding, the growth factor is endocytosed and degraded.[63] The receptor undergoes autophosphorylation[64] and a number of other physiologic changes, which include plasma membrane ruffling,[65] alteration in concentration of cytoplasmic cations,[66] induction of *c-myc* and *c-fos* expression,[67] and ultimately synthesis of DNA.[68] The number of M-CSF receptors increases as monocytes mature and become macrophages.[69]

M-CSF is biologically active in both normal and abnormal cell types in vitro. For example, monocyte-macrophages primed with M-CSF have exhibited cytolytic activity against melanoma and colon carcinoma tumor targets.[70] In addition, M-CSF–treated peripheral blood monocytes from primates demonstrated enhanced antibody-dependent cellular cytotoxicity against tumor targets in the presence of antibodies R24 and 3F8.[71] Finally, M-CSF reduced cellular proliferation and induced terminal differentiation of certain acute myeloid leukemia cells both in liquid suspension and clonogenic assays.[72]

M-CSF has been administered to nonhuman primates and other mammals in preclinical studies.[70,71,73] An increase in monocytes, frequently vacuolated, and the appearance of large unstained cells (LUC), felt to be promonocytes or macrophagelike cells, were the predominant hematologic changes. There was an associated decrease in the platelet count, despite an increase in marrow megakaryocytes, suggesting increased peripheral consumption. The platelet count returned to normal rapidly once the M-CSF infusion was discontinued. Of note, the circulating LUC from treated rabbits frequently demonstrated erythrocytosis and platelet phagocytosis.

The first clinical trials of M-CSF are underway. The early results of the first ongoing phase I trial in melanoma patients were recently reported.[74] Patients received doses of 10 to 80 μg/kg/day as a continuous infusion for two 7-day courses separated by a 1-week rest period. As was observed in the animal studies, there was an increase in the number of peripheral blood monocytes at doses ≥30 μg/kg/day. Enhanced in vitro antibody-dependent cellular cytotoxicity was observed with monocytes derived from patients receiving 50 and 80 μg/kg/day. One patient in this study demonstrated a delayed regression of skin lesions after what was felt to be an initial tumor progression. Thrombocytopenia that appeared to be dose dependent was also noted. As observed with other hematopoietic growth factors, M-CSF produced a dose-related decrease in serum cholesterol levels.[75] A recently initiated multicenter trial with M-CSF in the treatment of patients with MDS is in progress.

Interleukin-1

IL-1 is produced by activated monocytes and by a variety of other cell types (Table 4.2).[76] Two forms of human IL-1 have been described: IL-

TABLE 4.2. Sources of interleukin-1.[76]

Synovial fibroblast
Keratinocytes
Langerhans cells of the skin
Mesangial cells of the kidney
B lymphocytes
Natural killer cells
Astrocytes and microglial cells of the brain
Vascular endothelial and smooth muscle cells
Corneal, gingival, and thymic epithelial cells
T-lymphocyte cell lines

Reproduced with permission from ref. 25.

TABLE 4.3. Biological activities of interleukin-1.[76]

Inflammatory
    Human fibroblast proliferation
    Stimulation of prostaglandin E_2
    Increased collagen production
Systemic
    Decreased appetite
    Hypotension
    Somnolence
    Increased hepatic enzymes
    Induces synthesis of acute-phase proteins by hepatocytes
    Radioprotective effect in mice
Endocrine
    Stimulation of ACTH secretion
    Stimulation and inhibition of insulin production
    Destruction of pancreatic islet cells
    Inhibits Leydig cell testosterone production
Hematopoietic
    Neutrophilia induction
    Chemotaxis of neutrophils and monocytes
    Increases procoagulant activity

Reproduced with permission of ref. 25.

1α[77,78] and IL-1β.[79] Both have been purified and their cDNA has been cloned and expressed in *Escherichia coli*. The recombinant IL-1β protein is a glycoprotein with a molecular weight of 17,500 kDa. IL-1α is a protein consisting of 271 amino acids with only 27% homology to IL-1β. However, their biologic activities are similar (Table 4.3).[76]

IL-1 increases the transcription, production, and release of GM-CSF and G-CSF in response to fibroblasts and mononuclear phagocytes[80,81] and induces the production of GM-CSF by endothelial cells.[82] In vitro, IL-1 is a chemoattractant for neutrophils causing them to reduce nitro-blue tetrazolium and release their lysosomal enzymes. It also induces osteoblast proliferation and alkaline phosphatase production and stimulates osteoclasts to resorb bone. IL-1 is a chemoattractant for macrophages, and enhances prostaglandin production and tumoricidal activity.[76]

The administration of IL-1 has been shown to ameliorate leukopenia in animals given cytotoxic chemotherapy. Mice pretreated with high-dose cyclophosphamide were given a single dose of IL-1α intraperitoneally on day 0 and experienced higher peripheral blood granulocyte counts and increased bone marrow cellularity compared to controls.[83] Similar results were obtained with IL-1β.[84] Mice given 5-fluorouracil in combination with a 10-day course of intraperitoneal IL-1α experienced an accelerated recovery of neutrophil counts, especially in the mice who also received G-CSF.[85] Phase I studies in primates receiving IL-1β at doses of 0.2, 1.0, or 2.0 µg/kg/day for 14 days produced a dose-dependent increase in the absolute neutrophil counts. At a dose of 1 µg/kg/day Il-1β also accelerated the neutrophil recovery in animals previously treated with 5-fluorouracil (5-FU), reducing the recovery time from 14 to 2 to 7 days.[86] This study suggests that a short course of treatment is sufficient to enhance hematopoietic recovery and may be preferred since prolonged administration of IL-1β may induce hematopoietic inhibitors and delay recovery.[87]

The first human phase I–II study on the effects of IL-1β is underway in patients with metastatic colorectal cancer receiving 5-FU.[88] In the initial report, five of seven patients treated with IL-1β experienced a two to threefold increase in leukocyte counts and one of seven experienced an increase in platelet count. There were fewer neutropenic days after 5-FU and IL-1β than after 5-FU alone, suggesting that IL-1β had a myeloprotective effect. Toxic effects included fever, chills, and headaches in all patients. Nausea and vomiting, back pain, erythema at the infusion site, and blood pressure fluctuation were noted less frequently.

Interleukin-3

IL-3 is a pluripotent hematopoietic growth factor often referred to as "multi-CSF." The murine IL-3 gene was first identified and cloned in 1984.[89] Isolation of the equivalent human gene was not accomplished until 1986, when a gibbon cDNA clone was used as a hybridization probe.[90] Expression of the human gene is markedly restricted compared to those of other growth factors. Thus far only T lymphocytes[90] and natural killer cells[91] have been shown to have IL-3 mRNA. The human IL-3 gene, composed of 3200 base pairs, is located on the long arm of chromosome 5 at band q23-31 tandemly linked to the gene for GM-CSF.[11] It has been expressed in *Escherichia coli*, yeast, and COS7 cells.

This complex glycoprotein ranges in size from 14 to 28 kDa, depending on the degree of glycosylation. Purification of the protein to homogeneity revealed a 29% amino acid and a 45% nucleotide homology with

. the murine factor and a >90% amino acid homology with the gibbon protein.[90] The human protein does not stimulate hematopoiesis in mice but does have variable biologic activity in primates and dogs.[92] In addition to the name multi-CSF, this growth factor has also been referred to as CSF-2a and CSF-2B,[93] P-cell stimulating factor, hemopoietin-2,[94] thy-1–inducing activity,[95] colony forming unit–spleen stimulating activity,[96] and mast cell growth factor.[96–98] These terms reflect its various biologic activities. In early in vitro studies it was shown that murine IL-3 induced self-renewal of primitive stem cells,[99,100] promoted growth of inter-mediate and later stage progenitors[101] and mast cells,[102] and induced the production of histamine in hematopoietic precursor cells.[103]

As the recombinant human material became available, studies of bone marrow cells in semisolid media and suspension cultures revealed a similar spectrum of activity. Human blast cvells and uncommitted and committed myeloid colonies could be induced by rhIL-3. It stimulated the growth of day 14 myeloid colonies in a dose-dependent manner; however, it stimulated granulocyte colonies only weakly. Similarly, the response of 5-day cultures of promyelocytes and myelocytes was poor compared to other growth factors such as GM- or G-CSF.[104] In serum-free cultures, on the other hand, IL-3 demonstrated a superior ability compared to GM-CSF in promoting the formation of burst-forming unit (BFU)-Mega (derived from a more primitive hematopoietic progenitor) and colony-forming unit (CFU)-Mega colonies.[105,106] The response of BFU-E cultures to IL-3 or to GM-CSF was similar.[107] Results such as these led to the hypothesis that IL-3 exerts its effect on early progenitor cells, in contrast to factors such as G- and GM-CSF, which interact with intermediate and late progenitors.

The influence of IL-3 on effector cell function in vitro has also been examined. IL-3 prolongs survival of monocytes[108] and enhances the killing of WEHI 164 fibrosarcoma cells in response to endotoxin stimu-lation.[109] Pretreatment with IL-3 resulted in the appearance of eosinophils with the capacity to kill antibody-coated tumor cells. In contrast, IL-3–treated neutrophils lacked both antibody-dependent cellular cyto-toxicity (ADCC) and the ability to produce superoxide in response to f-met-leu-phe.[104] In basophils IL-3 increased the production and intra-cellular accumulation of histamine.[110] Biologic effects on lymphoid cells are less clearly defined. While IL-3 had no direct effect on peripheral blood T and B cells, it did enhance the IL-2–dependent growth of T cells,[111] and induced immunoglobulin G (IgG) secretion by activated B cells in the presence of IL-2.[112]

The first in vivo study of IL-3 in primates was carried out in normal macaque monkeys treated with continuous 7-day intravenous infusions of mammalian or recombinant derived IL-3 at doses of 5 to 50 µg/kg/day.[113] Blood studies revealed a gradual increase in white blood cell count in animals receiving doses of 20 µg/kg/day. A mean rise of 2- to 2.5-fold in the leukocyte count was due to increased neutrophils, eosinophils, lymphocytes, and atypical basophils. The greatest increases were an ~7-fold rise in mean eosinophil counts and an ~18-fold increase in mean atypical basophil count. There were also consistent increases in the corrected reticulocyte count and occasional increases in the platelet counts (3/4 monkeys). No significant systemic toxicity was noted with these doses. Notably, all animals receiving 20 μg/kg/day demonstrated an increase in some acute phase reactants: ~1.5- to 2-fold increase of alpha$_1$-antitrypsin and orosomucoid and a lesser rise in haptoglobin. These returned to normal by day 14. To investigate whether IL-3 expands an earlier progenitor population, a group of monkeys were treated for 7 days with 10 or 20 μg/kg/day of IL-3 followed by 4 days of treatment with a low dose of GM-CSF (2 μg/kg/day). A dramatic leukocytosis was observed that was higher than that seen with either growth factor alone, and that included a more impressive relative increase in the percentage of neutrophils and bands. This study was extended to treatment with GM-CSF for 14 days to determine if an IL-3 primed pool of progenitors would eventually be depleted. On the contrary, the leukocytosis was maintained at levels of >40,000 cells/mm^3 throughout the 14 days.

A similar study in normal rhesus monkeys using subcutaneous administration of recombinant IL-3 for 14 days (11–100 μg/kg/day) confirmed the marked increase in eosinophils and basophils with IL-3 alone and the improved neutrophil response when GM-CSF was added.[114]

Subsequently, the IL-3 primate studies were extended to the treatment of cynomologus monkeys receiving myelosuppressive chemotherapy with cyclophosphamide, or 5-FU (the latter is an agent that in mice spares only the primitive bone marrow progenitors with high proliferative potential).[115] Twenty-four hours after the last dose of chemotherapeutic agent, recombinant IL-3 was administered in doses of 20 μg/kg/day for 14 days by either continuous infusion or by subcutaneous injections twice a day. IL-3–treated animals demonstrated a marked enhancement of myeloid recovery following either cytotoxic agent. In general, their neutropenia was eliminated or less severe. Furthermore, clonal assays of the bone marrow revealed that IL-3 had accelerated the recovery of myeloid progenitor cells. Toxicity reactions included localized edema of the face and extremities and a pruritic rash. In addition, IL-3–treated animals developed mild splenomegaly.

A number of clinical trials are underway. Thus far, only one study on the use of yeast-derived rIL-3 in patients with advanced neoplasms or bone marrow failure has been reported.[116] Twenty-six patients were treated for 15 days with 30 to 500 μg/m^2 per day by subcutaneous administration. Nine patients had bone marrow failure. Those individuals who began treatment with adequate bone· marrow function demonstrated responses in all cell lineages. Substantial increases were noted in white blood cells, segmented neutrophils, eosinophils, and lymphocytes.

Interestingly, basophils increased only moderately. Reticulocytes increased in 12 of the 16 patients who could be evaluated. In the group with bone marrow failure, which included patients with MDS and aplastic anemia, platelets increased in five of eight evaluable patients, reticulocytes increased in three, and neutrophil counts rose threefold in eight evaluable patients. Only one patient was removed from the study because of toxicity reactions. These included fever, headache, flushing, and erythema at the injection site.

Future Clinical Uses of CSFs

Although the results obtained with growth factors in different clinical settings are promising and exciting, many issues need further investigation. A more thorough elucidation of the mechanism of action of the CSFs is required to maximize the therapeutic effects and to comprehend better the biology of the different disease states. For example, in the chronic neutropenias, the defect may be in the production of CSF, in its receptor binding, in the cytoplasmic transcription of its mRNA, or elsewhere. Preclinical studies that suggest that a combination of factors produce synergistic effects indicate the need for extensive clinical trials with combined agents. The search for additional hematopoietic growth factors, particularly factors that may affect the true stem cell or other lineages, such as the megakaryocyte, is crucial.

Although the administration of CSFs has ameliorated the neutropenia associated with cytotoxic chemotherapy or transplantation, the impact on long-term survival is still unknown. Whether incorporating the CSFs into future chemotherapeutic regimens will permit the use of larger doses ("dose intensification") and result in increased survival needs to be assessed. Furthermore, many investigators remain cautious in using these factors in the leukemic states fearing a possible stimulation of the malignant cells. As previously noted, blastic transformation has been reported in patients with MDS receiving growth factors, but it is difficult to distinguish between the natural course of the disease and a preferential stimulation of the malignant clone. On particular interest are the severe chronic neutropenia states, some of which have been considered to be preleukemic syndromes. Whether long-term administration of CSFs to these patients could increase the likelihood of leukemic transformation or marrow failure is unknown. And finally, the presence of hematopoietic growth factor receptors on tumors such as small cell carcinomas suggested the possibility that tumor growth may be enhanced. Although to date there have been no reports of growth stimulation by these factors in patients with solid tumors participating in clinical trials, this too bears careful monitoring.

Areas in which the CSFs may be beneficial that are just coming under study or have not yet been pursued include their adjuvant use in patients with extensive burns, diabetes mellitus, or other states associated with poor wound healing and frequent or chronic infections due to granulocyte dysfunction. These and other yet-to-be-discovered hematopoietic growth factors and inhibitors hold great promise for clinical manipulation of hematopoiesis, as tools for defining normal and abnormal hematopoiesis and for the correction of clinically significant hematologic disorders.

References

1. Bradley T, Metcalf D. The growth of mouse bone marrow cells in vitro. *Aust J Exp Biol Med Sci.* 1966;44:287–300.
2. Pluznik D, Sachs L. The cloning of normal "mast cells" in tissue culture. *J Cell Physiol.* 1965;66:319–324.
3. Clark S, Kamen R. The human hematopoietic colony stimulating factors. *Science.* 1987;236:1229–1237.
4. Old L. Polypeptide mediator network. *Nature.* 1987;326:330–331.
5. Groopman J, Molina J, Scadden D. Hematopoietic growth factors. *N Engl J Med.* 1989;321:1449–1459.
6. Walker F, Nicola N, Metcalf D. Hierarchical down modulation of hemopoietic growth factor receptors. *Cell.* 1985;43:269–276.
7. Gasson J, Weisbart R, Kaufman S, Clark S, Hewick R, Wong G, Golde D. Purified human granulocyte-macrophage colony-stimulating factor: direct action on neutrophils. *Science.* 1984;226:1339–1344.
8. Lee F, Yokata T, Otsuka T, et al. Isolation of cDNA for a human granulocyte-macrophage colony-stimulating factor by function expression in mammalian cells. *Proc Natl Acad Sci USA.* 1985;82:4360–4364.
9. Wong G, Witek J, Temple P, Wilkens K, Leary A, Luxenberg D, Jones S, Brown E, Kay R, Orr E, Shoemaker C, Golde D, Kaufman R, Hewick R, Wang E, Clark S. Human GM-CSF: molecular cloning of the complementary DNA and purification of the natural and recombinant proteins. *Science.* 1985;228:810–815.
10. Cantrell M, Anderson D, Ceretti D. Cloning, sequence, and expression of a human granulocyte/macrophage colony-stimulating factor. *Proc Natl Acad Sci USA.* 1985;82:6250–6264.
11. Yang Y, Koviac S, Kriz R. The human genes for GM-CSF and IL-3 are closely linked in tandem chromosome 5. *Blood.* 1989;71:958–961.
12. Fleishman J, Golde D, Weisbart R, Gasson J. Granulocyte-macrophage colony-stimulating factor enhances phagocytosis of bacteria by human neutrophils. *Blood.* 1986;68:708–711.
13. Weisbart R, Kwan L, Golde D, Gasson J. Human GM-CSF primes neutrophils for enhanced oxidative metabolism in response to the major physiological chemoattractants. *Blood.* 1987;69:18–21.
14. Lopez A, Williamson J, Gamble J, et al. Recombinant human granulocyte-macrophage colony stimulating factor stimulates in vitro human neutrophil and eosinophil function, surface receptor expression, and survival. *J Clin Invest.* 1986;78:1220–1228.

15. Arnaout M, Wang E, Clark S, Sieff C. Human recombinant granulocyte-macrophage colony-stimulating factor increases cell to cell adhesion and surface expression of adhesion-promoting surface glycoproteins on mature granulocytes. *J Clin Invest.* 1986;78:597–601.
16. Peters W, Stuart A, Affronti M, Kim C, Coleman R. Neutrophil migration is defective during recombinant human granulocyte-macrophage colony stimulating factor infusion after autologous bone marrow transplantation in humans. *Blood.* 1988;72:1310–1315.
17. Phillips N, Jacob S, Stoller R, Earle M, Przepiorka D, Shadduck R. Effect of recombinant human granulocyte colony stimulating factor on myelopoiesis in patients with refractory metastatic cancer. *Blood.* 1989;74:26–34.
18. Rifkin R, Hersh E, Salmon S. Continuous intravenous (IV) administration of recombinant human granulocyte-macrophage colony stimulating factor (rGM-CSF) is superior to IV bolus administration: a Phase I Study. *Proc Am Soc Clin Oncol.* 1988;7:165a.
19. Morstyn G, Campbell L, Souza L, et al. Effect of granulocyte colony stimulating factor on neutropenia induced by cytotoxic chemotherapy. *Lancet.* 1988;1:667–672.
20. Morstyn G, Lieschke G, Sheridan W, Layton J, Fox R. Clinical experience with recombinant human granulocyte colony stimulating factor and granulocyte-macrophage colony stimulating factor. *Semin Hematol.* 1989; 26(Suppl 1):9–13.
21. Kaplan S, Zdzarski V, Basford R, Wing E, Shadduck R. Effect of in vivo recombinant human granulocyte-macrophage colony stimulating factor on peripheral blood granulocyte function. *Clin Res.* 1988;36:566.
22. Groopman J, Mitsuyasu R, DeLeo M, Oette D, Golde D. Effect of recombinant human granulocyte-macrophage colony-stimulating factor on myelopoiesis in the acquired immunodeficiency syndrome. *N Engl J Med.* 1987;317:593–598.
23. Antman K, Griffin J, Elias A, Socinski M, Ryan L, Cannistra S, Oette D, Whitley M, Frei E, Shnipper L. Effect of recombinant human granulocyte-macrophage colony-stimulating factor on chemotherapy induced myelo-suppression. *N Engl J Med.* 1988;319:593–598.
24. Antin J, Smith B, Holmes W, Rosenthal A. Phase I/II study of recombinant human granulocyte-macrophage colony-stimulating factor in aplastic anemia and myelodysplastic syndrome. *Blood.* 1988;72:705–713.
25. Champlin R, Nimer S, Ireland P, Oette D, Golde D. Treatment of refractory aplastic anemia with recombinant human granulocyte-macrophage colony-stimulating factor. *Blood.* 1989;73:694–699.
26. Ganser A, Völkers B, Greher J, Ottman O, Walther F, Becher R, Betgmann L, Schulz G, Hoelzer D. Recombinant human granulocyte-macrophage colony-stimulating factor in patients with myelodysplastic syndromes: a phase I/II trial. *Blood.* 1989;73:31–37.
27. Vadhan-Raj S, Buescher S, Broxmeyer H, LeMaistre A, Lepe-Zuniga J, Ventura G, Jeha S, Horwitz L, Trujillo J, Gillis S, Hittelman W, Gutterman J. Stimulation of myelopoiesis in patients with aplastic anemia by recombinant human granulocyte-mactrophage colony-stimulating factor. *N Engl J Med.* 1988;319:1628–1634.

28. Nemunaitis J, Singer J, Buckner C, Hill R, Storb R, Thomas E, Appelbaum F. Use of recombinant human granulocyte-macrophage colony-stimulating factor in autologous marrow transplantation for lymphoid malignancies. *Blood*. 1988;72:834–836.

29. Brandt S, Peters W, Atwater S, Kurtzberg J, Borowitz M, Jones R, Shpall E, Bast R, Gilbert C, Oette D. Effect of recombinant human granulocyte-macrophage colony-stimulating factor on hematopoietic reconstitution after high-dose chemotherapy and autologous bone marrow transplantation. *N Engl J Med*. 1988;318:869–876.

30. Welte K, Platzer E, Lu L. Purification and biochemical characterization of human pluripotent hematopoietic colony-stimulating factor. *Proc Natl Acad Sci*. 1985;82:1526–1530.

31. Platzer E, Welte K, Gabrilove J. Biological activities of human pluripotent hematopoietic colony stimulating factor on normal and leukemic cells. *J Exp Med*. 1985;162:1788–1801.

32. Gabrilove J, Welte K, Harris P. Constitutive production of leukemia differentiation colony-stimulating erythroid burst-promoting and pluripoietic factors by a human hepatoma cell line: characterization of the leukemia differentiationg factor. *Blood*. 1985;66:407–416.

33. Simmers R, Webber L, Shannon M, Garson O, Wong G, Vadas M, Sutherland G. Localization of the G-CSF gene on chromosome 17 proximal to the breakpoint in the t(15;17) in acute promyelocytic leukemia. *Blood*. 1987;70:330–332.

34. Souza L, Boone T, Gabrilove J. Recombinant human granulocyte colony-stimulating factor: effects on normal and leukemic myeloid cells. *Science*. 1986;232:61–65.

35. Nagata S, Tsuchinya M, Asano S. Molecular cloning and expression of cDNA for human granulocyte colony stimulating factor. *Nature*. 1986; 319:415–418.

36. Burgess A, Metcalf D. The nature and action of granulocyte-macrophage colony stimulating factors. *Blood*. 1980;56:947–958.

37. McNiece I, McGrath H, PJ Q. Granulocyte colony-stimulating factor augments in vitro megakaryocyte colony formation by interleukin-3. *Exp Hematol*. 1988;16:807–810.

38. Nicola N, Metcalf D. Binding of the differentiation-inducer, granulocyte colony stimulating factor, to responsive but not unresponsive leukemic cell lines. *Proc Natl Acad Sci USA*. 1984;81:3765–3769.

39. Platzer E, Oez S, Welte K, Sendler A, Gabrilove J, Mertelsmann R, Moore M, Kalden J. Human pluripotent hematopoietic colony stimulating factor: activities on human and murine cells. *Immunobiology*. 1986;172:185–193.

40. Gabrilove J, Jakubowski A, Scher H, Sternberg C, Wong G. Grous J, Vagoda A, Fain K, Moore M, Clarkson B, Oettgen H, Alton K, Welte K, Souza L. Effect of granulocyte colony stimulating factor on neutropenia and associated morbidity due to chemotherapy for transitional cell carcinoma of the urothelium. *N Engl J Med*. 1988;318:1414–1422.

41. Bronchud M, Scarffe J, Thatcher N, Crowther D, Souza LM, Alton NK, Testa NG, Dezter TM. Phase I/II study of recombinant human granulocyte colony-stimulating factor after high-dose chemotherapy in patients receiving

intensive chemotherapy for small cell lung cancer. *Br J Cancer*. 1987; 56:809–813.

42. Taylor K, Jagannath S, Spitzer G, Spinolo J, Tucker S, Fogel B, Cabanillas F, Hagemeister F, Souza L. Recombinant human granulocyte colony-stimulating factor hastens granulocyte recovery after high dose chemotherapy and autologous bone marrow transplantation in Hodgkin's disease. *J Clin Oncol*. 1989;7:1791–1799.

43. Nagler A, Negrin R, Ginzton N, Donlon T, Souza L, Greenberg P. In vitro hemopoiesis in myelodysplastic syndrome (MDS) patients on maintenance therapy with recombinant human granulocyte colony-stimulating factor (G-CSF). *Blood*. 1989;74(Suppl 1):234a.

44. Hammond W, Price T, Souza L, Dale D. Treatment of cyclic neutropenia with granulocyte colony-stimulating factor. *N Engl J Med*. 1989;320: 1306–1311.

45. Bonilla M, Gillio A, Ruggiero M, Kernan N, Brochstein J, Abboud M, Fumagalli L, Vincent M, Gabrilove J, Welte K, Souza L, O'Reilly R. In vivo recombinant human granulocyte colony stimulating factor (rhG-CSF) corrects neutropenia in patients with congenital agranulocytosis. *Blood*. 1988;72(Suppl 1):110a.

46. Bonilla M, Gillio A, Ruggiero M, Kernan N, Brochstein J, Abboud M, Fumagalli L, Vincent M, Gabrilove J, Welte K, Souza L, O'Reilly R. Effects of recombinant human granulocyte colony-stimulating factor on neutropenia in patients with congenital agranulocytosis. *N Engl J Med*. 1989;320:1574–1580.

47. Welte K, Zeidler C, Reiter A, Odenwald E, Souza L, Riehm H. Differential effects of granulocyte-macrophage colony-stimulating factor and granulocyte colony-stimulating factor in children with severe congenital neutropenia. *Blood*. 1990;75:1056–1063.

48. Jakubowski A, Souza L, Kelly F, et al. Effects of human granulocyte colony-stimulating factor in a patient with idiopathic neutropenia. *N Engl J Med*. 1989;320:38–42.

49. Lanotte M, Metcalf D, Dexter T. Production of monocyte/macrophage colony-stimulating factor by preadipocyte cell lines derived from murine bone marrow stroma. *J Cell Physiol*. 1982;112:123–127.

50. Rambaldi A, Young D, Griffin J. Expression of the M-CSF (CSF-1) gene by human monocytes. *Blood*. 1987;69:1409–1413.

51. Bartocci A, Pollard J, Stanley E. Regulation of colony-stimulating factor-1 during pregnancy. *J Exp Med*. 1986;164:956–961.

52. Wong G, Temple P, Leary A, Witek-Giannotti JS, Yang Y, Carletta AB, Chung M, Murtha P, Krizm R, Kaufman R, Ferenz C, Sibley B, Turner K, Hewick R, Clark S, Yana N, Yokotz H, Tamada M, Saito M, Motoyoshi K, Takaku F. Human CSF-1: molecular cloning and expression of 4 Kb cDNA encoding the human urinary protein. *Science*. 1987;235:1504–1508.

53. Rettenmeir C, Roussel M. Differential processing of colony-stimulating factor-1 precursors encoded by two human cDNAs. *Mol Cell Biol*. 1988;8:5026–5034.

54. Rettenmeir C, Roussel M, Ashmun R, Ralph P, Price K, Scher CJ. Synthesis of membrane-bound colony-stimulating factor-1 (CSF-1) and

downmodulation of CSF-1 receptors in NIH3T3 cells transformed by cotransfection of the human CSF-1 and c-fyms (CSF-1 receptor) genes. *Mol Cell Biol.* 1987;7:2378–2387.

55. Das S, Stanley E. Structure-function studies of a colony-stimulating factor (CSF-1). *J Biol Chem.* 1982;257:13679–13684.

56. Ladner M, Martin G, Noble J, et al. Human CSF-1: gene structure and alternative splicing of mRNA precursors. *EMBO J.* 1987;6:2693–2698.

57. Pettenati M, Le Beau M, Lemons R, Shima EA, Kawasaki ES, Larson RA, Sherr CJ, Diaz MO, Riwkey JD. Assignment of CSF-1 to 5q33.1: evidence for clustering of genes regulating hematopoiesis and for their involvement in the deletion of the long arm of chromosome 5 in myeloid disorders. *Proc Natl Acad Sci USA.* 1987;84:2970–2974.

58. Das S, Stanley E. Human colony-stimulating factor (CSF-1) radio-immunoassay: resolution of human colony-stimulating factors. *Blood.* 1981; 58:630–641.

59. Gilbert H, Praloran V, Stanley E. Increased serum concentrations of colony stimulating factor-1 in myeloproliferative disease. *Blood.* 1987;70(Suppl 1): 135a.

60. Hanamura T, Motoyushi K, Yoshida K, et al. Quantification and identification of human monocytic colony-stimulating factor in human serum by enzyme-linked immunosorbent assay. *Blood.* 1988;72:886–892.

61. Sherr C, Rettenmeir C, Sacca R, Roussel M, Look A, Stanley E. The c-fms proto-oncogene product is related to the receptor for the mononuclear phagocyte growth factor CSF-1. *Cell.* 1985;41:665–676.

62. Rettenmeir C, Chen J, Roussel M, Sherr C. The product of the c-fms proto-oncogene: a glycoprotein with associated tyrosine-kinase activity. *Science.* 1985;228:320–322.

63. Bartocci A, Mastrogiannis D, Migliorate G, Stockert RJ, Wolkoff AW, Stanley ER. Macrophages specifically regulate the concentration of their own growth factor in the circulation. *Proc Natl Acad Sci USA.* 1987; 84:6179–6183.

64. Downing J, Rettenmeir C, Sherr C. Ligand-induced tyrosine kinase activity of the colony-stimulating factor-1 receptor in a murine macrophage cell line. *Mol Cell Biol.* 1988;8:1795–1799.

65. Tushinski R, Oliver I, Guilbert L. Survival of mononuclear phagocytes depends on a lineage-specific growth factor that the differentiated cells selectively destroy. *Cell.* 1982;28:71–81.

66. Vairo G, Hamilton J. Activation and proliferation signals in murine macrophages: stimulation of Na^+, K^+-ATPase activity by hematopoietic growth factors and other agents. *J Cell Physiol.* 1988;134:13–24.

67. Orlofsky S, Stanley E. CSF-1-induced gene expression in macrophages: dissociation from the mitogenic response. *EMBO J.* 1987;6:2947–2952.

68. Tushinski R, Stanley E. The regulation of mononuclear phagocyte entry into S phase by the colony-stimulating factor CSF-1. *J Cell Physiol.* 1985;122:221–228.

69. Guilbert L, Stanley E. Specific interaction of murine colony-stimulating factor with mononuclear phagocytic cells. *J Cell Biol.* 1980;85:153–159.

70. Garnick M, Stoudemire J, Donahue R, Metzger M, Bree A, Timony G, Wong G, Munn D, Cheung NKV. In vitro and in vivo biological effects of

recombinant human macrophage colony stimulating factor (rhM-CSF). *Proc ASCO*. 1989;8:184.

71. Munn D, Garnick M, Cheung N. Effects of parenteral macrophage colony-stimulating factor (M-CSF) on circulating monocyte number, immuno-phenotype and antitumor activity in cynomolgus monkeys. *Blood*. 1988; 72(Suppl 1):127a.

72. Miyauchi J, Wang C, Kelleher C, et al. The effects of recombinant CSF-1 on the blast cells of acute myeloblastic leukemia in suspension culture. *J Cell Physiol*. 1988;135:55–62.

73. Donahue R, Wong G, Metzger M, Seghal PK, Morris JP, Turner KJ, Morin SH, Sibley SB, Stoudemire J, Clark SC, Garnick M. In vivo effects of recombinant human macrophage-colony stimulating factor (M-CSF) in primates. *Blood*. 1988;72(Suppl 1):114a.

74. Bajorin D, Jakubowski A, Cody B, Munn D, Cheung N-K, Urmacher C, Dantis L, Templeton MA, Scheinberg DA, Chapman P, Toner G, Zakowski M, Haines C, Oettgen HF, Gabrilove J, Garnick MB, Houghton AN. Recombinant macrophage colony stimulating factor (rhM-CSF): a phase I trial in patients (pts) with metastatic melanoma. *Proc ASCO*. 1990;8: 183a.

75. Garnick M, Stoudemire J. Marked serum cholesterol (C) and low density lipoprotein C (LDLC) lowering activity induced by recombinant human macrophage colony-stimulating factor (rhM-CSF) and other hematopoietic growth factors (HGF) in primates. *Clin Res*. 1989;37:260A.

76. Dinarello C. Biology of interleukin 1. *FASEB J*. 1988;2:108–115.

77. Gubler U, Chua A, Stern A, Hellman P, Vitek MP, Dechiara TM, Benjamin WR, Collier KR, Dukovich M, Familletti PC, Fiedler-Nagy C, Jenson J, Kaffka K, Kilian PL, Stremlo D, Wittreich BH, Woehle D, Lomedico PT. Recombinant human interleukin 1: purification and biological characterization. *J Immunol*. 1986;136:2492–2497.

78. Lomedico P, Gubler U, Hellman C, et al. Cloning and expression of murine interleukin-1 in *Escherichia coli*. *Nature*. 1984;312:458–462.

79. Auron P, Webb A, Rosenwasser L, Mucci S, Rich A, Wolf S, Dinarello C. Nucleotide sequence of human monocyte interleukin 1 precursor cDNA. *Proc Natl Acad Sci USA*. 1984;81:7907–7911.

80. Fibbe W, Damme V, Billiau A, Voogt P, Duinkerken N, Kluck P, Falkenburg J. Interleukin-1 (22-K factor) induces release of granulocyte-macrophage colony-stimulating activity from human mononuclear phagocytes. *Blood*. 1986;68:1316–1321.

81. Kaushansky K, Lin N, Admanson A. Interleukin 1 stimulates fibroblasts to synthesize granulocyte-macrophage and granulocyte colony-stimulating factors: mechanism for the hematopoietic response to inflammation. *J Clin Invest*. 1988;81:92–97.

82. Sieff C, Tsai S, Faller D. Interleukin 1 induces cultured human endothelial cell production of granulocyte-macrophage colony-stimulating factor. *J Clin Invest*. 1987;79:48–51.

83. Fibbe W, van der Meer J, Falkenburg J, Hamilton M, Kluin P, Dinarello C, Willemze R. A single low dose of human recombinant interleukin-1 accelerates the reconstitution of neutrophils in mice with cyclophosphamide-induced neutropenia. *Blood*. 1988;72(Suppl 1):116a.

84. Stork L. IL-1 accelerates murine granulocyte recovery following treatment with cyclophosphamide. *Blood.* 1989;73:938–944.
85. Moore M, Warren D. Synergy of interleukin 1 and granulocyte colony-stimulating factor: in vivo stimulation of stem-cell recovery and hematopoietic regeneration following 5 fluorouracil treatment of mice. *Proc Natl Acad Sci USA.* 1987;84:7134–7138.
86. Gillio A, Laver J, Abboud M, Bonilla M, Gasparetto C, O'Reilly R. Short course of IL-1 accelerates hematopoietic recovery following 5-flourouraciil administration in cynomolgus primates. *Blood.* 1988;72(Suppl 1):118a.
87. Gasparetto C, Laver J, Abboud M, Gillio A, Smith C, O'Reilly R, Moore M. Effects of interleukin-1 on hematopoietic progenitors: evidence of stimulatory and inhibitory activities in a primate model. *Blood.* 1989; 74:547–550.
88. Crown J, Kemeny N, Jakubowski A, Sheridan C, Gasparetto C, Toner G, Meisenberg B, Abboud M, Sinha S, Gordon M, Houston C, Oettgen H, Buhles W, Cheney T, Moore M, Kelson M, Kelson D, Gabrilove J. Phase I–II trial of recombinant human interleukin-1β (IL-1) in patients (pts) with metastatic colorectal cancer (MCC) receiving 5-flourouracil (5FU). *Blood.* 1989;74(Suppl 1):15a.
89. Fung M, Hapel A, Ymer S, Cohen D, Johnson R, Campbell H, Young I. Molecular cloning of cDNA for murine interleukin-3. *Nature.* 1984;307: 233–237.
90. Yang Y, Ciarletta A, Temple P, Chung M, Kovacic S, Witek-Gianotti J, Leary A, Kriz R, Donahue R, Wong G, Clark S. Human IL-3 (multi-CSF): identification of expression cloning of a novel hematopoietic growth factor related to murine IL-3. *Cell.* 1986;47:3–10.
91. Cuturi M, Anegon I, Sherman F, Loudon R, Clark S, Perussia B, Trinchieri G. Production of hematopoietic colony-stimulating factors by human natural killer cells. *J Exp Med.* 1989;169:569–583.
92. Metcalf D. Multi-CSF dependent colony formation by cells of a murine hemopoietic cell line: specificity and action of multi-CSF. *Blood.* 1985; 65:357–362.
93. Prestidge R, Watson J, Urdal D, Mochizuki D, Conlon P, Gillis S. Biochemical comparison of murine colony-stimulating factors secreted by a T cell lymphoma and a myelomonocytic leukemia. *J Immunol.* 1984; 133:293–298.
94. Bartelmez S, Sacca R, Stanley E. Lineage specific receptors used to identify a growth factor for developmentally early hemopoitic cells: assay of hemopoietin-2. *J Cell Physiol.* 1985;122:362–369.
95. Schrader J, Crapper R. Autogenous production of hemopoietic growth factor, persisting-cell-stimulating factor, as a mechanism for transformation of bone marrow-derived cells. *Proc Natl Acad Sci USA.* 1983;80:6892–6896.
96. Schrader J, Clark-Lewis I. A T-cell derived factor stimulating multipotential hemopoietic stem cells: molecular weight and distinction from T cell growth factor and T-cell derived granulocyte-macrophage colony-stimulating factor. *J Immunol.* 1982;129:30–35.
97. Yung Y, Eger R, Tertian G, Moore M. Long-term in vitro culture of murine mast cells. II: Purification of a mast cell growth factor and its dissociation from TCGF. *J Immunol.* 1981;127:794–799.

98. Yung Y, Moore M. Long-term in vitro culture in murine mast cells. III: Discrimination of mast cell growth factor and granulocyte-CSF. *J Immunol.* 1982;129:1256–1261.

99. Goldwasser E, Ihle J, Prystowsky M, Rich I, Van Zant G. The effect of interlukin-3 on hematopoietic precursor cells. In: Golde, Marks, eds. *Normal and neoplastic hemopoisis.* New York: Alan R Liss; 1983.

100. Spivak J, Smith R, Ihle J. Interleukin-3 promotes the in vitro proliferation of murine pluripotent hemopoitic stem cells. *J Clin Invest.* 1985;76:1613–1621.

101. Kenichi K, Ogaswa M, Ihle J, et al. Recombinant murine granulocyte-macrophage (GM) colony-stimulating factor supports formation of GM and multipotential blast cell colonies in culture: comparison with the effects of interleukin-3. *J Cell Physiol.* 1987;131:458–464.

102. Razin E, Ihle J, Seldin B, Mencia-Huerta J, Katz HR, LeBlanc PA, Hein A, Caulfield JP. Interleukin-3: a differentiation and growth factor for the mouse mast cell that contains chondroitin sulfate E proteoglycan. *J Immunol.* 1984;132:1479–1486.

103. Dy M, Shneieder E, Guy-Grand D, Lekel B. Histamine-producing cell stimulating factor (H-CSF) and interleukin-3 (IL-3): their effects on hisitidine ornithine decarboxylase. In: Shrader, ed. *Interleukin-3: The panspecific hematopoietin.* San Diego: Academic Press; 1988.

104. Lopez A, Dyson P, LB T, et al. Recombinant human interleukin-3 stimulation of hematopoisis in humans: loss of responsiveness with differentiation in the neutrophilic myeloid series. *Blood.* 1988;72:1797–1804.

105. Briddell R, Brandt J, Staneva J, Srour E, Hoffman R. Characterization of the human burst-forming unit-megakaryocyte. *Blood.* 1989;74:145–151.

106. Bruno E, Miller M, Hoffman R. Interacting cytokines regulate in vitro human megakaryopoiesis. *Blood.* 1989;73:671–677.

107. Sonada Y, Yang Y, Wong G, Clark S, Ogawa M. Erythroid burst-promoting activity of purified recombinant human GM-CSF and interleukin-3: studies with anti-GM-CSF and anti-IL-3 sera and studies in serum-free cultures. *Blood.* 1988;72:1381–1386.

108. Yang Y, Clark S. *Interleukin-3: molecular biology and biologic activities.* Philadelphia: Saunders; 1989.

109. Cannistra S, Veilenga E, Groshek P, Rambaldi A, Griffin J. Human granulocyte-monocyte colony-stimulating factor and interleukin-3 stimulate monocyte cytotoxicty through a tumor necrosis factor-dependent mechanism. *Blood.* 1988;71:672–676.

110. Valent P, Schmidt G, Bessemer J, Mayer P, Zenke G, Liehl E, Hinterberger W, Lechner K, Maurer D, Bettelheim. Interleukin-3 is a differentiation factor for human basophils. *Blood.* 1989;73:1763–1769.

111. Santoli D, Clark S, Kreider B, Maslin PA, Rovera G. Amplification of IL-2-driven T cell proliferation by recombinant human IL-3 and granulocyte-macrophage colony-stimulating factor. *J Immunol.* 1988;141:519–526.

112. Tadmori W, Feingersh D, Clark S, Choi YS. Human recombinant interleukin-3 stimulates B cell differentiation. *J Immunol.* 1989;142:1950–1955.

113. Donahue R, Seehra J, Metzger M, Lefebvre D, Rodk B, Carbone S, Nathan D, Garnick M, Seghal P, Laston D, LaVallie E, Mckoy J, Schendel P,

Norton C, Turner K, Yang Y, Clark S. Human IL-3 and GM-CSF act synergistically in stimulating hematopoiesis in primates. *Science*. 1988;241:1820–1823.

114. Mayer P, Valent P, Schmidt G, Liehl E, Bettelheim P. The in vivo effects of recombinant human interleukin-3: demonstration of basophil differentiation factor, histamine-producing activity, and priming of GM-CSF-responsive progenitors in nonhuman primates. *Blood*. 1989;74:613–621.

115. Gillio A, Gasparetto C, Laver J, Abboud M, Bonilla M, Garnick M, O'Reilly R. Effects of interleukin-3 on hematopoietic recovery after 5-fluorouracil and cyclophosphamide treatment on cynomolgus primates. *J Clin Invest*. 1990;85:1560–1565.

116. Ganser A, Lindeman A, Seipelt G, Ottman O, Hermann F. Effect of recombinant human interleukin-3 (rhIL-3) in patients with bone marrow failure: a phase I/II trial. *Blood*. 1989;74(Suppl 1):50a.

5
Regulation of Lymphocyte Proliferation, Differentiation, and Functional Activity by Peptide Growth Factors

JOSEPH KAPLAN

During the past 20 years there has been an explosion of knowledge concerning the nature and mode of action of growth factors that control lymphocyte differentiation, proliferation, and functional activity. The rapid advances in this field have been made possible by the development of monoclonal antibodies that identify specific lymphocyte subsets and stages of differentiation, and by the cloning and large-scale production of growth factors using recombinant DNA technology.

Lymphocytes comprise functionally and phenotypically distinct subsets of T cells, B cells, and natural killer (NK) cells, all of which arise from bone marrow stem cell precursors. A large variety of peptide growth factors contribute to the differentiation from pluripotent stem cells of these diverse lymphocyte populations. In addition to promoting lymphocyte differentiation, they act in concert as positive and negative signals for regulating lymphocyte growth and functional activity. During the past 10 years there has been an increasing rate of discovery of new factors. Most were named according to the biologic activities leading to their detection. For example interleukin-2 (IL-2) and interleukin-4 (IL-4) were originally designated "T cell growth factor" and "B cell growth factor" based on their abilility to promote growth of T cells and B cells, respectively. However, it is now clear that most growth factors affect a much broader range of cells and have a wider range of biologic activities than implied by the assay sytems leading to their original detection. Thus, any given growth factor may exhibit growth inhibitory effects on some cells and growth-promoting effects on others, and may have effects on the functional activity of cells that are unrelated to its effects on cell growth. Moreover, the effect of a growth factor on a cell frequently depends on the state of activation or differentiation of that cell as influenced by other growth factors and hormones. Because of the multiple activities of individual growth factors, the original assay-dependent nomenclature for lymphohemopoietic growth factors led to designation by different investigators of different names for the same growth factors. To avoid further confusion their is now general agreement that lymphocyte growth factors will be designated "interleukins" and will be numbered interleukin-1, interleukin-2, etc. in order of the determination of their nucleotide sequences. Some interleukins characterized and named before the establishment of the interleukin nomenclature (e.g., interferons, transforming growth factors, and tumor necrosis factors) will retain their original designations.

Interleukin-1

In 1972 Gery et al.[1] described a macrophage-derived factor that enhanced the thymocyte proliferative response to mitogens. Originally designated "lymphocyte activating factor," it soon became apparent that the factor had multiple activities, and at the Second International Lymphokine Workshop it 1979 it was given the name interleukin-1.

The initial findings suggested that macrophages were the sole source of IL-1, and most of the evidence concerning the role of IL-1 in cellular interactions within the immune system has come from studies of macrophages and their products. However, it is now known that IL-1 is produced by a wide variety of hematopoietic and nonhematoietic cells including B cells, NK cells, and keratinocytes.

DNA cloning has revealed two species: IL-1α and IL-1β. These two molecules share only 26% homology by DNA sequence analysis[2] but share a common IL-1 receptor. The biologic activities of IL-1 are mediated by the secreted forms of both IL-1α and IL-1β as well as by a macrophage membrane-associated form of IL-1α.[3]

IL-1 production by macrophages occurs only after activation. A variety of stimuli result in macrophage activation. Some, like bacteria and endotoxin, act directly on macrophages. Others, like most soluble proteins, do so via T cells. Two T cell–dependent macrophage activation pathways have been identified.[3] One involves antigen-induced T cell production of macrophage-activating cytokines, including tumor necrosis factors. The other involves a direct cell-contact–dependent effect, which remains to be completely elucidated. IL-1 production is also regulated by the neuroendocrine system in a classical feedback loop. Thus IL-1, in addition to its well known activity in the central nervous system as a pyrogen, stimulates pituitary cells to release adrenocorticotropic hormone (ACTH), luteinizing hormone, growth hormone, thyroid-stimulating hormone and corticotropin-releasing factor. ACTH acts on adrenals to produce corticosteroids, which suppress expression of IL-1 mRNA.

Once IL-1 is produced its major immunologic effect is to activate T cells to produce other cytokines critical to specific immune responsiveness, including interleukin-2 (IL-2) and.interleukin-6 (IL-6). IL-1 triggering of T cell cytokine production is acomplished by binding to specific IL-1 receptors on the T cell membrane. The IL-1 receptor appears to consist of two peptide chains. The gene for one of these chains, an 80-kDa IL-1 binding protein, has been cloned and belongs to the immunoglobulin superfamily as do receptors for several other growth factors with which IL-1 shares biologic activities, including platelet-derived growth factor, colony-stimulating factors, and IL-6.[4] IL-1 shares stimulatory effects on smooth muscle cells with platelet-derived growth factor; like colony-stimulating factors, IL-1 stimulates hematopoietic cells proliferation and differention; like IL-6, IL-1 has pyrogenic activity, induces synthesis and release of acute-phase reactants, prostaglandins, and collagenase, and stimulates growth of T cells and B cells.

The role of IL-1 in T cell proliferation has been extensively investigated, and has recently been reconsidered. As discussed below, T cell proliferation is known to be IL-2 dependent. This and the IL-1 dependency of T cell IL-2 production demonstrated by Smith et al in 1980,[5] led to the development of the concept that IL-1 provides the essential signal to T cells which, together with antigen or mitogen, triggers IL-2 production by the T cell as well as T cell IL-2 receptor expression. However, in the past few years this attractively simple concept has required modification in light of evidence that antibody to IL-1 fails to inhibit T cell proliferation induced in unfractionated populations of spleen cells,[6] and that IL-1 requires the presence of accessory cells to cause T cell activation. The reason for this, it now appears, is that IL-1 activation of T cells is probably mediated by IL-6 produced by IL-1–stimulated macrophages.[7] (The role of IL-6 in T cell activation is discussed below.)

The role of IL-1 in B cell activation, differentiation, and proliferation is receiving increasing attention.[8] IL-1 acts as a "progression signal" driving activated B cells through the S and into the G2 phase of the cell cycle without necessarily further driving the cell to proliferate. The fact that B cells produce as well as respond to IL-1 suggests that IL-1 may be involved in an autocrine B cell regulatory loop.

Given the wide variety of cells that produce IL-1 and the pleiotropic effects of IL-1 on multiple organ systems, it is not surprising that IL-1 activity is affected by a wide variety of naturally occurring inhibitors.[9] Some act as competitive inhibitors of binding to IL-1 receptors. Others may act by blocking intracellular activation pathways induced by IL-1. Given the central role of IL-1 in all inflammatory reactions, the identification and potential clinical use of such inhibitors is a matter of considerable current interest.

Interleukin-2

In 1976 Morgan and colleagues reported that long-term in vitro growth of T cells could be accomplished using a T cell growth factor (TCGF) that was present in crude phytohemagglutinin-stimulated lymphocyte conditioned media.[10] Shortly thereafter Gillis and Smith reported that it was possible to create, maintain, and clone TCGF-dependent antigen-specific cytolytic T cell lines.[11] The ability to clone normal T cells using TCGF, now known as interleukin-2 (IL-2), made possible studies of the cellular and biochemical requirements for T cell activation and the development of monoclonal antibodies that identified individual species of T cell antigen receptors. Since many of the T cell clones themselves released lymphocyte-active growth factors, the discovery of IL-2 also led indirectly to the identification of such other lymphocyte growth factors as interleukin-4 and interleukin-5.

T cells are the main source of IL-2.[12] Two signals appear to be necessary for its release: lectin or antigen binding to the T cell membrane, and a signal from adherent cells that, as described above, appears to be IL-6 produced by IL-1–activated macrophages. Analysis of cloned and sequenced IL-2 cDNA indicates that IL-2 is a 15.5-kDa peptide that, unlike IL-1, contains a classic hydrophobic signal sequence.[13] T cells release IL-2 by the same pathway as that of other secretory proteins.

Although the initial studies suggested that IL-2 was a specific T cell growth factor, it is now evident that B cells and NK cells also respond to IL-2. Responsiveness is determined by expression of specific IL-2 cell surface receptors. Binding studies have shown that there are three types of IL-2 receptors. One, a low-affinity receptor termed the IL-2 alpha chain (IL-2Rα), is a 55-kDa glycoprotein expressed only on activated cells.[14] The second is a 70- to 75-kDa protein termed the IL-2 beta receptor chain (IL-2Rβ) that binds IL-2 with intermediate affinity[15] and is constitutively expressed by both resting T cells and NK cells. The third is a heterodimer formed by the noncovalent surface-membrane association of the two types of receptors that binds IL-2 with high affinity. The high-affinity binding of the heterodimer appears to be due to the fact that IL-2 binds rapidly to IL-2Rα and dissociates slowly from the IL-2Rβ.[16]

Binding of IL-2 to the p55 type IL-2 receptor, now termed the IL-2 receptor alpha chain (IL-1α), is not sufficient to trigger T cells. Binding of IL-2 to the p75 IL-2 receptor on NK cells triggers enhanced lytic activity, expression of IL-2Rα, and subsequent proliferation.[17] By contrast, binding of IL-2 to IL-2Rβ on resting T cells fails to induce a proliferating signal unless monocytes are present.[18] T cell surface expression of IL-2Rα molecules appears to require mitogenic or antigenic stimulation together with triggering by IL-6, which is released by IL-1–stimulated monocytes-macrophages. The monocyte-dependent expression of IL-2Rα by IL-2–stimulated resting T cells permits IL-2 to induce a proliferative signal by binding to the resulting IL-2Rα/IL-2Rβ heterodimer high-affinity receptor. This may explain why stimulation of resting peripheral blood lymphocytes with high doses of IL-2 preferentially stimulates proliferation and activation of NK cells rather than mature T cells. The resulting activated and proliferating NK cells have enhanced lytic activity with broader target specificity than resting NK cells and have been termed "lymphokine-activated killer" (LAK) cells.[19] The ability to generate large numbers of tumor-cytolytic LAK cells from peripheral blood using recombinant IL-2 has led to the use of such cells together with IL-2 in the experimental treatment of cancer patients. Administration of IL-2 alone with or without LAK cells has been associated with approximately a 20% rate of clinical responses in cancer patients who have failed other forms of therapy.[20]

Like NK cells, the earliest intrathymic T cell precursors constitutively express IL-2R[21] and proliferate in the presence of IL-2 in the absence of any other exogenous stimulating agent.[22] After IL-2 stimulation, these early precursor T cells differentiate into mature effector T cells. These same cells consitutively produce as well as bind IL-2, suggesting that the earliest event in intrathymic T cell differentiation may be autocrine IL-2 stimulation of proliferation and differentiation.

Resting T cells express only IL-2Rβ. However, binding of antigen to the T cell surface antigen receptor induces IL-2Rα expression, and this is upregulated by IL-2 itself.[23]

As with T cell regulation, IL-2 plays an important role in both up-regulating and down-regulating B cell activation, proliferation, and differentiation. IL-2 directly enhances initial B cell activation and promotes differentiation of activated B cells.[24] However, it also stimulates differentiation and proliferation of T suppressor cells of B cell function.[25] As is true in general for the complex effects on lymphocyes of all cytokines, which of these disparate effects predominates in any given situation appears to depend on the concentrations of IL-2 available, the presence or absence of other B cell–active cytokines, and state of differentiation and activation of the target B cells.

In addition to its direct effects on lymphocytes, IL-2 has multiple indirect effects by virtue of the fact that it stimulates cells to produce other lymphocyte-active cytokines including IL-1, IL-6, tumor necrosis factor, interferon, and transforming growth factor β.[26,27] That each of these in turn affect IL-2 production and responsiveness further illustrates the complexity of the cytokine network.

Interleukin-3

Interleukin-3 (IL-3), a T lymphocyte–derived 28-kDa glycoprotein, originally described as a T cell factor that induced the expression of the enzyme 20α-hydroxysteroid dehydrogenase in cultures of spleen cells from athymic nide mice,[28] promotes the development of a wide variety of hematopoietic cells including lymphocytes, myeloid cells, and erythrocytes. IL-3 is not thought to be essential for normal hematopoiesis since hematopoiesis is normal in animals and tissues that lack IL-3–producing cells. Rather, IL-3 appears to provide a stimulus for the production of additional hematopoietic cells that may be needed during an ongoing immunologic response.

Interleukin-4

As reviewed by Paul and Ohara,[29] interleukin-4, a 20-kDa glycoprotein, was originally described as a T cell– and mast cell–derived B cell stimulatory factor by virtue of its ability to act as a costimulator with antibody to B cell surface immunoglobulin of in vitro B cell growth. The original designation of IL-4 as B cell stimulatory factor is appropriate since IL-4 can act on B cells before they enter the cell cycle, rendering them susceptible to activation by other stimuli, particularly IL-2. It has been shown that IL-4 specifically induces the IL-2Rβ chain on B cells whereas IL-5 induces the IL-2Rα chain.[30] Therefore, IL-4 may render resting B cells responsive to relatively high local concentrations of IL-2, such as might occur during local cell–cell interaction between a B cell and an IL-2–producing T cell coactivated by the same antigen. Subsequent exposure to IL-5 may then render these cells responsive to the proliferative effects of lower concentrations of IL-2, as might be present in the milieu of B cells no longer in direct contact with IL-2–producing T cells.

In addition to rendering B cells responsive to activation signals, IL-4 increases their expression of class II major histocompatibility antigens that are known to play a role in B cell antigen presentation to T cells, and increases expression of receptors for the Fc portion of immunoglobulin E (IgE) molecules. Studies in mice suggest that IL-4 may influence the secretion of specific Ig isotypes by B cells. Thus, addition of IL-4 to lipopolysaccharide-stimulated murine B cells results in the preferential secretion of IgG_1 and IgE, the Ig isotype responsible for most acute allergic reactions. However, similar isotype-specific stimulation of human B lymphocytes by IL-4 has not been observed. Instead, IL-4 seems to stimulate production of all human Ig isotypes.[31]

IL-4 is also a growth factor for T cells. This effect is due in part to IL-4 stimulation of T cell production of IL-2 and expression of IL-2R, and in part to a direct IL-2–independent proliferative effect on activated but not resting T cells.[32] The ability of IL-4 to promote growth of activated T cells may explain why IL-4 enhances the proliferation-dependent generation of specific cytotoxic T cells. In contrast to its effects on T cells, IL-4 has only transient stimulatory effects on the lytic activity of NK cells,[33] and inhibits the IL-2–dependent generation of nonspecific cytotoxic T cells and LAK cells when added to resting cells at the same time as IL-2. IL-4 has no inhibitory effect when added subsequent to addition of IL-2.[34]

Moreover, its inhibitory effect on generation of LAK cells is partially reversed by IL-1, indicating that IL-1 and IL-4 have opposing activities on responsiveness of NK cells to IL-2.[35] The inhibitory effects of IL-4 on IL-2–mediated NK cell activation and proliferation is likely due to IL-4–mediated inhibition of IL-2 receptor expression.[36] In light of the fact that NK cells can nonspecifically suppress immunoglobulin production by B cells, it is possible that IL-4 production during T cell–B cell collaboration prevents unwanted suppression by activated natural killer cells (i.e., LAK cells) induced by IL-2.

In addition to its effects on lymphocytes IL-4 also acts as a regulator of growth of nonlymphoid hematopoietic cells. In most instances IL-4 acts on nonlymphoid hematopoietic cells as a costimulator with other growth factors. For example, IL-4 enhances the IL-3–dependent growth of mast cell lines, stimulates macrophage antitumor cytotoxic activity, and synergizes with granulocyte–colony stimulating factor (G-CSF) or erythropoietin to promote the in vitro growth of granulocyte or erythroid colonies.

Interleukin-5

As reviewed by Swain et al.[37] and Sanderson et al.,[38] the discovery of what is now termed interleukin-5 resulted from the convergence of separate lines of investigation into the nature of B cell and eosinophil growth factors. In one set of studies, lymphocyte-conditioned media were found to contain a factor distinct from IL-2 and IL-4 that could replace the T cell requirement for in vitro generation of antibody production. Cloning of cDNA for this factor, termed B cell growth factor II (BCGFII), showed that it was a glycoprotein with a molecular weight of 45,000—considerably greater than that of IL-2 or IL-4. In another set of studies Sanderson and coworkers found that T cell clones from parasite-infected mice produced a 45-kDa glycoprotein termed eosinophil differentiation factor (EDF) that enhanced eosinophil differentiation from immature precursors. In 1986 Sanderson and colleagues demonstrated that BCGFII and EDF were the same molecule that was designated interleukin-5 (IL-5). Cloning of IL-5[39] showed that the molecule is a dimer of two identical glycoprotein chains. In addition to stimulating eosinophil differentiation, recombinant IL-5 (a) stimulates B cell growth, particularly the $CD5^+$ B cell population thought to be the main source of autoantibodies,[40] (b) stimulates B cell immunoglobulin production, particularly IgA, which it selectively stimulates in conjunction with costimulation by transforming growth factor β,[41] (c) induces high-affinity IL-2 receptors on thymocytes by selective enhancement of expression of the p55 alpha chain of the IL-2 receptor, and (d) enhances generation of cytotoxic T lymphocytes.

Interleukin-6

Interleukin-6 (IL-6) was first identified as a B cell differentiation factor (BCDF) by its ability to induce antibody secretion by activated B cells without first inducing cellular proliferation.[42] As was the case with IL-5, a convergence of results from multiple laboratories using separately cloned genes led to the realization that a single molecule was responsible for multiple activities influencing a wide variety of different cell types and tissues. Thus, some investigators isolated what is now known to be IL-6 as the product of a cloned cDNA encoding a granulocyte macrophage—colony stimulating factor,[43] others isolated IL-6 by cloning a cDNA encoding a T cell activating factor (TAF),[44] and yet others identified IL-6 by examining the biologic activities of a protein encoded for by a cloned cDNA for interferon-β.[45] IL-6 is a 26-kDa glycoprotein that, upon appropriate stimulation, is produced by a wide variety of cells, including T cells, B cells, monoyctes-macrophages, fibroblasts, endothelial cells, mesangial cells, and keratinocytes. It binds to specific receptors that are distributed on various cell types including hepatocytes, activated B cells, and resting T cells. A single cDNA clone encoding both a high-affinity and low-affinity binding site for IL-6 has been isolated, and like the IL-1 receptor, it belongs to the Ig gene superfamily.[46] It is produced constitutively by some tumor cells including human myeloma cells which, in turn, respond to IL-6 by proliferation. This suggests that IL-6 may be an autocrine growth factor in the development of multiple myeloma. Support for this notion has come from findings of Yamasaki and coworkers,[47] who showed that transgenic mice that constitutively overexpressed IL-6 develop plasmacytomas.

IL-6 is produced during the course of an immune response and has multiple effects on growth and differentiation of both T and B lymphocytes. As described above, IL-6, released by IL-1 or tumor necrosis factor–activated monocytes-macrophages, induces antigen or mitogen-activated T cells to express IL-2Rα molecules that then combine with constitutively produced IL-2Rβ molecules to form the high-affinity IL-2R required for ongoing IL-2–dependent proliferation. Together with gamma interferon and IL-2, IL-6 is also required for differentiation of cytotoxic T cells.[48] IL-6 acts synergistically with IL-1 to promote IgM and with IL-5 to promote IgA synthesis by activated B cells.[49]

In addition to its effects on lymphocytes, IL-6, like IL-1, regulates the acute inflammatory response by inducing hepatocyte production of acute-phase proteins, and acts on the central nervous system as a pyrogen. The striking parallel between many of the biologic activities of IL-6 and IL-1 and the requirement of IL-1 for IL-6 production raises the possibility that many of the biologic effects previously attributable to IL-1 may in fact be indirect effects of IL-1 directly mediated by IL-1–indiced IL-6. Studies using specific neutralizing antibodies specific for each of these cytokines should help clarify this issue.

Interleukin-7

Interleukin-7 (IL-7) is a 25-kDa protein originally detected as a constitutive product of bone marrow stromal cells transfected with the transforming sequences of SV40.[50] The supernatants of these cells supported the in vitro growth of B cell precursors and could replace the need for stromal cells in long-term bone marrow cultures. Murine and human IL-7 cDNAs have been cloned and show a strong degree of homology (60%).

IL-7 induces proliferation of immature B cells and both immature and mature T cells. In the B cell lineage IL-7 induces proliferation of both pro–B cells, the earliest identified cells committed to the B cell lineage that contain immunoglobulin light and heavy chain genes in unrearranged "germ line" configuration, and pre–B cells that rearrange and express cytoplasmically the immunoglobulin heavy chain genes, but does not stimulate mature B cells, which express surface but not cytoplasmic immunoglobulin. By contrast, IL-7 stimulates both immature and mature T cells to proliferate, perhaps by enhancing T cell IL-2 receptor expression.[51]

Interleukin-8

Interleukin-8 (IL-8) was originally detected as a monocyte-derived neutrophil chemotactic factor.[52,53] Recent studies indicate that IL-8 is also chemotatic for T cells.[54] Its expression pattern is most closely related to that of IL-6. Like IL-6, IL-8 is induced by IL-1 and tumor necrosis factor, and inhibited by glucocorticoids. IL-8 is produced by a wide variety of cells including T cells, fibroblasts, keratinocytes, and endothelial cells[55] and appears to be a key component of the inflammatory response.

Interleukin-9

Interleukin-9 (IL-9) was originally isolated by functional expression cloning of a T cell–derived factor'that stimulated growth of a human IL-3–dependent leukemic cell line M-07E. This human factor was shown to be the equivalent of a murine T cell growth factor P40.[56] Recent findings indicate that its primary hematopoietic effect is to stimulate erythropoiesis selectively.[57] The range of cells that produce IL-9, stimuli for its production, the nature of its receptor, and its role in lymphocyte proliferation and differentiation remain to be determined.

Transforming Growth Factor–β

Transforming growth factor–beta (TGFβ) is a polypeptide first identified by its ability to transform certain normal cells in vitro.[58] However, it is now clear that the main importance of TGFβ is its activity as a ubiquitous mediator of growth, differentiation, and functional activity of normal cells. Almost all normal cells both produce TGFβ and have receptors for TGFβ. As discussed below, TGFβ has pleiotropic effects on normal lymphocytes at extremely low (femtomolar) concentrations. In fact, the immunosuppressive effects of TGFβ make it molecule-for-molecule the most potent endogenous immunosuppressive molecule identified to date.

Originally purified from human platelets, the richest source in the body, TGFβ is a stable polypeptide consisting of two identical chains each containing 112 amino acids. Highly conserved in evolution, its amino acid sequence is identical in humans, monkeys, pigs, and chickens. In addition to the initially characterized form (TGFβ1), several other forms have been isolated. One of these, TGFβ2, is a close homolog of TGFβ1 found primarily in bone that appears to play an important role in cartilage formation by inducing synthesis of collagen and proteoglycans.

Although it was originally discovered as a growth factor, TGFβ is a strong growth inhibitor for many normal cells including liver cells, fibroblasts, keratinocytes, epithelial cells, and lymphocytes. In general, TGFβ stimulates growth of mesenchymal cells and inhibits growth of epithelial cells.

TGFβ appears to play an important role at both the onset and resolution stages of the inflammatory response.[59] Thus, after any local injury platelet aggregation and degranulation results in release of numerous mediators of inflammation including TGFβ, which has a potent chemotactic factor for circulating monocytes. TGFβ then activates the locally recruited monocytes to produce more TGFβ as well as other inflammatory mediators including IL-1, platelet-derived growth factor, fibroblast growth factor, and tumor necrosis factor. Together, these factors act to promote local vascular growth and fibrosis. IL-1 and tumor necrosis factor activate local cells to produce IL-6 and IL-8. IL-6 with IL-1 induces fever and hepatic synthesis of acute-phase proteins and activates IL-2 and IL-2 receptor expression by locally accumulated lymphocytes recruited by IL-8. IL-2 promotes growth and activation of T cells and NK cells, which in turn further amplify the reaction. As discussed below, TGFβ, perhaps released by activated T cells and NK cells themselves, then acts to reduce further progression of the reaction by inhibiting both T cell and NK cell growth and functional activity.

Given the potent and multiple effects of TGFβ on the inflammatory response and the fact that TGFβ is produced by nearly all cells, it is obvious that there must be strong mechanisms for regulating TGFβ activity. Most cells release TGFβ in a latent form that does not bind to TGFβ receptors unless cleaved to its active form by proteolytic enzymes (in vitro, TGFβ can be activated by heat or acidification). Monocytes constitutively produce and release TGFβ in its latent form, but only activated monocytes are able to cleave TGFβ to its active form, perhaps by release of such proteolytic enzymes as plasmin and cathepsin D. Therefore, processing of the latent form of TGFβ is probably an important mechanism for regulating the activity of locally released TGFβ.

Another important immunoregulatory mechanism for controlling TGFβ inflammatory activity involves its effects on lymphocytes. Thus, activation of T cells, B cells, and NK cells causes them to increase their production and secretion of TGFβ which, in turn, acts as a potent inhibitor of T cell, B cell, and NK cell proliferation and activation induced by IL-1 and IL-2. Thus, TGFβ acts as an autocrine lymphocyte regulator, thereby dampening the lymphocyte component of the inflammatory response.

The mechanisms of inhibition of lymphocyte growth and activation by TGFβ are currently matters of intense interest. Although TGFβ inhibits IL-1-induced expression of IL-2 and IL-2 receptors by some cells,[60] in other instances TGFβ does not inhibit IL-2 production by IL-1-stimulated cells but rather appears to inhibit IL-2-dependent S-phase progression and mitosis by suppressing growth factor-dependent transcriptional activation of the proto-oncogene c-myc.[61] As a result, T cells accumulate in a nonproliferative (G_0/G_1) phase of the cell cycle. This is overcome by high concentrations of IL-2, suggesting that TGFβ has a modulatory rather than blocking effect on T cell growth. TGFβ probably acts in a similar way to inhibit B cell proliferation since activated B cells accumlate at the G_1/S interphase in the presence of TGFβ.[62] In contrast to its inhibitory effects on B cell proliferation and production of IgG and IgM, TGFβ enhances B cell production of IgA, apparently by promoting a switch in Ig heavy chain synthesis to the production of the alpha heavy chain.[63] It also enhances B cell expression of class II major histocompatibility antigens, which are known to be critical to T cell-B cell interaction.

Given the potent immunosuppressive effects of TGFβ and its production by lymphocytes, it is not surprising that a variety of different cell-mediated immunosuppressive phenomena appear to be mediated by TGFβ. Thus, Clark et al. have shown that nonspecific suppression of maternal rejection of allogeneic fetuses is mediated by TGFβ released by uterine decidual cells.[64] Recently we have obtained evidence (J. Kaplan, *unpublished observations*) that TGFβ mediates the natural suppressor activity of neonatal lymphocytes and appears to be responsible for the failure of infants to mourt immune responses to certain antigens.

Summary

A large number of different growth factors regulate lymphocyte proliferation, differentiation, and functional activity. The multiple cellular sources and multiple acitivities of each of these factors indicates that they have been used for many purposes during the course of evolution. Perhaps the most important lesson from consideration of the multiple biologic effects of each factor is that the activity of any given lymphocyte growth factor on any given lymphocyte depends on the state of that cell as determined by other growth factors.

Biomedical Perspective

The identification and cloning of lymphokines and their cell-surface receptors, and the development of monoclonal antibodies to these moieties, has spurred a rapid increase in our understanding of the molecular basis for cell–cell interactions in the lymphohematopoietic system. The knowledge gained in this field has broad implications for our understanding of the molecular control of cell growth and differentiation in general. In addition, the ability to produce on an industrial scale large quantities of these growth factors, receptors, and corresponding monoclonal antibodies has set the stage for increasing application of "biological response modifier" (BRM) therapy in clinical medicine. IL-2, with and without coadministration of LAK cells, is already being used with some success in the treatment of advanced metastatic cancers.[20] The use of IL-2 to enhance the immune responses to vaccines and the use of toxin-conjugated IL-2 and monoclonal antibodies to IL-2 receptors as immunosuppressive agents for control of transplant rejection and auto-immune diseases is under active investigation. In the near future we are likely to see development of potent immunosuppressive drugs that act by blocking the binding of specific lymphokines to their cell-surface receptors. More detailed knowledge of the molecular requirements for triggering cells via cell-surface receptors for lymphokines will likely also lead to development of synthetic lymphokine agonists useful as immunopotentiating agents.

References

1. Gery I, Gershon R, Waksman BH. Potentiation of the thymocyte response to mitogens. I. The responding cells. *J Exp Med.* 1972;136:128–136.

2. March CJ, Mosley B, Larsen A, Cerretti DP, Braedt G, Price V, Gillis S, Henney CS, Krohnheim SR, Grabstein K, Conlon PJ, Hopp TP, Cosman D. Cloning, sequence, and expression of two distinct human interleukin-I complementary DNAs. *Nature.* 1985;315:641–647.

3. Weaver CT, Unanue ER. The costimulatory function of antigen presenting cells. *Immunol Today.* 1990;11:49–55.

4. Sims JE, March CJ, Cosman D, Widmer MB, MacDonald HR, McMahan CJ, Grubin CE, Wignall JM, Jackson JL, Call SM, Friend D, Alpert AR, Gillis S, Urdal DL, Dower SK. cDNA expression cloning of the IL-1 receptor, a member of the immunoglobulin superfamily. *Science.* 1988; 241:585–589.

5. Smith K, Gilbride KJ, Favata MF. Lymphocyte activating factor promotes T-cell growth factor production by cloned murine lymphoma cells. *Nature.* 1980;287:853–855.

6. Mizel SB. Interleukin 1 and T cell activation. *Immunol Today.* 1987;8: 330–332.

7. Wong GG, Clark SC. Multiple actions of interleukin 6 within a cytokine network. *Immunol Today.* 1988;9:137–139.

8. Gordon J, Guy GR. The molecules controlling B lymphocytes. *Immunol Today.* 1987;8:339–344.

9. Larrick JW. Native interleukin 1 inhibitors. *Immunol Today.* 1989;10:61–66.

10. Morgan DA, Ruscetti FW, Gallo RC. Selective in vitro growth of T lymphocytes from normal human bone marrow. *Science.* 1976;193: 1007–1008.

11. Gillis S, Smith KA. Long-term culture of tumor-specific cytotoxic T cells. *Nature.* 1977;268:154–156.

12. Smith KA, Baker PE, Gillis S, Ruscetti FW. Functional and molecular characteristics of T cell growth factor. *Mol Immunol.* 1980;17:579–589.

13. Taniguchi T, Matsui H, Fujita T, Takaoka C, Kashima N, Yoshimoto R, Hamuro J. Structure and expression of a cloned cDNA for human interleukin-2. *Nature.* 1983;302:305–310.

14. Uchiyama T, Broder S, Waldmann TA. A monoclonal antibody (anti-Tac) reactive with activated and functionally mature human T cells. 1. Induction of anti-Tac monoclonal antibody and destribution of Tac+ cells. *J Immunol.* 1981;126:1393–1397.

15. Siegel JP, Sharon M, Smith PL, Leonard WJ. The IL-2 receptor beta chain (p70): role in mediating signals for LAK, NK, and proliferative activities. *Science.* 1987;238:75–77.

16. Smith K. Interleukin-2: inception, impact, and implications. *Science.* 1988;240:1169–1176.

17. London L, Perussia B, Trinchieri G. Induction of proliferation in vitro of resting human natural killer cells: expression of surface activation antigens. *J Immunol.* 1985;134:718–727.

18. Ben Aribia M-H, Moire N, Metivier D, Vaquero C, Lantz O, Olive D, Charpentier B, Senik A. IL-2 receptors on circulating natural killer cells and T lymphocytes. Similarity in number and affinity but difference in transmission of the proliferation signal. *J Immunol.* 1989;142:490–499.

19. Herberman RB, Heberman RB, Hiserodt J, Vujanovic N, Balch C, Lotzova E, Bolhuis R, Golub S, Lanier LL, Phillips JH, Riccardic, Ritz J, Santoni A, Schmidt RE, Uchida A. Lymphokine-activated killer cell activity. Charac-

teristics of effector cells and their progenitors in blood and spleen. *Immunol Today*. 1987;8:178–181.

20. Rosenberg SA, Lotze MT, Mule LM, Leitman S, Chang AE, Ettinghausen SE, Matori YL, Skibber JM, Shiloni E, Vetto JT, Seipp CA, Simpson C, Reichert CM. Observations on the systemic administration of autologous lymphokine-activated killer cells and recombinant interleukin-2 to patients with metastatic cancer. *N Eng J Med*. 1986;313:1485–1489.

21. Haynes BF, Martin ME, Kay HH, Kurtzberg J. Early events in human T cell ontogeny. Phenotypic characterization and immunohistologic localization of T cell precursors in early human fetal tissues. *J Exp Med*. 1988;168:1061–1080.

22. De la Hera A, Toribio MS, Marcos MAR, Marquez C, Martinez AC. Interleukin-2 pathway is autonomously activated in human T11$^+$3$^-$4$^-$8$^-$ thymocytes. *Eur J Immunol*. 1987;17:683–687.

23. Jankovic DL, Gibert M, Baran D, Ohara J, Paul Wm E, These J. Activation by IL-2 but not IL-4, up-regulates the expression of the p55 subunit of the Il-2 receptor on IL-2 and IL-4-dependent T cell lines. *J Immunol*. 1989;142:3113–3120.

24. Jelinek DR, Lipsky PE. Regulation of human B lymphocyte activation, proliferation, and differentiation. *Adv Immunol*. 1987;40:1–59.

25. Hirohata S, Lipsky PE. T cell regulation of human B cell proliferation and differentiation. Regulatory influences of CD45R$^+$ and CD45R$^-$ subsets. *J Immunol*. 1989;142:2597–2607.

26. Kovacs EJ, Beckner SK, Longo DL, Varesio L, Young HA. Cytokine gene expression during the generation of human lymphokine-activated killer cells: early induction of interleukin-1β by interleukin-2. *Cancer Res*. 1989;49:940–943.

27. Kasid A, Director EP, Rosenberg SA. Induction of endogenous cytokine-mRNA in circulating peripheral blood mononuclear cells by IL-2 administration to cancer patients. *J Immunol*. 1989;143:736–739.

28. Ihle JN, Weinstein Y. Immunological regulation of hematopoietic/lymphoid stem cell differentiation by interleukin 3. *Adv Immunol*. 1986;39:1–50.

29. Paul WE, Ohara J. B-cell stimulatory factor-1/interleukin 4. *Annu Rev Immunol*. 1987;5:429–460.

30. Loughnan MS, Nossal GJV. Interleukins 4 and 5 control expression of IL-2 receptor on murine B cells through independent induction of its two chains. *Nature*. 1989;340:76–79.

31. Splawski JB, Jelinek DF, Lipsky PE. Immunolmodulatory role of IL-4 on the secretion of Ig by human B cells. *J Immunol*. 1989;142:1569–1575.

32. Mitchell LC, Davis LS, Lipsky PE. Promotion of human T lymphocyte proliferation by IL-4. *J Immunol*. 1989;142:1548–1557.

33. Peace DJ, Kern DE, Schultz KR, Greenberg PD, Cheever MA. IL-4-induced lymphokine-activated killer cells. Lytic activity is mediated by phenotypically distinct natural killer-like and T cell-like large granular lymphocytes. *J Immunol*. 1988;140:3679–3685.

34. Spits H, Yssel H, Paliard X, Kastelein R, Figdor C, De Vries JE. IL-4 inhibits IL-2 mediated induction of human lymphokine-activated killer cells, but not the generation of antigen-specific cytotoxic T lymphocytes in mixed leukocyte cultures. *J Immunol*. 1988;141:29–36.

35. Keever CA, Pekle K, Gazzola MV, Collins NH, Bourhis JH, Gillio A. Natural killer and lymphokine-activated killer cell activities from human

marrow precursors. II. The effects of IL-3 and IL-4. *J Immunol.* 1989;143:3241–3249.

36. Martinez OML, Gibbons RS, Gadovoy MR, Aronson FR. IL-4 inhibits IL-2 receptor expression and IL-2-dependent proliferation of human T cells. *J Immunol.* 1990;144:2211–2215.

37. Swain SL, McKenzie DT, Dutton RW, Tongonogy SL, English M. The role of IL-4 and IL-5: characterization of adistinct helper T cell subset that makes IL-4 and IL-5 (Th2) and requires priming before induction of lymphokine secretion. *Immunol Rev.* 1988;102:77–106.

38. Sanderson CJ, Campbell HD, Young IG. Molecular and cellular biology of eosinophil differentiation factor (interleukin 5) and its effects on human and mouse B cells. *Immunol Rev.* 1988;102:29–50.

39. Yokota T, Coffman RL, Hagiwara H, Rennick DM, Takebe Y, Yokoda K, Gemmell L, Shrader B, Yang G, Meyerson P, Luh J, Hoy P, Pène J, Brière F, Spitz H, Banchereau J, De Vries J, Lee FD, Arai N, Arai K. Isolation and characterization of lymphokine cDNA clones encoding mouse and human IgA-enhancing factor and eosinophil colony-stimulating factor activities: Relationship to interleukin 5. *Proc Nat Acad Sci USA.* 1987;84:7388–7392.

40. Hitoshi Y, Yamaguchi N, Mita S, Sonoda E, Takaki S, Tominaga A, Takatsu K. Distribution of IL-5 receptor-positive B cells. Expression of IL-5 receptor on LY-1 (CD5$^+$) B cells. *J Immunol.* 1990;144:4218–4225.

41. Kim P-H, Kagnoff MF. Transforming growth factor β1 is a costimulator for IgA production. *J Immunol.* 1990;144:3411–3416.

42. Hirano T, Yasukawa K, Harada H, Taga T, Watanabe Y, Matsuda T, Kashiwamura S, Nakajima K, Koyama K, Iwamatsu A, Tsunasawa S, Sakiyama F, Matsui H, Takahara Y, Taniguchi T, Kishimoto T. Complementary DNA for a novel human interleukin (BSF-2) that induces B lymphocytes to produce immunoglobulin. *Nature.* 1986;324:73–76.

43. Wong GG, Witek-Giannotti JS, Temple PA, Kriz R, Ferenz C, Hewick RM, Clark SC, Ikebuchi K, Ogawa M. Stimulation of murine hemopoietic colony formation by human IL-6, *J Immunol.* 1988;140:3040–3044.

44. Garman RD, Jacobs KA, Clark SC, Raulet DH. B cell-stimulatory factor$_2$ (β$_2$ interferon) fuctions as a second signal for interleukin-2 production by mature mureine T cells. *Proc Natl Acad Sci USA.* 1987;84:7629–7633.

45. Poupart P, Vadenabeele P, Cayphas S, Van Snick J, Haegeman G, Kruys V, Fiers W, Content J. B cell growth modulating and differentiating activity of recombinant human 26 kd protein (BSF-2, HuIFN-β$_2$). *Eur Mol Biol Organ J.* 1987;6:1219–1224.

46. Kishimoto T, Hirano T. Molecular regulation of B lymphocyte response. *Annu Rev Immunol.* 1988;6:485–512.

47. Yamasaki K, Taga T, Matsuda T, Snematsu S, Hirano T, Kishimoto T. Interleukin 6 and its receptor. In: Fradelizi D, Bertoglio J, eds. *Lymphokine receptor interactions.* London: John Libbey Eurotext; 1989:143–150.

48. Takai Y, Wong GG, Clark SC, Burokoff SJ, Herrmann SH. B cell stimulatory factor-2 is involved in the differentiation of cytotoxic T lymphocytes. *J Immunol.* 1988;140:508–512.

49. Kunimoto DY, Nordan RP, Strober W. IL-6 is a potent cofactor of IL-1 in IgM synthesis and of IL-5 in IgA synthesis. *J Immunol.* ˹89;143:2230–2235.

50. Henney CS. Interleukin 7: effects on early events in lymphopoiesis. *Immunol Today*. 1989;10:170–173.

51. Welch PA, Namen AE, Goodwin RG, Human IL-7: a novel T cell growth factor. *J Immunol*. 1989;143:3562–3567.

52. Yoshimura T, Matsushima K, Oppenheim JJ, Leonard EJ. Neutrophil chemotactic factor produced by lipopolysaccharide (LPS)-stimuated human blood mononuclear leukocytes: partial characterization and searation from interleukin 1 (IL-1). *J Immunol*. 1987;139:788–793.

53. Matsushima K, Morishita K, Yoshimura T, Lavu S, Kobayashi Y, Lew W, Appella E, Kung HF, Leonard EJ, Oppenheim JJ. Molecular cloning of a human monocyte-derived neutrophil chemotactic factor (MDNCF) and the induction of NDNCF mRNA by interleukin 1 and tumor necrosis factor. *J Exp Med*. 1988;167:1883–1888.

54. Larsen CG, Anderson AO, Apella E, Oppenheim JJ, Matsushima K. The neutrophil activating activity protein (NAP-1) is also chemotoactic for T lymphocytes. *Science*. 1989;243:1464–1465.

55. Schroder JM, Mrowietz U, Christophers E. Purification and partial characterization of a human lymphocyte-derived peptide with potent neutrophil-stimulating activity. *J Immunol*. 1988;140:3534–3540.

56. Yang Y-C, Ricciardi S, Ciarletta A, Calvetti J, Kelleher K, Clark SC. Expression cloning of a cDNA encoding a novel human hematopoietic growth factor: human homologue of murine T-cell growth factor p40. *Blood*. 1989;74:1880–1883.

57. Donahure RE, Yang Y-C, Clark SC. Human P40 T-cell growth factor (interleukin-9) supports erythroid colony formation. *Blood*. 1990;75: 2271–2275.

58. Roberts AB, Sporn MB. The transforming growth factor-betas. In: Sporn MB, Roberts AB, eds. *Peptide growth factors and their receptors: handbook of experimental pharmacology*. New York: Springer-Verlag; 1989.

59. Wahl SM, McCartney-Francis N, Mergenhagen SE. Inflammatory and immunomodulatory roles of TGF-β. *Immunol Today*. 1989;10:258–261.

60. Espevik T, Waage A, Faxvaag A, Shalaby MR. Regulation of interleukin-2 and interleukin-6 production from T cells: involvement of interleukin-1 beta and transforming growth factor-beta. *Cell Immunol*. 1990;126:47–56.

61. Ruegemer JJ, Ho SN, Augustine JA, Schlager JW, Bell MP, McKean DJ, Abraham RT. Regulatory effects of transforming growth factor-beta on IL-2- and IL-4-dependent T cell-cycle progression. *J Immunol*. 1990;144:1767–1776.

62. Cross D, Cambier JC. Transforming growth factor β1 has differential effects on cell proliferation and activation antigen expression. *J Immunol*. 1990; 144:432–439.

63. Lebman DA, Lee FD, Coffman RL. Mechanism of transforming growth factor beta and IL-2 enhancement of IgA expression in lipopolysaccharide-stimulated B cell cultures. *J Immunol*. 1990;144:952–959.

64. Clark DA, Falbo M, Rowley RB, Banwaff D, Stedronska-Clark J. Active suppression of host-vs-graft reaction in pregnant mice. IX. Soluble suppressor activity obtained from allopregnant mouse decidua that blocks they cytolytic effector response to IL-2 is related to transforming growth factor β. *J Immunol*. 1988;141:3833–40.

6
The Inhibitory Effects of Growth Factors and Cytokines on Cell Proliferation

DAVID GOLDSTEIN and GEORGE WILDING

Our understanding of the field of growth factors and their influence on cell proliferation has provided insight into the growth control mechanisms of cancer cells. Various lines of evidence have suggested that cancer cells not only produce peptide growth factors that normally control cell proliferation during processes such as organogenesis and wound healing, but they also have receptors for those factors and can respond to these factors in a positive or negative fashion.[1] This is called autocrine growth control. A typical example of autocrine growth control is the production of bombesin by small cell lung cancer cell lines.[2] These cells not only produce, secrete, and bind bombesin but they are stimulated to grow by bombesin. Monoclonal antibodies to the C-terminal region of bombesin inhibit bombesin binding and cell growth both in vitro and in xenografts.[2] A variety of tumor cell types have now been shown to secrete and respond to a wide array of growth factors in an autocrine manner.[1]

As the number of identified and characterized oncogenes has grown, some of these gene products have been shown to represent abnormal forms of growth factor or cell-surface receptor cellular genes, proto-oncogenes.[3,4] For example, a number of transmembrane receptors such as the epidermal growth factor (EGF) receptor, platelet-derived growth factor (PDGF) receptor, the insulin receptor, the colony-stimulating factor 1 (CSF-1) receptor, and others have protein kinase activity.[5] The binding of ligand to the receptor molecule results in phosphorylation of the receptor itself and other proteins. Alterations in the receptor proteins by mutation can alter their activity and regulation leading to uncontrolled growth. Two of the viral oncogenes are homologous to genes encoding growth factor receptors. The gene product of v-erb B represents a truncated and constitutively activated form of the EGF receptor,[6] whereas the gene product of v-fms is homologous to the CSF-1 receptor.[7] Another oncogene, neu, codes for a 185,000 molecular weight protein with the characteristics of a cell-surface receptor.[8] It is homologous but distinct from the EGF receptor and it is activated by a point mutation. Amplification and high levels of this gene have been found in many breast

TABLE 6.1. Families of inhibitory factors.

TGFβ:
  Human   1
  Human   2
  Human   3
  Chicken   4
  Amphibian 5
TGFβ related:
  Vg1 *Xenopus laevis*
  Decapentaplegic complex *Drosphila* (DPP-C)
  Mammalian Müllerian inhibitory substance (MIS)
  Inhibins
  Activins
Tumor necrosis factor:
  TNFα
  TNFβ
Interleukin:
  IL-1
  IL-6
Interferons:
  IFNα
  IFNβ
  IFNγ

cancers.[9] Alternatively, other cancer cell lines have shown high levels of receptor expression with the presence of many receptors with high affinities to circulating factors.[10-12]

To date, the gene product of only one viral oncogene has been found to be homologous to a gene whose protein product is a growth factor. The *sis* oncogene is homologous to the gene for the α-chain of PDGF.[13] In addition to oncogenes whose number or expression is increased in malignancy, several genes whose absence is associated with malignancy have been identified. These genes (e.g., the retinoblastoma gene) presumably secrete products that inhibit uncontrolled cell proliferation.[14] Despite the apparent importance of growth factors in controlling cell growth, there is little evidence that in vivo production of growth factors induces malignancy. Rather, malignant transformation of cells may be accompanied by altered production of and response to growth factors and their receptors.

In this chapter we examine the role of growth factors and cytokines in the growth control mechanisms of cancer cells. In particular, we focus on the growth inhibitory effects of these factors and their potential use as a therapeutic modality (Table 6.1). In the first section we address the inhibitory growth factors focusing on the transforming growth factor–beta (TGFβ) family. These factors not only represent some of the most potent cell inhibitors known but also influence a myriad of cell types in a paracrine fashion. Among their other activities, members of the TGFβ

TABLE 6.2. Biologic effects of TGFβ.
_____

Cell growth in vitro
  Inhibits the growth of many normal and malignant epithelial cells
  Stimulates the growth of many mitogens
  Stimulates anchorage-independent growth of nonmalignant fibroblasts
Differentiation
  Inhibits adipogenic differentiation of 3T3 cells
  Inhibits differentiation of cultured myoblasts
  Stimulates terminal differentiation of bronchial epithelial cells
Wound healing
  Enhances would healing
  Induces cartilage formation
  Induces fibroblast chemotaxis
  Modulates plasminogen activator activity and induces endothelial-type plasminogen
    activator inhibitor
  Enhances production of connective tissue components
  Regulates acute-phase protein expression and secretion
Immune response
  Impairs immune surveillance
  Modifies B and T cell growth
  Regulates NK cell cytotoxic activity
  Affects B cell function and immunoglobulin production
In vivo
  Inhibits human lung tumor xenografs
  Reversibly inhibits mammary gland growth
_____

Modified from Keski-Oja J, Postlethwaite AE, Moses HL., Transforming growth factors in the regulation of malignant cell growth and invasion. *Cancer Invest.* 1988;6:705–724.

family can induce differentiation and cell motility, stimulate fibroblasts and stroma formation, and alter bone metabolism. All of these actions may play important roles in tumor establishment, growth, invasion, and metastases. In the second part of this chapter we examine the effects of a variety of cytokines on cell proliferation. The interferons, tumor necrosis factor, and interleukins 1 and 6 have been shown to exert antiproliferative effects in vitro and in animal models. We review these cytokines as negative growth factors and discuss their potential in the autocrine models of neoplastic growth.

Inhibitory Growth Factors

Control of epithelial cell growth is poorly understood. Most epithelial cells exist in a tightly controlled state maintained by a balance between the effects of stimulatory and inhibitory factors. Holley proposed that transformed cells escape this control and require fewer exogenous growth factors in culture than their normal counterparts.[15] This observation formed the basis of the autocrine growth control hypothesis whereby transformed cells produce and respond to their own growth factors.[1,16]

Transforming growth factors (TGF) that have the ability to support anchorage-independent growth of indicator cell lines include the TGFα and the TGFβ family of growth factors. TGFα generally supports growth in a wide variety of cell types.[17] The TGFβs, on the other hand, exhibit both stimulatory and inhibitory properties, depending on the cell type examined[18] (Table 6.2). For example, normal rat kidney (NRK) cells require both TGFα and a member of the TGFβ family to form colonies optimally in soft agar medium, whereas mouse embryo AKR 2B cells require only TGFβ to grow under anchorage-independent conditions. Overall, the effects of TGFβ on epithelial cells are complex and are probably a manifestation of growth regulatory factors and their receptors that are present in the cell at a given time.[19]

The initial observation that density arrested BSC-1 cells secrete a growth inhibitor identical to or closely related to TGFβ raised the possibility that autocrine inhibitory factors may be contributory factors in growth control.[20] However, just as important may be the effects these factors have on neighboring cells and surrounding tissue. For example, TGFβ seems to promote chemotaxis and pericellular proteolysis, as well as stimulate stromal development.[21] This portion of our chapter on the effects of cytokines and growth factors focuses not only on the negative effects of the TGFβ family in epithelial proliferation but also examines the paracrine effects of TGFβ on the surrounding cellular and stromal elements. In the end, the net effect of these factors on tumor growth will represent the summation of these disparate influences.

TGFβ was initially discovered as one of the components of conditioned medium of murine sarcoma virus transformed fibroblasts that supported anchorage-independent growth of nontumorigenic fibroblasts.[22] In addition, TGFβ was discovered as a stimulatory factor for AKR2B mouse embryo cells and NRK 49F cells.[19,23] As mentioned above, Holley had described an inhibitor of BSC-1 cells that was identical to TGFβ.[20] Thus, from the beginning it was apparent that TGFβ was a pluripotent molecule with multifunctional effects on a wide variety of cells.

TGFβ1 is ubiquitous; platelets, however, comprise one of the richest sources of TGFβ1.[24] The active growth factor has a molecular weight of 23,000 and is a homodimer of two 12,500 molecular weight subunits with 112 amino acids each. The dissociated subunits are inactive.[24] TGFβ1 is derived from a larger precursor and is secreted from the cell in a latent form.[25] In addition to acid activation, the latent form of the molecule undergoes proteolytic activation.[26] It binds to a cell-surface receptor that is a dimeric glycoprotein of 560,000 molecular weight with lower molecular weight forms also described.[27-29] No enzymatic activity has been demonstrated, although receptor binding activity can be found in most cells.[30,31]

TABLE 6.3. Common characteristics of the TGFβ family.

Characteristic	TGFβ				
	1	2	3	4	5
Signal peptide: first methionine preceded by about 22 hydrophaobic a.a.	Y	Y	Y	N	Y
Conserved a.a. #39–49	Y	Y	Y	N	Y
Conserved glycosylation sites	Y	Y	Y	N	Y
Putative integrin binding site (ca. 233–235 β5)	Y	N	Y	Y	Y
Tetrabasic amino acid site (RKKR—processing of latent form—ca. 267–270 β5)	Y	Y	Y	Y	Y
No. of a.a. in C-terminal 112 protein	112	112	112	112	Mature
Conservation of 9 cysteines in mature protein	Y	Y	Y	Y	Y
Percent identity to TGFβ$_1$	100	71	72	82	76

Y, yes; N, no; a.a., amino acids.

The widespread distribution of TGFβ1 receptors and the release of latent forms of TGFβ1 by most cells seem to indicate that the influence of TGFβ1 on cells may be controlled by the activation process or via post-receptor mechanisms. This allows one to postulate that other growth factors may exert control or at least modulate the activation process of TGFβ1 and, therefore, its biologic effects.

This is consistent with the hypothesis that normal growth control reflects a balance between positive and negative mitogens. The very tight homology between TGFβ1 of human and murine origin showing only one amino acid difference[25] emphasizes the importance of this molecule in normal growth and development. In recent years, multiple forms of TGFβ have been identified. The members of the TGFβ family share characteristics which are outlined in Table 6.3. The prototypical form, TGFβ1, was the first molecule isolated. Its gene is located on chromosome 19q subbands q13.1–q13.3,[32] and Northern blot analysis shows a 2.5-kilobase (kb) transcript.

TGFβ type 2 (TGFβ2) was initially isolated from a tamoxifen-treated human prostatic adenocarcinoma cell line (PC3).[33,34] The cDNA sequence indicates that TGFβ2 is synthesized as a 442–amino acid polypeptide precursor from which the mature 112–amino acid TGFβ2 subunit is derived by proteolytic cleavage. The proteins coded for by the human TGFβ1 and β2 cDNAs show an overall homology of 41%. The mature and aminoterminal precursor regions show 71% and 31% homology, respectively. Northern blot analysis identifies TGFβ2 transcripts of 4.1, 5.1, and 6.5 kb. Heterodimers of TGFβ1 and β2 subunits can form and are designated TGFβ1 and TGFβ2. TGFβ1 and β2 correspond to factors previously described as bone-derived cartilage-inducing factor A (CIF-A)[35] and CIF-B,[36] respectively.

Functionally, the biologic response of most cells to TGFβ1 and TGFβ2 appear to be very similar. For example, Zugmaier et al. demonstrated equipotent inhibitory effects for TGFβ1 and β2 in a number of breast cancer cell lines.[37] On the other hand, Rosa and coworkers showed that TGFβ2 but not TGFβ1 induced mesoderm in the amphibian embryo.[38] Differential regulation of expression of three forms of TGFβ was observed in several human breast cancer cell lines by estradiol.[39] Thus, although the effects of the TGFβ species seem very similar, differential expression of each gene can be induced by steroid hormones such as estradiol. The TGFβ type 3 gene codes for a protein that includes a signal peptide of 20 to 30 amino acids such as in human TGFβ1 and 2 and a precursor protein consisting of 412 amino acids, which can be cleaved at a lys-arg site to produce a 112–amino acid processed peptide containing nine cysteine residues in the same positions as in human TGFβ1 and 2. At the nucleotide level, the processed coding region of TGFβ3 shows 72% and 76% identity with the processed coding regions of human TGFβ1 and β2, respectively; at the amino acid level, TGFβ3 shows 76% identity with TGFβ1 and 79% with TGFβ2. RNA analysis shows expression of a 3.0-kb mRNA in primary chick embryo chondrocytes. TGFβ3 mRNA is mainly expressed in cell lines from mesenchymal origin, suggesting a biologic role different from the other species of TGFβ.[40-42]

Unlike other described TGFβ species that are 390 to 414 amino acids long, the predicted precursor protein of TGFβ4 is only 304 amino acids and does not appear to contain a signal peptide. Also unique to TGFβ4 is an insertion of two amino acids near the N-terminus of the processed peptide, which results in a 114–amino acid mature protein from the precursor and contains the nine cysteine residues characteristic of all TGFβ species. TGFβ4 shows 82%, 64%, and 71% identity with the amino acid sequences of processed TGFβ1, 2, and 3, respectively.[43] A role for TGFβ4 distinct from other TGFβ species has not been demonstrated. Most recently, TGFβ5 was isolated. The TGFβ5 mRNA is developmentally regulated and highly expressed beginning at the early neurul stage and in many adult tissues in *Xenopus laevis*.[44]

Besides the five TGFβ species described above, there is a growing list of less homologous but related proteins that should be considered part of this family of growth factors. Characteristics common to this family of proteins are (a) their multifunctional nature, (b) their function as growth regulators, both positive and negative, and (c) their ability to induce differentiation. For example, homology has been shown between TGFβ and the βA and βB chains of inhibin, an inhibitor of follicle-stimulating hormone secretion that is produced by both the ovaries and testes. The inhibin molecule is comprised of one α chain and one β chain.[45] However, when the β-subunits form homodimers or heterodimers activins are formed and they have effects opposite those of the inhibins.[46,47] Similar homology exists between TGFβ and the Müllerian inhibiting substance, which has a molecular weight of 150,000[48] and sequences in the decapentaplegic gene complex of *Drosophila*.[49]

One inhibitory protein, mammastatin, derived from the conditioned media of normal mammary cells, is not related to TGFβ. This growth inhibitor consists of 47- and 65-kDa polypeptides and inhibits the growth of five transformed human mammary cell lines derived from nonmammary tissue.[50]

As noted previously, the manner in which TGFβ expression, secretion, activation, and function are controlled is unclear. Massague and Kelly[51] have demonstrated that, in BALB/c 3T3 mouse fibroblasts, the level of TGFβ receptors at the cell surface is not depleted by sustained exposure to ligand. This is due to rapid recycling of the receptor after ligand-induced internalization or rapid replenishment of surface receptors from a large intracellular pool. Wakefield et al.[31] confirmed that TGFβ causes only minimal down-regulation of its receptor in a wide variety of cell types. In addition, they showed that the degree of TGFβ binding in any given cell type was unaffected by agents that affect the biologic action of TGFβ in that cell type, such as other growth factors, retinoic acid, phorbol esters, and epinephrine. They suggested that modulation of binding may not be an important control mechanism in the action of this growth factor; rather, the cellular response to TGFβ may be modulated at subsequent steps.

The inhibitory nature of TGFβ on cancer cells is exemplified by the influence of TGFβ on human breast and prostate cancer cells. Human breast cancer cells secrete TGFβ, specifically TGFβ1. Using the MCF7 cell line, Knabbe et al. demonstrated that whereas TGFβ1 secretion was inhibited by the addition of mitogens such as estrogen and insulin, growth inhibitory antiestrogens and glucocorticoids strongly stimulated its production in a clone of MCF7 breast cancer cells.[52] The mechanism of TGFβ induction is not yet fully defined; it is not at the steady state mRNA level. Control may involve both protein synthesis and conversion of a latent form to an active form of TGFβ. TGFβ1 from antiestrogen-treated MCF7 cells inhibited the growth of an estrogen receptor negative cell line, MDA-MB-231. This growth inhibition was reversed by the addition of a polyclonal antibody directed against native TGFβ1. To the contrary, Arteaga and coworkers demonstrated that other clones of MCF-7 cells do not respond to antiestrogens in this fashion and that most estrogen-receptor negative breast cancer cell lines are inhibited by TGFβ1.[53] Zugmaier et al. have recently shown TGFβ2 to be equipotent in its inhibition of human breast cancer cells, whereas Müllerian inhibitory substance had no effect.[37] Differential regulation of expression of three forms of TGFβ was observed in several human breast cancer cell lines by estradiol.[39] Thus, steroid hormones can differentially affect the TGFβ genes while the growth inhibitory effects of the TGFβ species seem very similar.

Of interest, TGFβ2 was initially isolated from the human prostate epithelial cancer cell line PC3 after treatment with tamoxifen.[33] These cells do not contain either estrogen or androgen receptors. With regard to TGFβ1, the androgen-independent PC3 and DU145 human prostate cancer cell lines produce, secrete, and are inhibited by TGFβ whereas the androgen-responsive LNCaP human prostate cancer cells neither produce nor respond to TGFβ1.[54] Interestingly, both the PC3 and the DU145 cells have high-affinity TGFβ receptors on their cell surface whereas none were found on the LNCaP cells. Schuurmans, however, has reported that TGFβ1 decreases the growth response of LNCaP cells to epidermal growth factors.[55]

The role of TGFβ in cancer biology is not limited to negative growth regulation. Its paracrine effects on stromal cells, the immune system, and bone metabolism may be even more important. One of the primary targets for TGFβ are fibroblasts. TGFβ has been shown to enhance the proliferative effect of epidermal growth factor on fibroblasts and is mitogenic alone toward many cells of mesenchymal origin.[56] It also enhances the production of collagen[56,57] and fibronectin[58] by fibroblasts. Increased production of these proteins by fibroblasts exposed to TGFβ suggests that TGFβ may play an important regulatory role in the synthesis and deposition of extracellular matrices during wound healing and tumor formation.[59,60-63] In addition, TGFβ appears to modulate the action of other growth factors such as basic fibroblast growth factor, as well as the expression of collagenase and protease genes.[64] Induction of proteases by TGFβ may, in turn, foster increased activation of latent forms of TGFβ.[26]

Immune surveillance may also be altered by TGFβ. In its activated form TGFβ suppresses activation/differentiation of T and B lymphocytes as well as natural killer cells.[65-67] However, TGFβ has little effect on lymphocyte proliferation once an immune response has been initiated, and it does not suppress functions of activated lymphocytes such as cytolysis. Torre-Amione and coworkers demonstrated the ability of TGFβ to promote the escape of tumor cells from immune surveillance.[68] Using the highly immunogenic C3H-derived UV-induced fibrosarcoma, they transfected the tumor cells with a TGFβ1 gene under the control of an SV40 promoter. Stable clones were isolated and examined for the effects of endogenously produced TGFβ1 on cytolytic T-lymphocyte responses. Although the transfected tumor cells continued to express class I major histocompability complex molecules and a tumor-specific antigen, they did not stimulate primary cytolytic T-lymphocyte responses in vitro or in vivo. In addition, TGFβ has been shown to induce monocyte chemotaxis and the production of growth factors such as interleukin-1 (IL-1).[69,70] As a modulator of inflammation and tissue repair, TGFβ1 increases the secretion of positive acute-phase proteins such as α_1-protease inhibitor and α_1-antichymotrypsin while it decreases secretion of the negative acute-phase protein albumin. When TGFβ1 and interleukin-6 (IL-6) are combined, additive induction of the above two positive acute-phase proteins is observed along with a decrease in albumin and α-fetoprotein secretion. TGFβ inhibits the induction of fibrinogen caused by IL-6.[71] Thus, TGFβ not only mediates, in part, the inflammatory response but it can also modulate the actions of other factors involved in inflammation and tissue repair.

Although endothelial cell proliferation and motility are inhibited by TGFβ,[72,73] osteoblasts are mitogenically and metabolically stimulated by TGFβ,[74,75] an indication that TGFβ may be important in the regulation of bone remodeling. In fact, demineralized bone is a rich source of TGFβ. In cancers such as prostate cancer, which commonly metastasize to bone, production and secretion of TGFβ by tumor cells may play an important role in the establishment of metastases in the bone. One could hypothesize that the blastic characteristics of prostate cancer bone metastases are the result of TGFβ production by the prostate cancer cells.

The pathway of signal transduction for TGFβ is unknown. No enzymatic activity for the TGFβ receptor has been noted. Pertovaara and coworkers reported that an early effect of TGFβ is an enhancement of the expression of two genes encoding serum and phorbol ester tumor promoter-regulated transcription factors: the *jun B* gene and the *c-jun* proto-oncogene, respectively.[76] This stimulation was observed in human lung adenocarcinoma A549 cells whose growth is inhibited by TGFβ, AKR-2B mouse embryo fibroblasts whose growth is stimulated by TGFβ, and K562 human erythroleukemia cells, whose growth is not appreciably affected by TGFβ. The enhanced mRNA levels of *jun B*, *c-jun*, and the nuclear proto-oncogene *c-fos* were noted within 1 hr of treatment of A549 cells with picomolar quantities of TGFβ. However, differential and cell type–specific regulation appeared to determine the timing and magnitude of the response of each *jun* gene in a given cell. The investigators concluded that one of the earliest genomic responses to TGFβ may involve nuclear signal transduction and amplification by the *jun B* and *c-jun* transcription factors in concert with *c-fos*. The differential activation of the *jun* genes may explain some of the pleiotropic effects of TGFβ.[76] Of interest, in addition to early induction of transcription factors and nuclear proto-oncogenes, TGFβ can enhance its own expression in normal as well as transformed cells.[77]

Cytokines

The colony-stimulating factors and interleukin-2 have been dealt with in another chapter. We will concentrate on those cytokines that have demonstrated in vitro antiproliferative activity. The best studied of these are the interferons. There are three broad categories of interferons. Type I interferons are divided into two groups that share the same receptor and both mediate potent antiviral activity. At least 20 subtypes of alpha interferon[78] and at least two beta-interferons have been recognized.[78] All have a molecular weight between 18,000 and 22,000[78] and in addition to their antiviral and immune-modulating activities have antiproliferative effects against a wide variety of tumor cell lines in vitro as well as in nude mice. They have also been shown to have activity against several tumors in humans,[79] but whether this is an immunomodulatory or direct antiproliferative effect remains unresolved. Although interferons are pleiotropic agents with a wide variety of biologic effects particularly on the immune system, in this chapter we shall examine only their direct antiproliferative effects.

In vitro the alpha interferons can be shown to inhibit the proliferation of many different murine and human cell lines including adenocarcinomas of colon,[80] breast,[81-83] bladder,[84] ovary,[85] lung,[86,87] kidney,[88] sarcomas,[89,90] and leukemias.[91,92] Similar effects on colony formation have been seen with cultured cell lines grown in soft agar, including both myeloid and carcinoma cell lines.[93] The effect on primary colony formation in soft agar for a variety of human tumor types has also been studied. An inhibition of 50% or more has been noted in 25% to 50% of the tumors tested. Nearly 20% were inhibited >70%.[94] Tumor types tested included lung, breast, ovary, melanoma, sarcoma, myeloma, acute leukemia, and renal cancer.[94] These antiproliferative effects have been replicated in other studies of primary ovarian tumors,[95] primary melanoma,[96] leukemias and myelomas,[97] and adenocarcinomas.[98] The effects are in general dose responsive in vitro.[95,96] Similarly, there are reports of activity in murine tumor models, including the L1210, Ehrlich ascites, and Lewis lung carcinoma models[99-101] and the B16 melanoma.[102] These studies have suggested that efficacy is proportional to tumor burden, that prolonged daily treatment is required for maximal efficacy, and that the effect is proportional to dose. Prevention of metastasis has also been documented in osteogenic sarcoma[103] and B16 melanoma.[104] Human xenograft models of breast cancer,[81,105,106] osteosarcoma,[107] and bowel cancer[106] have shown that interferons do possess direct antiproliferative effects on tumors in vivo. Conversely, work on murine models has also shown that interferon-resistant tumors in vitro can be interferon sensitive in vivo,[101] implying that a host effect is also responsible for the antitumor effects of interferons.

When examining reports on the antiproliferative effects of inteferons, it is necessary to note several important points. First, the antiproliferative effects are in general cytostatic; this is, when the drug is removed the cells will usually recover their former growth rate. Second, growth suppression is rarely complete. Third, in vivo models show that very high dosage and daily administration are required for effective growth suppression.[106] These effects may explain why the antitumor effects of interferons in humans have been relatively modest in all but a few tumors.[79]

Interferon beta has also been studied in vitro using both natural and recombinant materials. The antiproliferative effects are against a broad range of tumor cells similar to those of the alpha-interferons,[84,108,109] but with increasd sensitivity. However, these encouraging effects in vitro have not yet been reflected in greater clinical efficacy in the tumor types tested.[79]

Interferon gamma was originally reported to be more active than the type I interferons but the preparations used were impure and probably contained other cytokines.[110] Although significant antiproliferative activity has been reported for gamma-interferon in vitro,[111,112] in vivo models using recombinant preparations have not shown much activity in xenograft models of lung[86] or breast cancer.[106] Rodent models show activity with very low tumor burdens.[113,114] This activity is lost after total body irradiation, suggesting that a host effect is the primary antitumor effect of interferon gamma.

Clearly, the optimal use of the antiproliferative potential of the interferons is in situations of low tumor burden such as in adjuvant therapy; a series of studies are currently testing this hypothesis in interferon-sensitive tumors (renal carcinomas and melanoma). Unfortunately, not knowing whether a direct antitumor effect or a host modulatory effect is more important has made the choice of doses empirical. There is a need to explore both high antiproliferative doses as well as the much lower immune modulating doses. An alternative avenue is the integration of the interferons with other modalities such as chemotherapy or radiotherapy. A variety of chemotherapeutic agents have been shown to synergize with interferons including vinblastine,[115] 5-fluorouracil,[116] Adriamycin,[117,118] cyclophosphamide,[119] and cisplatin.[120] This synergy has also been reproduced in vivo.[120–122] Beta-interferon has shown synergy with radiation[123] in bronchogenic carcinoma and is currently under study in human trials.[79] Similarly, impressive synergy has been shown in vitro and in vivo between type I and type II interferons.[124,125] Unfortunately, early clinical studies in advanced disease did not show clinical efficacy, possibly because due to the toxicity of the combination the doses of both interferons have been quite low.[126,127] Alternatively, concentrating on their other properties such as tumor antigen up-regulation and immune modulation may prove to be more productive.

In addition, there is evidence from a number of in vitro systems that interferons may cause differentiation of both normal and malignant cells,[128] including human skeletal muscle cells, erythroid cells, adipocytes, and a variety of myeloid cell lines.[129] This is another mechanism through which interferons could influence cell proliferation.

Of most interest from a biologic standpoint, however, is the increasing evidence for an interaction between interferons and growth factors. These and other studies showing that interferons can suppress oncogene expression as well as induce differentiation suggest that they may be among the putative negative growth factors, the absence of which may play a role in tumor formation and progression.

Initial data came from experiments where interferon alpha (IFNα) inhibited serum-induced stimulation of resting fibroblasts, suggesting antagonism of mitogen-stimulated growth.[130-132] Subsequent work suggested specific inhibition of platelet-derived growth factor (PDGF), epidermal growth factor (EGF), and insulin. Thus, IFNs inhibited the growth-stimulating effect of PDGF in 3T3 cells,[133-136] of EGF in 3T3 cells[137] and A549 lung cancer cells,[138] acidic fibroblast growth factor (FGF) in endothelial cells,[139] of insulin and transferrin in a colon carcinoma cell line Colo 205,[80] and of insulin in Daudi cells.[140] Similar inhibitory effects of IFNα were observed on the growth response of murine sarcoma virus tumors in mice in response to PDGF[138] and TGFα.[138] In the former case, the inhibitory effect of IFNα alone was greater than when given with PDGF, suggesting a common pathway. Similarly, EGF has decreased the antiproliferative effect of IFNα on A549 cells[138] and 3T3 cells.[136]

We have investigated also the interaction of IFNs and growth factors on the estrogen receptor positive MCF-7 breast cancer cell line. This cell line is sensitive to the mitogenic effects of both estradiol and EGF[141] and to the antiproliferative effects of IFNs.[83] We found that the mitogenic effects of estradiol (E_2) and EGF on MCF-7 cells were decreased in the presence of IFNβ$_{ser}$ in phenol red free media (E. Cormier, D. Goldstein, *unpublished data*).

Several pathways have been proposed to explain the observed antagonism between IFNs and growth factors. One is the inhibition of specific genes associated with growth stimulation. In 3T3 cells, IFNα has been shown to decrease the levels of mRNA of several genes associated with growth stimulation such as *c-myc*, *c-fos*, ornithine decarboxylase, and β-actin that are turned on by competence growth factors such as PDGF.[142] Decreases in *c-myc* mRNA have also been shown in IFNα-treated Daudi cells[143] and proteins induced by PDGF have been decreased by IFNα in 3T3 cells.[135,144] PDGF has been shown to stimulate IFNβ and 2'5'A synthetase, an IFN-induced enzyme, providing evidence for feedback inhibition of mitogen-induced growth stimulation in fibroblasts by reciprocal induction of growth-inhibiting IFN.[145,146] Treatment of 3T3 cells with IFN after transfection by the *H-ras* oncogene prevents malignant transformation[147] and is associated with decreased expression of *ras* mRNA. Other examples of reversion of the neoplastic phenotype have been reported.[148-150] The persistent presence of the oncogene DNA after phenotypic reversion has been noted,[147,150,151]

Studies on the effect of IFNα on the stimulation of vascular smooth muscle and endothelium by PDGF and acidic FGF showed that IFNα could still be inhibitory even if added up to 8 hr after the mitogen and could not be overcome by addition of more mitogen.[139] Similar results have been obtained in 3T3 cells[152,153] and COLO 205 cells.[80] Other studies have shown inhibition of proliferation without a decrease in total mRNA levels.[80,135,152,154,155] Thus, general nuclear events, such as changes in total mRNA levels, are not sufficient to account for all the antiproliferative effect of IFNs. Effects later in the cell cycle have been implicated.[156]

The absence of interferon gene expression has been noted in a variety of leukemic cell lines and primary leukemia cells.[157] In addition, both interferon gamma and alpha have potent inhibitory effects on myeloid colony formation,[129] which further supports an important role for IFNs as negative growth factors in hematopoietic cells. We speculate that one of the mechanisms by which tumors may escape growth regulation is through the development of mitogenic pathways resistant to interferon blockade.

Tumor necrosis factor (TNF) is another cytokine that has received much attention for its antiproliferative properties. It was originally described on the basis of its ability to cause necrosis when injected directly into tumors.[158] Subsequently, it became clear that it was identical to another factor, cachectin,[159] which has been shown to have pleiotropic effects on the body. In particular, the effects on metabolism are associated with chronic conditions that cause wasting. Also, potent inflammatory effects suggest that TNF is the mediator of endotoxic shock. Subsequently, numerous immunologic effects have been described. Monocytes are the primary source of TNF in vivo and it is the cytokine that mediates most of the antiproliferative effects of monocytes.[160,161]

TNF has been shown to be growth inhibitory for murine and human tumor cells in vitro and in vivo. A variety of tumor types can be growth inhibited in vitro including melanoma,[162,163] lung,[163] cervix,[163,164] breast,[163-165] and colon[163,165-167] tumor cell lines. Differentiation of human myeloid cell lines[168] and inhibition of primary leukemic cell colony formation has also been noted.[169]

Like interferon gamma, TNF has been shown to inhibit formation of normal hematopoietic and leukemic cell colonies in vitro.[168] It also has activity in vivo against murine tumors[170,171] and human xenografts of both tumor cell lines and primary cultures.[170,172] These in vivo antitumor effects have been shown to be associated with host effects including both T-cell stimulation[173,174] and direct effects on tumor endothelium.[175] Indeed, the growth of cells resistant in vitro to TNF is inhibited when injected in vivo.[176] TNF has been shown to be synergistic with interferon gamma in vitro[164,177-179] and in vivo.[172] This may be related to the up-regulation of TNF receptors by interferon gamma.[180] Synergy with IL-2 in vivo in a murine tumor model has also been shown.[181] By contrast, like TGFβ it is growth stimulatory for human fibroblasts.[146]

TNF antagonizes the positive mitogenic effects of EGF and TGFα in an epithelial cell line and of TGFβ in a fibroblast cell line. This suggests that the presence of a variety of growth factors may alter the antiproliferative effects of TNF in vivo.[164] Similarly, with regard to its potential therapeutic use in hematopoietic tumors, TNF has been reported to stimulate GCSF[182] and GMCSF[183] production. Thus, its in vivo effects may be far more complex than its in vitro antiproliferative and differentiating effects might suggest. This may explain the lack of efficacy of TNF in clinical trials. Other important problems include its toxicity,[184] which has prevented the administration of the high levels used in animal experiments to humans.

Of particular interest has been the observation that the anatomic compartment into which the cytokine is injected influences its antitumor effects. Intraperitoneal (IP) injection of TNF is effective against IP tumor growth but subcutaneous (SC) injection is not.[172] Similarly, IP injection is less effective than SC against SC tumors.[171,172] Future development of TNF as an antitumor agent will clearly require more innovative strategies, such as localized therapy or use in combination with other cytokines.

In terms of its role in the biology of tumor development, TNF appears to be one of the cytokines available to cells of the immune system that recognize tumors as foreign and is used by them to inhibit cell proliferation. For example, TNF decreases c-myc expression in HeLa cells. This may provide insight into TNF's mechanism of action.[185] Tumor cells, on the other hand, might overcome TNF's inhibitory actions by producing stimulatory growth factors. Likewise, production of endogenous TNF by tumors appears to protect them from the effects of exogenous TNF.[186]

Two interleukins (IL) have demonstrated antiproliferative activity in preclinical studies. Interleukin-1 is a cytokine produced by monocytes and its primary action is to stimulate T helper cells to produce IL-2, as part of the cellular immune response. However, it shares a great variety of other systemic effects with TNF, including antiproliferative activity in vitro. IL-1, for example, inhibits human breast cancer and melanoma cell lines.[187–189] It also stimulates differentiation and inhibits the growth of human myeloid leukemia cell lines.[190,191,192]

The antineoplastic effects of IL-1 may also be mediated immunologically since it augments the production of other lymphokines[193,194] and modulates monocyte cytocidal activity.[193] Its use in cancer therapy is, however, linked to its influence on hematopoiesis rather than on its antiproliferative effects. IL-1 enhances the early recruitment of stem cells and synergizes with other hematopoietic growth factors to protect against the lethal effects of both chemotherapy and radiotherapy in vivo models.

Interleukin-6 is another pleiotropic cytokine that has had a variety of names related to its many different activities, including hepatocyte-stimulating factor, hybridoma growth factor, interferon β2, and B cell differentiating factor (BSF-2). Although its many other properties are beyond the scope of this chapter, its effects on cell growth suggest two major roles in cancer biology and therapy. IL-6 has been shown to synergize with other colony-stimulating factors, such as IL-3, in hematopoietic stem cell proliferation.[194] By contrast, IL-6 inhibits cell proliferation and colony formation of human breast[195] and HeLa cervical carcinoma cells.[195] It also stimulates differentiation of human and mouse myeloid leukemic cell lines.[190]

The potential importance of IL-6 as an autocrine inhibitory factor is demonstrated by three studies. In studies on human fibroblasts, the growth stimulation induced by TNF was limited by the subsequent autocrine production of IFNβ2 (IL-6). Conversely, antibodies to IL-6 increased growth.[196] A similar inhibitory effect of IL-6 was noted with regard to stimulation by PDGF or IL-1.[146] In a study on the effect of CSF on myeloid differentiation of a leukemic cell line, the interferon beta–induced enzyme 2'5'oligoadenylate synthetase was substantially increased and interferon beta antibodies prevented growth arrest.[197] A similar effect has been seen during induction of differentiation of U937 cells to monocytes.[198] Finally, in a study of human endometrial stromal cells, a variety of cytokines induced IL-6 production which, in turn, can inhibit endometrial cell proliferation. By contrast, estradiol, which stimulates endometrial cell proliferation, inhibits IL-6 production.[199] The bifunctional nature of this cytokine on cell proliferation is demonstrated by its identification as an autocrine growth factor for human multiple myeloma cells in vitro.[200]

Conclusions

The control of epithelial cell growth appears to involve a balance between the effects of inhibitory and stimulatory influences. As evident from a variety of in vitro studies, one mechanism by which tumors may escape normal growth control is by constitutively activating autocrine growth stimulatory pathways. The inhibitory growth factors, as exemplified by the TGFβ, IFN, and IL families, may offer an opportunity to restore that balance and thus gain control of cancer cell growth. Promotion of differentiation by these factors could render a therapeutic advantage also.[62,63] Of concern is the possibility that the paracrine effects of these factors may convey an advantage to tumor cells in terms of immune suppression, proteolysis, and stromal development, which may outweigh the growth inhibitory effects. Regardless, these factors and some of the mechanisms behind their inhibitory influences will without doubt open new therapeutic avenues.

The discovery of the interferons sparked a decade of effort directed toward synthesis and clinical application of them as anticancer agents because of their growth inhibitory properties. It has become clear that they will only have limited effectiveness as single agents. However, the enormous amount of research stimulated by their discovery led to the identification of numerous other cytokines with a broad range of effects and potential applications in the management of human disease. The concurrent identification and study of somatic growth factors has now allowed us to begin to visualize the complexity of growth control in normal human cells and some of the mechanisms by which this control is regulated. It has also given us insight into the concept of tumor progression through autocrine growth stimulation. In addition to autocrine production of positive growth factors, it seems that cells are capable of producing negative growth factors that limit their own growth. Thus, one of the mechanisms of uncontrolled growth may be a loss or suppression of such factors or the development of mitogenic pathways that are not blocked by these factors.

We have reviewed a number of cytokines such as the interferons and interleukins 1 and 6 and TNF that are used both by cells involved in tumor surveillance and, in an autocrine fashion, in the prevention of uncontrolled cell growth. Excess production of growth factors would appear to overcome this growth suppression. The response of cells to this variety of factors appears to depend on the sequence and number of opposing stimuli, since many of these factors can be bifunctional. An understanding of this signalling "language"[198] is crucial to our understanding of the biology of tumor progression and will undoubtedly lead to new antitumor strategies we cannot even conceive of today. We look forward with anticipation to the next few decades of research.

Acknowledgments. George Wilding is a recipient of an American Cancer Society Career Development Award. Supported in part by National Cancer Institute grant R29-CA50590 and the Veterans Administration.

References

1. Sporn MB, Roberts AB. Autocrine growth factors and cancer. *Nature.* 1984;313:745–747.
2. Cuttitta F, Carney DN, Mulshine J, Moody TW, Fedorki J, Fischler A, Minna JD. Bombesin-like peptides can function as autocrine growth factors in human small cell lung cancer. *Nature.* 1985;316:823–826.
3. Bishop JM. Cellular oncogenes and retroviruses. *Annu Rev Biochem.* 1983;52:301–354.
4. Varmus HE. The molecular genetics of cellular oncogenes. *Annu Rev Genet* 1984;18:553–612.
5. Hunter T, Cooper JA. Protein-tyrosine kinases. *Annu Rev Biochem.* 1985;54:897–930.
6. Downward J, Yarden Y, Mayes E, Scrace G, Totty N, Stockwell P, Ullrich A, Schlessinger J, Waterfield MD. Close similarity of epidermal growth factor receptor and v-erb B oncogene protein sequences. *Nature.* 1984; 307:521–527.

7. Sherr CJ, Rettenmier CW, Sacca R, Roussel MF, Look AT, Stanley ER. The c-fms proto-oncogene product is related to the receptor for the mononuclear phagocyte growth factor CSF-1. *Cell*. 1985;41:665–676.

8. Bargmann CI, Hung MC, Weinberg RA. The *neu* oncogene encodes an epidermal growth factor receptor related protein. *Nature*. 1986;319:226–230.

9. Slamon DJ, Clark GM, Wong SG, Levin WJ, Ullrich A, McGvire Wm L. Human breast cancer: correlation of relapse and survival with amplification of the HER2/neu oncogene. *Science*. 1987;235:177–182.

10. Wheeler EF, Rettenmier CW, Look AT, Sherr CJ. The v-fms oncogene induces factor independence and tumorigenicity in CSF-1 dependent macrophage cell line. *Nature*. 1986;324:377–379.

11. Bradley S, Garfinkle G, Walker E, Salem R, Chen LB, Steele GJ. Increased expression of the epidermal growth factor on human colon carcinoma cells. *Arch Surg*. 1986;121:1242–1247.

12. Liberman TA, Nusbaum HR, Razon N, Kris R, Lax I, Soreq H, Whittle N, Waterfield MD, Ullrich A, Schlessinger J. Amplification, enhanced expression and possible rearrangement of EGF receptor gene in primary human brain tumors of glial origin. *Nature*. 1985;313:144–147.

13. Waterfield MD, Serace GT, Whittle N, Stroobant P, Johnsson A, Wasteson Å, Westermark B, Heldin C-H, Huang JS, Devel TF. Platelet derived growth factor is structurally related to putative transforming protein p28^s is of simian sarcoma virus. *Nature*. 1983;304:35–39.

14. Huang HJS, Lee JK, Shew JY, Chen P-L, Bookstein R, Friedmann T, Lee Ey H-P, Lee W-H. Suppression of the neoplastic phenotype by replacement of the RB gene in human cancer cells. *Science*. 1988;242:1563–1566.

15. Holley RW. Control of growth of mammalian cells in cell culture. *Nature*. 1975;258:487–490.

16. DeLarco JE, Todaro GJ. Growth factors from murine sarcoma virus transformed cells. *Proc Natl Acad Sci USA*. 1978;75:4001–4005.

17. Goustin AS, Leof EB, Shipley GD, Moses HL. Growth factors and cancer. *Cancer Res*. 1896;46:1015–1029.

18. Massague J. The TGFβ family of growth and differentiation factors. *Cell*. 1987;49:437–438.

19. Roberts AB, Anzano MA, Wakefield LM, Roche NS, Stern DF, Sporn MD. Type β transforming growth factor: a bifunctional regulator of cell growth. *Proc Natl Acad Sci USA*. 1985;82:119–123.

20. Tucker RF, Shipley GD, Moses HL, Holley RW. Growth inhibitor from BSC-1 cells closely related to platelet type beta transforming growth factor. *Science*. 1984;226:705–709.

21. Keski-Oja J, Postlethwaite AE, Moses HL. Transforming growth factors in the regulation of malignant cell growth and invasion. *Cancer Invest*. 1988;6:705–724.

22. Anzano MA, Roberts AB, Smith JM, Sporn MD, De Larco JE. Sarcoma growth factors from conditioned medium of virally transformed cells is composed of both type a and type b transforming growth factors. *Proc Natl Acad Sci USA*. 1983;80:6264–6268.

23. Moses HL, Branum EB, Proper JA, Robinson RA. Transforming growth factor production by chemically transformed cells. *Cancer Res*. 1981;41:2842–2848.

24. Assoian RK, Komoriya A, Meyers CA, Miller DM, Sporn MB. Transforming growth factor β in human platelets. Identification of a major storage site, purification and characterization. *J Biol Chem*. 1983;258:7155–7160.

25. Derynck R, Jarrett JA, Chen EY, Eaton DH, Bell JR, Assoian RK, Roberts AB, Sporn AB, Sporn MB, Goeddel DV. Human transforming growth factor β complementary DNA sequence and expression in normal and transformed cells. *Nature*. 1985;316:701–705.

26. Lyon RM, Keski-Oja J, Moses HL. Proteolytic activation of latent transforming growth factor β from fibroblast conditioned medium. *J Cell Biol*. 1988;106:1597–1605.

27. Massague J, Like B. Cellular receptors for type β transforming growth factor. Ligand binding and affinity labelling in human and rodent cell lines. *J Biol Chem*. 1985;260:2636–2645.

28. Fanger BO, Wakefield LM, Sporn MB. Structure and properties of the cellular receptor for transforming growth factor type β. *Biochemistry*. 1986;25:3083–3091.

29. Cheifetz S, Like B, Massague J. Cellular distribution of type I and type II receptors for transforming growth factor β. *J Biol Chem*. 1896;261:9972–9978.

30. Tucker RF, Barnum EL, Shipley GD, Ryan RJ, Moses HL. Specific binding to cultured cells of ^{125}I-labeled transforming growth factor β from human platelets. *Proc Natl Acad Sci USA*. 1984;81:6757–6761.

31. Wakefield LM, Smith DM, Masui T, Harris CC, Sporn MB. Distribution and modulation of the cellular receptor for transforming growth factor beta. *J Cell Biol*. 1987;105:965–975.

32. Fujii D, Brissenden JE, Derynck R, Francke U. Transforming growth factor β gene maps to human chromosome 19 long arm and to mouse chromosome 7. *Somatic Cell Mol Genet*. 1986;12:281–288.

33. Ikeda T, Lioubin MN, Marquardt H. Human transforming growth factor type $β_2$: production by a prostatic adenocarcinoma cell line, purification and initial characterization. *Biochemistry*. 1987;26:2406–2410.

34. Madisen L, Webb NR, Rose TM, Marquadt H, Ikeda T, Twadzik D, Seyedin S, Purchio AF. Transforming growth factor $β_2$: cDNA cloning and sequence analysis. *DNA*. 1988;7:1–8.

35. Seyedin SM, Thomas TC, Thompson AY, Rosen DM, Piez K-A. Purification and characterization of two cartilage inducing factors from bovine demineralized bone. *Proc Natl Acad Sci USA*. 1985;82:2267–2271.

36. Seyedin SM, Segarini PR, Rosen DM, Thompson AY, Bentz H, Graycar J. Cartilage-inducing factor β is a unique protein structurally and functionally related to transforming growth factor β. *J Biol Chem*. 1987;262:1946–1949.

37. Zugmaier G, Ennis BW, Deschauer B, Katz D, Knabbe C, Wilding G, Daly P, Lippman ME, Dickson RB. Transforming growth factor β1 and β2 but not Mullerian inhibiting substance are equipotent growth inhibitors of human breast cancer cell lines. *J Cell Physiol*. 1989;141:353–361.

38. Rosa F, Roberts AB, Danielpour D, Dart LL, Sporn MB, Dawid IB. Mesoderm induction in amphibians: the role of TGFB2-like factors. *Science*. 1988;239:783–785.

39. Arrick BA, Korc M, Derynck R. Differential regulation of expression of three transforming growth factor β species in human breast cancer cell lines by estradiol. *Cancer Res*. 1990;50:299–303.

128   David Goldstein, George Wilding

40. TenDijke PT, Hanson P, Iwata KK, Piela C, Foulkes JG. Identification of another member of the transforming growth factor type β gene family. *Proc Natl Acad Sci USA*. 1988;85:4715–4719.
41. Derynck R, Lindquist PB, Lee A, Wen D, Tamm J, Graycar JL, Rhee L, Mason AJ, Miller DA, Coffey RJ. A new type of transforming growth factor β, TGFβ₃. *EMBO J*. 1988;7:3737–3743.
42. Jakowlew SB, Dillard PJ, Kondaiah P, Sporn MB, Roberts AB. Complementary deoxyribonucleic acid cloning of a novel transforming growth factor β messenger ribonucleic acid from chick embryo chondrocytes. *Mol Endocrinol*. 1988;2:747–755.
43. Jakowlew SB, Dillard PJ, Sporn MB, Roberts AB. Complementary deoxyribonucleic acid encoding transforming growth factor β4 from chicken embryo chondroycytes. *Mol Endocrinol*. 1988;2:1186–1195.
44. Kondaian P, Sands MJ, Smith JM, Fields A, Roberts AB, Sporn MB, Melton DA. Identification of a novel transforming growth factor-β (TGF-β5) mRNA in *Xenopus laevis*. *J Biol Chem*. 1990;265:1089–1093.
45. Mason AJ, Hayflick JS, Ling N, Esch F, Ueno N, Ying S-Y, Guillemin R, Niall H, Seeburg PH. Complementary DNA sequences of ovarian follicular fluid inhibin show precursor structure and homology with transforming growth factor β. *Nature*. 1985;318:659–663.
46. Vale W, Rivier J, Vaughan J, McClintock R, Corrigan A, Woo W, Karr D, Spiess J. Purification and characterization of an FSH releasing protein from porcine ovarian follicular fluid. *Nature*. 1986;321:776–779.
47. Ling N, Ying S, Ueno N, Shimasaki S, Esch F, Hotta M, Guillemin R. Pituitary FSH is released by a heterodimer of β subunits from the two forms of inhibin. *Nature*. 1986;321:779–782.
48. Cate RL, Mattaliano RJ, Hession C, Tizard R, Farber NM, Cheung A, Ninfa EG, Frey AZ, Gash DJ, Chow EP. Isolation of the bovine and human genes for Mullerian inhibiting substance and expression of the human gene in animal cells. *Cell*. 1986;45:685–698.
49. Padgett RW, St Johnston RD, Gelbart WM. A transcript from a Drosophila pattern gene predicts a protein homologous to the transforming growth factor β family. *Nature*. 1987;325:81–84.
50. Ervin PR, Kaminski MS, Cody RL, Wicha MS. Production of mammastatin, a tissue-specific growth inhibitor, by normal human mammary cells. *Science*. 1989;244:1585–1587.
51. Massague J, Kelly B. Transforming growth factor β stimulates the expression of fibronectin and collagen and their incorporation into the extracellular matrix. *J Cell Physiol*. 1986;128:216–222.
52. Knabbe C, Lippman ME, Wakefield LM, Flanders KC, Kasid A, Derynck R, Dickson RB. Evidence that transforming growth factor β is a hormonally regulated negative growth factor in human breast cancer cells. *Cell*. 1987; 48:417–428.
53. Arteaga CL, Tandon AK, VonHoff DD, Osborne CK. Transforming growth factor β: potential autocrine growth inhibition of estrogen receptor negative human breast cancer cells. *Cancer Res*. 1988;48:3898–3904.
54. Wilding G, Zugmeier G, Knabbé C, Flanders KC, Gelmann E. Differential effects of transforming growth factor β on human prostate cancer cells in vitro. *Mol Cell Endocrinol*. 1989;62:79–87.

55. Schuurmans ALG, Bolt J, Mulder E. Androgens and transforming growth factor β modulate the growth response to epidermal growth factor in human prostatic tumor cells (LNCaP). *Mol Cell Endocrinol.* 1988;60:101–104.
56. Fine A, Goldstein RH. The effect of transforming growth factor beta on cell proliferation and collagen formation by lung fibroblasts. *J Cell Biol Chem.* 1987;262:3897–3902.
57. Roberts AB, Sporn MB, Assoian RK, Smith JM, Roche NS, Wakefield LM, Heine UI, Liotta LA, Falanga V, Kehrl JH, Fauci AS. Transforming growth factor beta: rapid induction of fibrosis and angiogenesis in vivo and stimulation of collagen formation in vitro. *Proc Natl Acad Sci USA.* 1986;83: 4167–4174.
58. Ignotz RA, Massague J. Transforming growth factor beta stimulates the expression of fibronectin and collagen and their incorporation into the extracellolae matrix. *J Biol Chem.* 1986;261:4337–4345.
59. Sporn MB, Roberts AB. Peptide growth factors and inflamation, tissue repair and cancer. *J Clin Invest.* 1986;78:329–332.
60. Sporn MB, Roberts AB, Shull JH, Smith JM, Ward JM, Sodek J. Polypeptide transforming growth factors isolated from bovine sources and used for wound healing in vivo. *Science.* 1983;219:1329–1331.
61. Roberts AB, Sporn MB, Assoian RK, Smith, JM, Roch NS, Wakefield LM, Heine UI, Liotta LA, Falanga V, Kehrl JH, Fauci AS. Transforming growth factor type β: rapid induction of fibrosis and angiogenesis in vivo and stimulation of collagen formation in vitro. *Proc Natl Acad Sci USA.* 1986; 83:4167–4171.
62. Twardzik DR, Ranchalis JE, McPherson JM, Ogawa Y, Gentry L, Purchio A, Plata E, Todaro, GJ. Inhibition and promotion of differentiated-like phenotype of human lung carcinoma in athymic mice by natural and recombinant forms of transforming growth factor β. *J Natl Cancer Inst.* 1989;81:1182–1185.
63. Masui T, Wakefield LM, Lechner JF, LaVeck MA, Sporn MB, Herris CC. Type beta transforming growth factor is the primary differentiation inducing serum factor for normal human bronchial epithelial cells. *Proc Natl Acad Sci USA.* 1986;83:2438–2442.
64. Edwards DR, Murphy G, Reynolds JJ, Whitman SE, Doherty AJP, Angel P, Heath JK. Transforming growth factor beta modulates the expression of collagenase and metalloproteinase inhibitor. *EMBO J.* 1989;6:1899–1904.
65. Kehrl JH, Roberts AB, Wakefield LM, Jakowlew S, Sporn MB, Fauci AS. Transforming growth factor beta is an important immunomodulatory protein for human B-lymphocytes. *J Immunol.* 1986;137:3855–3860.
66. Wrann M, Bodmer S, deMartin A, Siepl C, Hofer-Warbinek R, Frei K, Hofer E, Fontant A. T Cell suppressor factor from human glioblastoma cells is a 125 KD protein closely related to transforming growth factor beta. *EMBO J.* 1987;6:1633–1636.
67. Rook AH, Kehrl JH, Wakefield LM, Roberts AB, Sporn MB, Burlington DB, Land HC, Fauci AS. Effects of transforming growth factor beta on the functions of normal killer cells: depressed cytolytic activity and blunting of interferon responsiveness. *J Immunol.* 1986;136:3916–3920.
68. Torre-Amione G, Beauchamp RD, Koeppen H, Park BH, Schreiber H, Moses HL, Rowley DA. A highly immunogenic tumor transfected with a

murine transforming growth factor type β_1 cDNA escapes immune surveillance. *Proc Natl Acad Sci USA*. 1990;87:1486–1490.

69. Wahl SM, Hunt DA, Wakefield LM, McCartney-Francis N, Wahl LM, Reberts AB, Sporn MB. Transforming growth factor beta induces monocyte chemotaxis and growth factor production. *Proc Natl Acad Sci USA*. 1987; 84:5788–5792.

70. Assoian RK, Fleurdelys BS, Stevenson HD, Miller PJ, Madtes DK, Raines EW, Ross R, Sporn MD. Expression and secretion of type beta transforming growth factors by activated human macrophages. *Proc Natl Acad Sci USA*. 1987;84:6020–6024.

71. Mackiewicz A, Ganapathi MK, Schultz D, Brabenec A, Weinstein J, Kelley MF, Kushner I. Transforming growth factor β_1 regulates production of acute phase proteins. *Proc Natl Acad Sci USA*. 1990;87:1491–1495.

72. Muller G, Behrens J, Nussbaumer U, Bohlen P, Birchmeier W. Inhibitory action of transforming growth factor β on endothelial cells. *Proc Natl Acad Sci USA*. 1987;84:5600–5604.

73. Takehara K, LeRoy EC, Grotendorst GR. TGFβ inhibition of endothelial cell proliferation: alteration of EGF binding and EGF induced growth regulatory gene expression. *Cell*. 1987;49:415–422.

74. Pfeilschifter J, Mundy GR. Modulation of type beta transforming growth factor activity in bone cultures by osteotropic hormones. *Proc Natl Acad Sci USA*. 1987;84:2024–2028.

75. Centrella M, McCarthy TL, Canalis E. Transforming growth factor beta is a bifunctional regulator of replication and collagen synthesis in osteoblast-enriched cell cultures from fetal rat bone. *J Biol Chem*. 1987;262:2869–2874.

76. Pertovaara L, Sistonen L, Bos TJ, Vozt PK, Keski-Oja J, Alitalo K. Enhances *jun* gene expression is an early genomic response to transforming growth factor β stimulation. *Mol Cell Biol*. 1989;9:1255–1262.

77. Van Obberghen-Schilling E, Roch NS, Flanders K, Sporn MB, Roberts AB. Transforming growth factor $\beta 1$ positively regulates its own expression in normal and transformed cells. *J Biol Chem*. 1988;263:7741–7746.

78. Pestka S, Langer JA, Zoon KC, Sammuel CE. Interferons and their actions. *Annu Rev Biochem*. 1987;56:727–777.

79. Goldstein D, Laszlo J. Interferon; a current perspective. *Cancer Res*. 1988;38:258–290.

80. Hamburger AW, Condon ME, O'Donnell K. Inhibition of mitogen stimulated growth of human colon cancer cells by interferon. *Br J Cancer*. 1988;58:147–151.

81. Taylor-Papdimitriou J, Shearer M, Balkwill FR. Effects of Hu IFN alpha 2 and Hu IFN alpha (Namalwa) on breast cancer cells grown in culture and as xenografts in the nude mouse *J Interferon Res*. 1982;2:479–485.

82. Balkwill F, Watling D, Taylor-Papadmitiriou J. Inhibition by lymphoblastiod interferon of growth of cells derived from the human breast. *Int J Cancer*. 1978;22:258–265.

83. Goldstein D, Bushmeyer SM, Witt PL, Jordan VC, Borden EC. Effects of type I and II interferons on cultured human breast cells: interaction with estrogen receptors and with tamoxifen. *Cancer Res*. 1989;49:2698–2702.

84. Borden EC, Groveman DS, Nasu T, Reznikoff C, Bryan GT. Antiproliferative activities of interferons against human bladder carcinoma cell lines in vitro. *J Urol*. 1984;132:800–802.

85. Bradley EC, Ruscetti FW. Effect of fibroblast, lymphoid and myeloid interferons on human tumor colony formation in vitro. *Cancer Res.* 41:244–249.

86. Twentyman PR, Workman P, Wright KA, Bleehan NM. The effects of alpha and gamma interferons onhuman lung cancer cells grown in vitro or as xenografts in nude mice. *Br J Cancer.* 1985;52:21–29.

87. Munker M, Munker R, Saxton RE, Koeffler P. Effect of recombinant monokines. lymphokines and other agents on clonal proliferation of human lung cancer cell lines. *Cancer Res.* 1987;47:4081–4085.

88. Schmid SM, Borden EC, Bryan GT, Trump DL, Cummings KB. Antiproliferative efects of recombinant interferon alpha, beta, and gamma on renal carcinoma cells. *Surg Forum.* 1984;35:655–657.

89. Strander H, Einhorn S. Effect of human leukocyte interferon on the growth of human osteosarcoma cells in tissue culture. *Int J Cancer.* 1977;19:468–473.

90. Crane JL, Glasgow LA, Kerr ER, Younger JS. Inhibition of murine osteogenic sarcomas by treatment with Type I or Type II interferons. *J Natl Cancer Inst.* 1978;61:871–874.

91. Verma DS, Spitzer G, Gutterman J, Zander AR, McCredie KB, Dicke KA. Human leukocyte preparation blocks granulocyte differentization. *Blood.* 1979;43:1423–1427.

92. Williams CK, Svet-Moldavskaya I, Vilcek J. Inhibitory effects of human leukocyte and fibroblast interferons on normal and chronic myelogenous leukemic granulocytic progenitor cells. *Oncology.* 1981;38:356–360.

93. Bradley EC, Ruscetti FW. Effects of fibroblast, lymphoid and myeloid interferons on human tumor colony formation in vitro. *Cancer Res.* 1982; 41:244.

94. Salmon SE, Durie GM, Young L, Liu RM, Trown P, Stebbing N. Effects of cloned human leukocyte interferons in the human tumor stem cell assay. *J Clin Oncol.* 1983;1:217–225.

95. Willson JKV, Bittner G, Borden EC. Antiproliferative activity of human interferons against ovarian cancer cells grown in human tumor stem cell assay. *J Interferon Res.* 1984;4:441–447.

96. Schiller JH, Willson JKV, Bittner G, Wolberg WH, Wm H, Hawkins MJ, Borden EC. Antiproliferative effects of interferons on human melanoma cells in the human tumor colony-forming assay. *J Interferon Res.* 1986; 6:615–625.

97. Ludwig H, Swetley P. In vitro inhibitory effects of interferon on colony formation of myeloma stem cells. *Cancer Immunol Immunother.* 1980; 9(3):139–143.

98. Von Hoff DD, Gutterman JU, Portnoy B, Coltman CA Jr. Activity of human leukocyte interferon in a human tumor cloning system. *Cancer Chemother Pharmacol.* 1982;8:99–103.

99. Gresser I, Coppey I, Bourai C. Interferon and murine leukemia IV. *J Natl Cancer Inst.* 1970;45:365–374.

100. Gresser I, Bourai-Maury C. Inhibition by interferon preparations of a solid malignant tumor and pulmonary metastasis in mice. *Nature New Biol.* 1972;236:78–79.

101. Gresser I, Maury C, Brouty-Boye D. Mechanism of the antitumor effect of interferon in mice. *Nature (Lond).* 1972;239:167–168.

102. Bart RS, Porzio NR, Kopf AW. Inhibition of growth of B16 murine melanoma by exogenous interferons. *Cancer Res.* 1980;40:614–619.
103. Glasgow LA, Kern ER, Effect of interferon administration on pulmonary osteogenic sarcomas in an excperimental murine model. *J Natl Cancer Inst.* 1981;67:207–212.
104. Nishimura J, Mitsui K, Ishikawa T, Tanaka Y, Yamamoto R, Suhara Y, Ishitsuka H. Restoration by recombinant interferon alpha A/D of host defence sytens against tumor in immunosuppressed mice. *Clin Exp. Metastasis.* 1985;3:295–304.
105. Balkwill FR, Moodie EM, Freedman V, Fantes KH. Human interferon inhibits the growth of established human breast tumours in the nude mouse. *Int J Cancer.* 1982;30:231–235.
106. Balkwill FR, Goldstein L, Stebbing N. Differential action of six human interferons against two human carcinomas in the nude mouse. *Int J Cancer.* 1985;35:613–617.
107. Brosjo O, Bauer HCF, Brostrom L-A, Nilsonne U, Nilsson OS, Reinholt FP, Strander H, Tribukait B. Influence of human alpha interferons on four human osteosarcoma xenografts in nude mice. *Cancer Res.* 1985;45:5598–5602.
108. Borden EC, Hogan TF, Voelkel JC. Comparative antiproliferative activity in vitro of natural interferons alpha and beta for diploid and transformed human cells. *Cancer Res.* 1982;42:4948–4953.
109. Tanaka N, Nagao S, Tohgo A, Sekiguchi F, Kohno M, Ogawa H, Matsui T, Matsotani M. Effects of human fibroblast interferon on human gliomas transplanted into nude mice. *Japanese Journal of Cancer Research* (GANN) 1983;74:308–316.
110. Tyring S, Klimpel GR, Fleischmann R, Barons S. Direct cytolysis by partially pure preparations of immune interferon. *Int J Cancer.* 1982;30:59–64.
111. Saito T, Berens ME, Welander CE. Direct and indirect effects of human recombinant gamma-interferon on tumor cells in a clonogenic assay. *Cancer Res.* 1986;46:1142–1147.
112. Richtsmeier WJ. Interferon gamma induced oncolysis—an effect on head and neck squamous carcinoma cultures. *Arch Otolaryngol Head Neck Surg.* 1988;114:432–437.
113. Nooter K, Van der Meide PH, Deurloo J, Van Bekkum DW, Schellekens H. Antitumor effects of rat recombinant IFN-γ in rats bearing transplantable rat tumors. In: Cantell K, Schellekens H, eds. *The biology of the interferon system.* Dordrecht: Martinus Nigjhoff; 1987:287–292.
114. Trown PW, Brunda MJ, Sim IS, Truitt GA. Antitumor antiviral and immonomodulatory activities of combinations of recombinant interferons alpha and gamma in mice. In: Stewart II WE, Schellekens H, eds. *The biology of the interferon system.* Amsterdam: Elsevier Science Publishers; 1986:371–377.
115. Aapro MS, Alberts DS, Salmon SE. Interactions of human leukocyte interferon with vinca alkaloids and other chemotherapeutic agents against human tumors in clonogenic assay. *Cancer Chemother Pharmacol.* 1983;10:161–166.
116. Miyoshi T, Ogawa S, Kanamori T, Nobuhara M, Namba M. Interferon potentiates cytotoxic effects of 5FU on cell proliferation of established

human cell liners originating from neoplastic tissue. *Cancer Lett.* 1983; 17:239–247.

117. Welander CE, Morgan TM, Homesley HD, Troita PP, Spiegel RJ. Combined recombinant human interferon alpha2 and cytotoxic agents studied in a clonogenic assay. *Int J Cancer.* 1985;35:721–729.

118. Kataoka T, Fusiko OH, Sakura Y. Enhancement of antiproliferative activity of vincristine and Adriamycin by interferon. *Gann.* 1984;75:548–556.

119. Balkwill FR, Monshowitz S, Seilman S. Positive interactions between interferon and chemotherapy due to tumour action rather than effects on host drug metabolizing enzymes. *Cancer Res.* 1984;44:5249–5255.

120. Mowshowitz SL, Chin-Bow ST, Smith GD. Interferon and cis-DDP: combination chemotherapy for P388 leukemia in CFI mice *J Interferon Res.* 1982;2:587–591.

121. Namba M, Yamamoto S, Tanaka H, Kanamori T, Nobuhara M, Kimoto T. In vitro and in vivo studies on potentiation of cytotoxic effects of anticancer drugs or cobalt 60 gamma ray by interferon on human neoplastic cells. *Cancer.* 1984;54:2262–2267.

122. Carmichael J, Fergusson RJ, Wolf CR, Balkwill FR, Smyth JF. Augmentation of cytotoxicity of chempotherapy by human alpha interferons in human non small cell cancer xenografts. *Cancer Res.* 1986;46:4916–4920.

123. Gould MN, Kakria RC, Olson S, Borden EC. Rdaiosensitivity of human bronchogenmic carcinomacells by interferon beta. *J Interferon Res.* 1984; 4:123–128.

124. Schiller JH, Groveman DS, Schmid SM, Wilson JKV, Cummings KB, Borden EC. Synergistic antiproliferative effects of human recombinant alpha 54 or beta ser interferon with gamma interferon on human cell lines of various histogenesis. *Cancer Res.* 1986;46:483–488.

125. Ratliff TL, Kadmon D, Shapiro A, Jacobs AJ, Heston WDW. Inhibition of mouse bladder tumor proliferation by murine interferon-gamma and its synergism with interferon-beta. *Cancer Res.* 1984:44:4377–4281.

126. Creagan E, Loprinzi CL, Ahmann DC, Schaid DJ. A phase I-II trial of the combination of recombinant leukocyte A interferon and recombinant human interferon-gamma in patients with metastatic malignant melanoma. *Cancer.* 1988;62:2472–75.

127. Osanto S, Jansen R, Naipal A, Gratama JW, van Leeuwen A, Cleton FJ. In vivo effects of combination treatment with recombinant interferon gamma and alpha in metastatic melanoma. *Int J Cancer.* 1989;43:1001–1006.

128. Fisher PB, Grant S. Effects of interferon on differentation of normal and tumor cells. *Pharmacol Ther.* 1985;27:143–166.

129. Raefsky EL, Platanias LC, Zoumbos NC, Young NS. Studies of interferon as a regulator of hematopoietic cell proliferation. *J Immunol.* 1985:135(4): 2507–2511.

130. Sokawa Y, Watanabe Y, Watanabe Y, Kawade Y. Interferon suppresses the transition of quiescent 3T3 cells to a growing state. *Nature.* 1977;260: 236–238.

131. Balkwill F, Taylor-Papadimitriou J. Interferons affect both G1 and S + G2 in cells stimulated from quiescence to growth. *Nature.* 1978;274:798–800.

132. Lundgren E, Larrson I, Miorner H. Effects of leukocyte and fibroblast interferon on events in the fibroblast cell cycle. *J Gen Virol.* 1979;42:589–595.

133. Oleszak E, Inglot AD. Platelet derived growth factor inhibits antiviral and anticellular action of interferon in synchronized mouse or human cells. *J Interferon Res*. 1980;1:37–48.

134. Inglot A, Pajtasz E. Measurements of anatagonistic effects of growth factors and interferons. *Methods Enzymol*. 1986;119:657–660.

135. Lin SL, Kikiuchi WJ, Pledger WJ, Tamm I. Interferon inhibits the establishment of competence in G0/S phase transition. *Science*. 1986;233:336–338.

136. Taylor-Ppadimitriou J, Shearer M, Rozengurt E. Inhibitory effect of interferon on cellular DNA synthesis: modultation by pure mitogenic factors. *J Interferon Res*. 1981;1:401–409.

137. Lin LS. Epidermal growth factor-urogastrone action: induction of 2'5'-oligoadenylate synthetase activity and enhancement of the mitogenic effect of ant-interferon antibody. *Life Sci*. 1983;32:1479–1488.

138. Popik W, Inglot AD. Interactions of interferons and transforming growth factors during clonal growth of mouse or human cells in soft agar and in mice. *Int J Cancer*. 1987;40:108–113.

139. Heyns A, Eldor A, Vlodasky I, Kaiser N, Fridman R, Pamet A. The antiproliferative effect of interferon amnd the mitogenic activity of growth factors are independent of cell cycle events. *Exp Cell Res*. 1985;161:297–306.

140. Pfeffer L, Donner DB, Tamm I. Interferon alpha down regulates insulin receptors in lymphoblastoid cells. *J Biol Chem*. 1987;262:3665–3670.

141. Cormier E, Jordan VC. Contrasting ability of antiestrogens to inhibit MCF-7 growth stimulated by estradiol òr epidermal growth factor. *Eur J Cancer Clin Oncol*. 1989;25:57–63.

142. Einat M, Resnitzky D, Kimchi A. Close link between reduction of c-myc expression by interferon and G0/G1 arrest. *Nature*. 1985;313:597–600.

143. Jonak GJ, Knight E. Selective reduction of c-myc mRNA in Daudi cells by human beta interferon. *Proc Natl Acad Sci*. 1984;81:1747–1750.

144. Tominaga S, Lengyel P. Beta-Interferon alters the pattern of proteins secreted from quiescent and platelet derived growth factor-treated BALB/c-3T3 cells. *J Biol Chem*. 1985;260:1975–1978.

145. Zullo JN, Cochran BH, Huang AS, Stiles CD. Platelet derived growth factor and double stranded ribonucleic acids stimulate expression of the same genes in 3T3 cells. *Cell*. 1985;43:793–800.

146. Kohase M, May LT, Tamm I, Vilček J, Sehgal PB. A cytokine network in human diploid fibroblasts: Interactions of beta interferons, tumor necrosis factor, platelet derived growth factor and interleukin-1. *Mol Cell Biol*. 1987;7:273–280.

147. Smid D, Schaft Z, Chang EH, Friedman RM. Interferon induced modulation of human ras oncogene expression. *Prog Clin Biol Res*. 1985;192:259–264.

148. Brouty-Boye D, Gresser I. Progressive reversion of the phenotype of x-ray transformed C3H/10T1/2 cells under prolonged treatment with Interferon. *Int J Cancer*. 1981;28:165–173.

149. Hicks NJ, Morris AG, Burke DC. Partial reversion of the transformed phenotype of murine sarcoma virus transformed cells in the presence of

interferon: a possible mechanism for the anti-tumour effect of interferon. *J Cell Sci.* 1981;49:225–236.

150. Sergiescu D, Gerfaux J, Joret AM, Chany C. Persistent expression of v-mos oncogen in transformed cells that revert to non malignancy after prolonged treatment with interferon. *Proc Nat Acad Sci.* 1986;83:5764–5768.

151. Dani C, Machti N, Piechaczyk M, Blanchard JM. Increased rate of degradation of c-myc mRna in Interferon treated Daudi cells. *Proc Nat Acad Sci.* 1985;82:4896–4899.

152. Ebsworth N, Rozengurt E, Taylor-Papadimitriou J. Measurement of the antiproliferative effect of interferon: influence of growth factors. *Methods Enzymol.* 1986;119:643–649.

153. Cheung HS, Mitchell PG, Pledger WJ. Induction of expression of c-fos and c-myc protooncogenes by basic calcium phosphate crystal: effect of beta interferon. *Cancer Res.* 1989;49:134–138.

154. Dron M, Modjtahedi N, Brison O, Tovey MG. Interferon modulation of c-myc expression in cloned Daudi cells: relationship to the phenotype of interferon resistance. *Molecular Cellular Biology.* 1986;6(5):1374–1378.

155. Landreth GE, Williams LK, Reiser GP. Association of the growth factor kinase with the detergent insoluble cytoskeleton of A431 cells. *J Cell Biol.* 1985;101:1341–1350.

156. Mehmet H, Taylor-Papadimitriou J, Rozengurt E. Interferon inhibition of Bombesin-stimulated mitogenesis in Swiss 3T3 cells occurs without blocking c-fos and c-myc expression. *J Interferon Res.* 1989;9:205–213.

157. Diaz MO, Zieman S, Le Beau MM, Pitha P, Smith SD, Chilcote RR, Rowley JD. Homozygous deletion of the alpha and beta 1-interferon genes in human leukemia and derived cell lines. *Proc Natl Acad Sci USA.* 1988; 85:5259–5263.

158. Carswell EA, Old LJ, Kassel RL, Green S, Fiore N, Williamson B. An endotoxin-induced serum factor that causes necrosis of tumors. *Proc Natl Acad Sci USA.* 1975;72:3666–3670.

159. Beutler B, Greenwald D, Hulmes JD, Beutler B, Greewald D, Hulmes JD, Chang M, Pan Y-CE, Mathison J, Vlevitch R, Cerami A. Identity of tumour necrosis factor and the macrophage secreted factor cachectin. *Nature.* 1985;316:552–554.

160. Urban JL, Shepard HM, Rothstein JL, Sugarman BJ, Schreiber H. Tumor necrosis factor: a potent effector molecule for tum, or cell killing by activated macrophages. *Proc Natl Acad Sci USA.* 1986;83:5233–5237.

161. Phillip R, Epstein LB. A dual role of TNF: immunomodulator and mediator of monocyte cytotoxicity induced by IFN-gamma and IL-1. *Nature.* 1986; 423:86–89.

162. Carswell EA, Green S, Everson TC, Old LJ. Effect of tumor necrosis factor on cultured human melanoma cells. *Nature (Lond).* 1975;258:731–732.

163. Sugarman BJ, Aggarwal BB, Hass PE, Figari IS, Palladino MA Jr, Shepard MJ. Recombinant human tumor necrosis factor-alpha: effect on proliferation of normal and transformed cells in vitro. *Science.* 1985;230:943–945.

164. Sugarman BJ, Lewis GD, Eessalu, TE, Aggarwal BB, Shepard HM. Effects of growth factors on the antiproliferative activity of tumor necrosis factors. *Cancer Res.* 1987;47:780–786.

165. Creasey AA, Doyle LV, Reynolds MT, Jung T, Lin LS, Vitt CR. Biological effects of recombinant human tumour necrosis factor and its novel muteins on tumour and normal cell lines. *Cancer Res*. 1987;47:145–149.

166. Schiller JH, Bittner G, Storer B, Willson JKV. Syenrgistic antitumor effects of tumour necrosis Factro and gamma interferon on human colon carcinoma cell lines. *Cancer Res*. 1987;47:2809–2813.

167. Fransen L, Ruysschaert M, Vann Der Heyeden J. Recombinant tumor necrosis factor: species specificity for a variety of human and murine transformed cell lines. *Cell Immunol*. 1986;100:260–267.

168. Peetre C, Gulberg U, Nilsson E, Olsson I. Effects of recombinant tumor necrosis factor in proliferation and differentiation of leukemic and normal hemopoietic cells in vitro. *J Clin Invest*. 1986;78:1674–1700.

169. Beran M, McCredie KB, Keating MJ, Gutterman JU. Antileukemic effect of recombinant tumor necrosis factor alpha in vitro and its modulation by alpha and gamma interferons. *Blood*. 1988;72:728–738.

170. Haranaka KH, Satomi N, Sakurai A. Antitumor activity of murine tumor necrosis factor (TNF) against transplanted murine tumors and heterotransplanted human tumors in nude mice. *Int J Cancer*. 1984;34:263–267.

171. Creasey AA, Reynolds T, Laird W. Cures and partial regression of murine and human tumors by recombinant human tumor necrosis factor. *Cancer Res*. 1986;46:5687–5690.

172. Balkwill FR, Ward BG, Moodie E, Fiers W. Therapeutic potential of tumour necrosis factor alpha and gamma interferon in experimental human ovarian cancer. *Cancer Res*. 1987;47:4755–4758.

173. Palladino JR, Refaat Shalaby M, Kramer SM, Ferraiolo BL, Baughman RA, Deleo AB, Crasé D, Marafino B, Aggarwal BB, Figarils, Liggitt D, Patton JS. Characterization of the antitumour activities of human necrosis factor alpha and the comparison with other cytokines: induction of tumour specific immunity. *J Immunol*. 1987;138:4023–4032.

174. Havell EA, Fiers W, North RJ. The antitumour function of tumour necrosis factor. (TNF). *J Exp Med*. 1988;167:1067–1085.

175. Shimomura K, Manda T, Mukomoto S, Katsumasa K, Nakano K, Mori J. Recombinant human tumor necrosis factor-alpha: thrombus formation is a cause of anti-human activity. *Int J Cancer*. 1988;41:243–247.

176. Gresser I, Beladelli F, Tavernier J, Fiers W, Podo F, Federico M, Carpinelli G, Dovillard P, Prade M, Maury C. Antitumor effects of interferon in mice injected with interferon-sensitive and interferon-resistant friend leukemia cells. V. comparisons with the action of the tumor necrosis factor. *Int J Cancer*. 1986;38:771–778.

177. Lee SH, Aggarawal BB, Rinderknecht E, Assisi F, Chiu H. The synergistic antiproliferative effect of gamma interferon and human lymphotoxin. *J Immunol*. 1984;133:1083–1086.

178. Stone-Wolff DS, Yip YK, Kelker H, Le J, Henriksen-Destefano D, Rubin BY, Rinderknecht E, Aggerwal BB, Vilćek J. Interrelationships of human interferon gamma with lymphotoxin and monocyte cytotoxin. *J Exp Med*. 1984;159:828–843.

179. Lewis GD, Aggarwal BB, Eessalu TE, Sugarman BJ, Shepard HM. Modulation of the growth of transformed cells by human tumor necrosis factor and interferon gamma. *Cancer Res*. 1987;47:5382–5385.

180. Aggarwal BB, Eessalu TE, Hass PE. Characterization of receptors for human tumor necrosis factor and their regulation by gamma interferon. *Nature.* 1985;318:665–667.
181. Winkelhake JL, Stampfl A, Zimmerman RJ. Synergistic effects of combination therapy with human recombinant interleukin-2 and tumor necrosis factor in murine tumor models. *Cancer Res.* 1987;47:3948–3953.
182. Broudy VC, Kaushansky K, Segal GM, Harlan JM, Adamson JW. Tumor necrosis factor type alpha stimulates human endothelial cells to produce granulocyte/macrophage colony-stimulating factor. *Proc Natl Acad Sci USA.* 1986;83:7467–7471.
183. Koeffler HP, Gasson J, Ranyard J, Souza L, Shepard M, Munker R. Recombinant human TNF alpha stimulates production of granulocyte colony-stimulating factor. *Blood.* 1987;70(1):55–59.
184. Blick M, Sherwin SA, Rosenblum K, Gutterman J. Phase I study of recombinant tumor necrosis factor in cancer patients. *Cancer Res.* 1987; 47:2986–2989.
185. Yarden A, Kimichi A. Tumor necrosis factor reduced c-myc expression and co-operates with IFN-gamma in HeLa cells. *Science.* 1986;234:1419–1422.
186. Spriggs DR, Imamura K, Rodriguez C, Sariban E, Kufe DW. Tumor necrosis factor expression in human epithelial tumor cell lines. *J Clin Invest.* 1988;81:455–460.
187. Onozaki K, Matsushima K, Aggarawal BB, Oppenheim JJ. Interleukin 1 is a cytocidal factor for several tumour cell lines. *J Immunol.* 1985;135:3962–3968.
188. Gaffney EV, Tsai SC. Lymphocyte-activating and growth-inhibitory activities for several sources of native and recombinant interleukin. *Cancer Res.* 1986;46:3834–3837.
189. Tsai SC, Gaffney EV. Inhibitition of cell proliferation by interleukin-1 derived from monocytic leukaemia cells. *Cancer Res.* 1986;46:1471–1477.
190. Onozaki K, Akiyama Y, Okana A, Hirano T, Hirano T, Kishimoto T, Hashimoto T, Yoshigawa K, Taniyama T. Synergistic regulatory effects of interleukin 6 and interleukin 1 on the growth and differentiation of human and mouse myeloid leukemic cell lines. *Cancer Res.* 1989;49:3602–3607.
191. Smith KA, Lachman LB, Oppenheim JJ, Favata MF. The functional relationships of the interleukins. *J Exp Med.* 1980;151:1551–1556.
192. Kasahara T, Mukaida N, Hatake K, Motoyoshi K, Kawai T, Shiori-Nakano K. Interleukin 1 dependent lymphokine production by human leukemic T cell line HSB. 2 subclones. *J Immunol.* 1985;134:1682–1689.
193. Onozaki K, Matsushima K, Kleinerman ES, Saito T, Oppenheim JJ. Role of interleukin 1 in promoting human monocyte mediated tumor cytotoxicity. *J Immunol.* 1985;135:314–320.
194. Leary AG, Ikebuchi K, Hirai Y, Wong GG, Yank YC, Clark SC, Ogawa M. Synergism between interleukin-6 and interleukin-3 in supporting proliferation of human hematopoietic stem cells: comparison with interleukin-1 alpha. *Blood.* 1988;7:1759–1763.
195. Chen L, Mory Y, Zilberstein A, Revel M. Growth inhibition of human breast carcinoma and leukemia/lymphoma cell lines by recombinant interferon-beta 2. *Proc Natl Acad Sci USA.* 1988;85:8037–8041.

196. Kohasse M, Henriksen-Destefano D, May LT, Vilcek J, Seghal PB. Induction of B2-interferon by tumor necrosis factor: homeostatic mechanism in the control of cell proliferation. *Cell.* 1986;45:659–666.
197. Resnitzky D, Yarden A, Zipori D, Kimchi A. Autocrine beta related interferon controls c-myc suppression and growth arrest during hemotopoietic cell differentation. *Cell.* 1986;46:31–40.
198. Ferbus D, Testa U, Titeux M, Louache F, Thang MN. Induction of (2'–5') oligoadenylate synthetase, activity during granulocyte and monocyte differentation. *Mol Cell Biochem.* 1985;67:125–133.
199. Tabibzadeh SS, Santhanam U, Seghal PB, May LT. Cytokine-induced production of IFN-beta 2/IL-6 by freshly explanted human endometrial stromal cells. *J Immunol.* 1989;142:3134–3139.
200. Kawano M, Hirano T, Matsuda T, Taga Y, Taga T, Horii Y, Iwato K, Asaoku H, Tang B, Tanabe O, Tanaka H, Kuramoto A, Kishimoto T. generation and essential requirement of BSF-2/IL6 for human multiple myelomas. *Nature.* 1988;332:83.

7
Cytokine Regulation of Hypothalamic and Pituitary Hormone Secretion

Michael D. Lumpkin

Mammalian organisms respond to microbial infections, trauma, inflammatory processes, and a number of other physical and psychological stressors by modulating the activity of the immune system. As a part of this response, the brain and its neural inputs can activate certain immune components either by direct innervation of immune system tissues such as the thymus gland and spleen, or the central nervous system (CNS) can act through neuroendocrine mechanisms to stimulate the secretion of certain pituitary hormones that regulate lymphoproliferative activity. In turn, macrophages, monocytes, and lymphocytes may produce hormone-like substances and cytokines that, in addition to providing a level of paracrine regulation, can also transmit long-loop feedback signals to the hypothalamic–pituitary unit to adjust further hormonal output. Through the many studies that have delineated these interactions, it has now become clear that bidirectional communication exists between immune components and the neuroendocrine system. In this chapter, the regulatory relationship between a number of the cytokines, including interleukin-1 (IL-1), IL-2, IL-6, tumor necrosis factor (TNF), gamma-interferon (IFN), the hypothalamic releasing factors, and pituitary hormones will be considered.

Among the actions of the cytokines, those of IL-1 in controlling the secretion of the neuropeptide corticotropin-releasing factor (CRF) and the pituitary hormone adrenocorticotropin (ACTH) have been the most highly characterized. There also exists some information on IL-1's regulation of growth hormone (GH) secretion via hypothalamic somatostatin (SRIF) and growth hormone-releasing hormone (GHRH). Thus, the regulatory relationship between IL-1 and the controllers of ACTH and GH is discussed first as prototypes of cytokine-directed hormone secretion. This is followed by a description of IL-1's effects on luteinizing hormone (LH) and thyroid-stimulating hormone (TSH) secretion. Finally, the actions of additional cytokines on pituitary hormone secretions are discussed.

Background

The host response to an episode of infection or inflammation, during which the acute-phase response is elicited, will result in macrophages, monocytes, and glial cells producing the polypeptide cytokine known as interleukin-1 (IL-1).[1] Human IL-1 is a polypeptide monokine whose propeptides consist of 269 to 271 amino acids and that exists predominantly in two forms known as IL-1α and IL-1β.[2] March et al.[2] showed that the positions of 70 of the 271 amino acids (26%) of human IL-1β are identical to those of human IL-1α. Human IL-1β shares 80 of 270 amino acids with murine IL-1 (30%), whereas human IL-1α demonstrates a homology of 167 of the 271 amino acids (62%) with murine IL-1.[1] Mature forms of IL-1α and IL-1β are represented by C-terminal cleavage products of 159 and 153 amino acids, respectively.

The mediation of the acute-phase response by IL-1 provokes certain catabolic changes leading to the mobilization of metabolic substrates, the production of specific immune substances by the liver, an increase in the number and immaturity of circulating neutrophils, and fever.[1] IL-1 also activates T- and B-cell function, promotes IL-2 production, and synergizes with lymphokines to enhance natural killer cell activity directed against certain tumor target cells.[1] With such a substantial immune reaction being driven by IL-1, it is desirable for the body to have a mechanism by which this aggressive response to infection and inflammation can ultimately be reduced in order to prevent the induction of pathologic processes.[3] It has been observed that when a reduction in stimulated IL-1 activity is finally needed, this cytokine can promote the production of glucocorticoid hormones such as corticosterone in the rat or cortisol in humans[5,6] which, in turn, suppress the further production of IL-1.[7-9]

Not only can immune cell products such as IL-1 promote immune functions, but it is now apparent that certain hormones of the anterior pituitary gland can also stimulate immune system activity. Growth hormone (GH) and, to a lesser extent prolactin (PRL), which shares some homology with GH,[4] can reverse the inhibitions on lymphoproliferation, antibody synthesis, and delayed skin graft survival produced either by stress or glucocorticoids.[10-13] GH has also been shown to stimulate the growth and activity of the thymus gland, lymphoid cells, phagocytic cells, and stem cells.[14]

Furthermore, GH has recently been described to be a macrophage activating factor,[15] whereas prolactin was observed to enhance the tumoricidal activity of macrophages and the synthesis of interferon gamma.[16] The mechanisms by which IL-1 ultimately down-regulates its own activity through the stimulation of corticosteroid synthesis may also pertain to the coordinated and integrated control of GH secretion and perhaps to the subsequent regulation of immune cell function.

Relationships Between IL-1, Stress, ACTH, and GH Secretion

Since the immunostimulating action of GH opposes many of the immunosuppressive effects of corticosteroids,[10-13] the secretions of which are stimulated by IL-1 in the latter phases of an immune response, it would be advantageous for the organism to possess a coordinated signal for the suppression of GH-driven immune stimulation when the overall desired effect becomes a reduction or moderation of a prior, vigorous immunopromoting event. As will be seen below, IL-1 itself is positioned to provide such a signal for the integrated control between the CRF–ACTH–corticosteroid and the GHRH–SRIF–GH secretion axes and their effects on immune cell activity. That there is indeed coordination between the control of ACTH and GH release is seen during various types of stress. In a number of stressful conditions, ACTH levels are elevated whereas those of GH are suppressed in the rat.[17,18] Infection and inflammation represent two types of stressful events and during such episodes, immune cells secrete higher levels of IL-1 while the pituitary increases ACTH but decreases GH secretion. Additional work has shown that injections of the bacterial endotoxin known as lipopolysaccharide (which mimics bacterial infection) will increase the release of endogenous IL-1.[19] Separate studies have shown that endotoxin injection will increase ACTH[20] but decrease GH[21] secretion in the rat. Similarly, administration of the Newcastle disease virus (NDV) to mice elevated plasma ACTH and corticosterone levels.[5,22,23] Considered together, these results suggest a causal relationship between production of endogenous IL-1 and the secretion of pituitary hormones. A number of other studies have investigated the possibility that IL-1 is the driving force of the CRF–ACTH–glucocorticoid axis during stressful episodes,[4,5,6,16,24-28] but fewer studies have examined the equally interesting possibility that IL-1 may also influence the GHRH–SRIF–GH axis.

IL-1, the Hypothalamus, and Releasing Factor Hormones

There is some doubt that peripherally produced IL-1, a large polypeptide, can penetrate the hypothalamus to modulate the secretion of neurohormones such as CRF and GHRH.[29] The reason for this is that large peptides do not cross the blood–brain barrier in significant amounts.[29] However, some studies have indicated that peripherally produced IL-1 from monocytes might reach the brain through the circumventricular organs such as the organum vasculosum of the lamina terminalis (OVLT).[30,31] Since most of the hypothalamic sites mediating the neural responses to plasma IL-1 are located at some distance from the OVLT, some investigators believe it is unlikely that these far-removed hypothalamic neurons are affected directly by circulating IL-1.[32] However, with the discovery by Breder et al.[32] that IL-1β is located in neuronal elements of the human hypothalamus, including the median eminence (ME), the concern over IL-1 access to intrahypothalamic brain sites is now circumvented. Therefore, it is necessary to begin elucidating the function of hypothalamically produced IL-1.

Interestingly, Breder et al.[32] have reported that IL-1β is contained in nerve fibers that terminate in the paraventricular, the periventricular, and the arcuate nuclei of the human hypothalamus. The paraventricular nucleus is also the site of the parvicellular CRF cell bodies,[33] while the anterior periventricular region contains most of the hypothalamic SRIF perikarya[34,35] and the arcuate nucleus is where most GHRH somata arise.[36,37] Furthermore, it was reported that IL-1β cell bodies are located in the periventricular and arcuate areas of the human hypo-thalamus.[32] Another study in rats has found immunoreactive IL-1β in the cell bodies and nerve fibers of the same hypothalamic nuclei that contain hypophysiotropic hormones.[38] In addition, it has been observed that administration of convulsant agents to rats will increase the content of IL-1β mRNA in the hypothalamus, as well as in cerebral cortex and thalamus.[39] This places IL-1 in a favorable anatomic position to control hypothalamic neurohormone secretion, including the regulation of CRF release by IL-1 (which has been much studied as of late[6,24,25]) and, potentially, the regulation of GHRH and SRIF synthesis and release. The governance of these neuropeptides could ultimately influence the GH secretion pattern and thereby alter somatic growth. The finding of IL-1 receptors in hypothalamic slices[40] and on hypothalamic membrane preparations[41] also support the likelihood that this cytokine regulates hypothalamic hormone activity.

In investigating the actual regulation of CRF secretion by IL-1, it was found that IL-1 injections would cause depletion of CRF from the median eminences of colchicine-treated rats.[24] Suda et al.[42] observed increases in hypothalamic CRF mRNA after giving intraperitoneal injections of IL-1 to rats. Furthermore, intravenous injections of IL-1 were found to stimulate the release of CRF into the portal blood of the pituitary vascu-lature of the rat.[6] This finding was consistent with the abolition of IL-1–induced ACTH release after administration of CRF antiserum (i.e., in vivo immunoneutralization) to rats.[6,24,25]

In vitro experiments have shown that IL-1 will stimulate the release of CRF from whole rat hypothalamic and dispersed cells into incubation culture media.[43,44]

IL-1 has also been shown to modulate the neurosecretion of hypo-thalamic factors controlling GH secretion. Our laboratory has reported[45] that injection of recombinant human IL-1β into the third ventricle of the

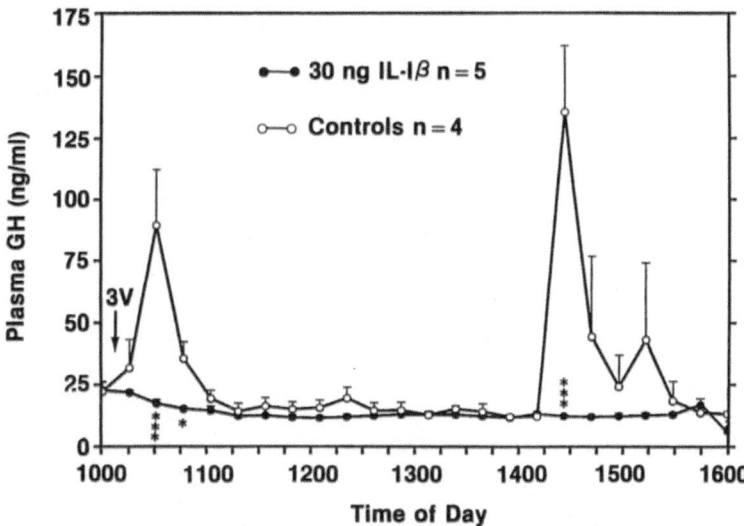

FIGURE 7.1. Effect of IL-1β (30 ng in 2 μl of sterile, nonpyrogenic saline) or saline alone (controls) on pulsatile GH release when microinjected into the third ventricle (3V) of the hypothalamus of conscious, undisturbed adult male rats. Shown are the mean ±SEM; ***$p < .005$, *$p < .05$.

hypothalamus of conscious rats will completely eliminate the spontaneous pulses of plasma GH (Fig. 7.1), while having no effect on tonic prolactin secretion (Fig. 7.2). To define better the site of action of IL-1β, this cytokine was next microinfused into the sites corresponding to either the location of GHRH neurons in the hypothalamic arcuate nucleus or that of SRIF neurons in the hypothalamic periventricular nucleus. Administration of IL-1 into the arcuate nucleus[45] produced a significant inhibition on the GHRH-driven peaks of pulsatile GH secretion,[46-48] as seen in Figure 7.3.[45] However, injection of IL-1 into the periventricular region of SRIF cells did not alter the pulsatile pattern of GH secretion.[45]

In vitro studies by Scarborough et al.[49] have shown that exposure of cultured hypothalamic cells from immature rats to IL-1β for 24 hr will significantly increase the synthesis of SRIF mRNA. After 4 days of chronic exposure to IL-1 (10^{-8} M), the amount of SRIF peptide secreted into media also was significantly elevated. The overall conclusion from these in vivo and in vitro approaches is that IL-1β acts acutely to decrease GH secretion by inhibiting GHRH release, but functions chronically to lower GH levels by stimulating SRIF synthesis and secretion. It is also possible that IL-1 acts for longer periods to maintain a suppression of GHRH secretion, but this has not been examined.

In understanding the neuroendocrine physiology of IL-1, it would be useful to know whether certain endogenous neural substances might

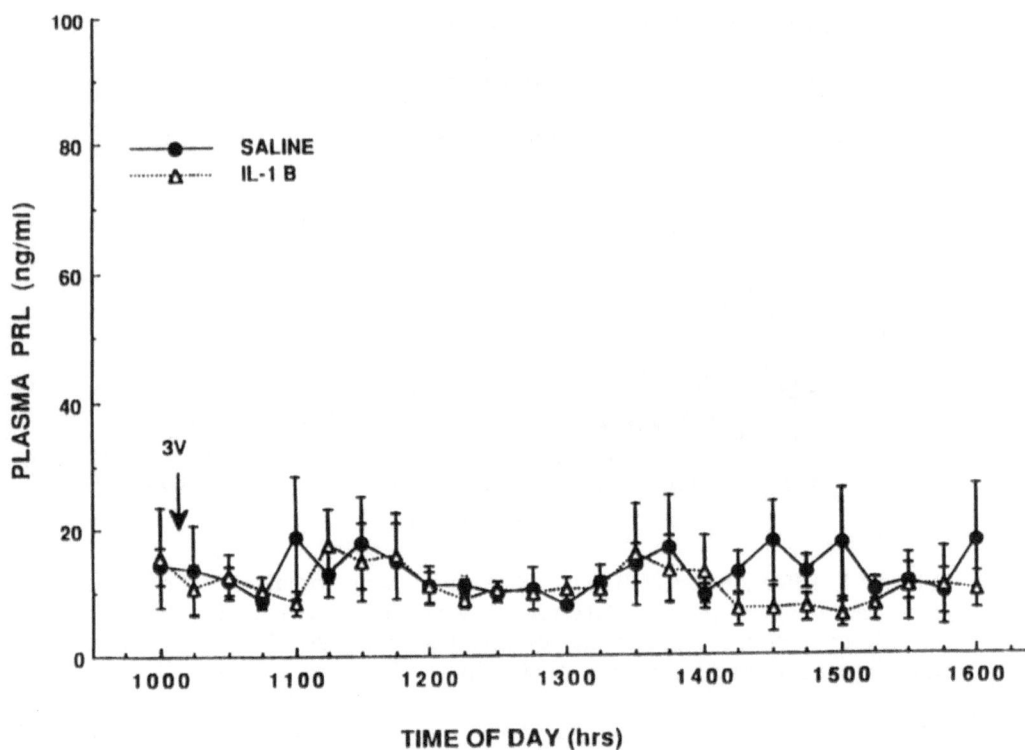

FIGURE 7.2. Effect of IL-1β (30 ng in 2 μl of sterile, nonpyrogenic saline) or saline alone (2 μl) on plasma PRL levels after microinjection into the third ventricle (3V) of the hypothalamus of conscious, undisturbed adult male rats. Shown are the mean ±SEM. No significant differences were detected.

antagonize the central effects of IL-1, particularly if excessive neural IL-1 activity is ever shown to produce pathologic conditions that require treatment. One candidate substance that this laboratory has examined is alpha melanocyte-stimulating hormone (α-MSH). This 13–amino acid fragment of ACTH is found in the arcuate nucleus,[50,51] as are GHRH and IL-1, and has already been observed to antagonize IL-1–mediated fever induction in rabbits[52] and IL-1–induced elevation of ACTH secretion in mice.[53] Studies by the author[54] have shown that prior administration of 200 ng of α-MSH into the bilateral arcuate nucleus will completely block the inhibitory effect of IL-1 on pulsatile GH secretion, when injected into the arcuate nucleus 15 min after α-MSH administration by the same route (Fig. 7.3). Lesser doses of α-MSH were only effective for a shorter period.[45,54] In explaining the mechanism of α-MSH antagonism of IL-1's effect, Cannon et al.[55] have shown that increasing doses of IL-1 could overcome the inhibition produced by α-MSH on thymocyte pro-

PLASMA GH IN ADULT MALE RATS RECIEVING a-MSH AND IL-1B

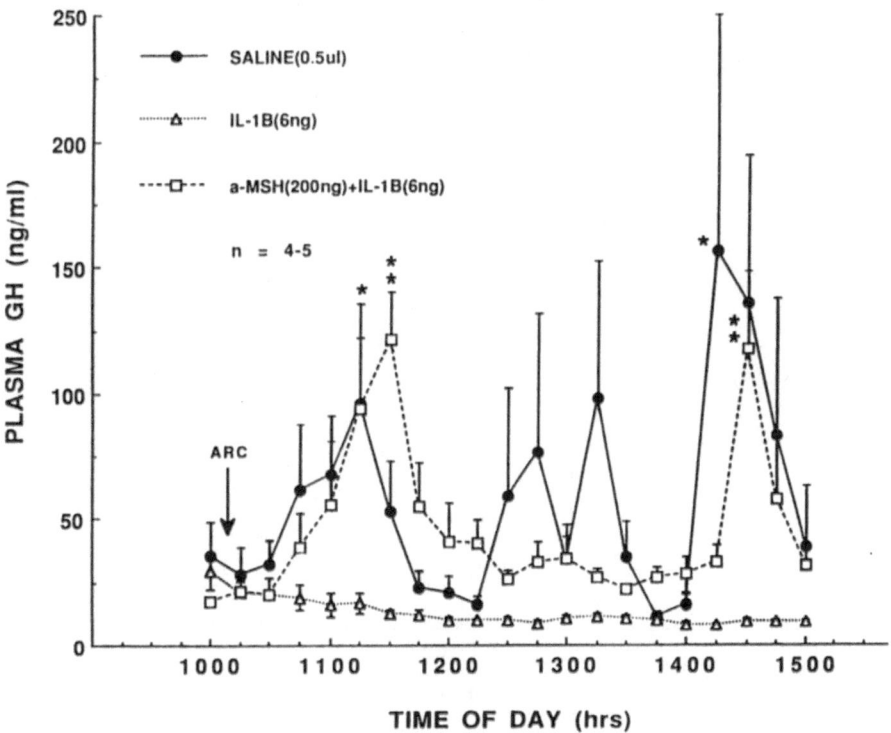

FIGURE 7.3. Effect of microinjection of sterile, nonpyrogenic saline (controls) or IL-1β (6 ng in 0.5 μl sterile saline) into the bilateral arcuate (*ARC*) nucleus of the hypothalamus on the pulsatile pattern of GH secretion (mean ±SEM) in conscious, undisturbed adult male rats. A third group of rats (*open boxes*) first received 200 ng of α-MSH into the bilateral ARC followed 15 min later by 6 ng of IL-1β into the ARC. Shown are mean ±SEM. Mean peaks of GH release were statistically compared to the IL-1β group: **$p < .025$; *$p < .05$. The mean peak values between the saline-treated group and the α-MSH plus IL-1β group are not significantly different.

liferation; therefore, α-MSH may simply interfere with the binding of IL-1 on its target receptors. A second possibility is that since IL-1 stimulates CRF secretion[6,24,25] and also activates the gene of proopiomelanocortin (POMC),[56] both of which ultimately result in the production of α-MSH via the processing of POMC either in the brain[57] or in leukocytes,[58] the increasing levels of α-MSH may operate as a negative feedback signal to the further action of either IL-1 or CRF in stimulating production of POMC derivatives.

PLASMA GH IN ADULT MALE RATS RECEIVING CRF ANTAGONIST

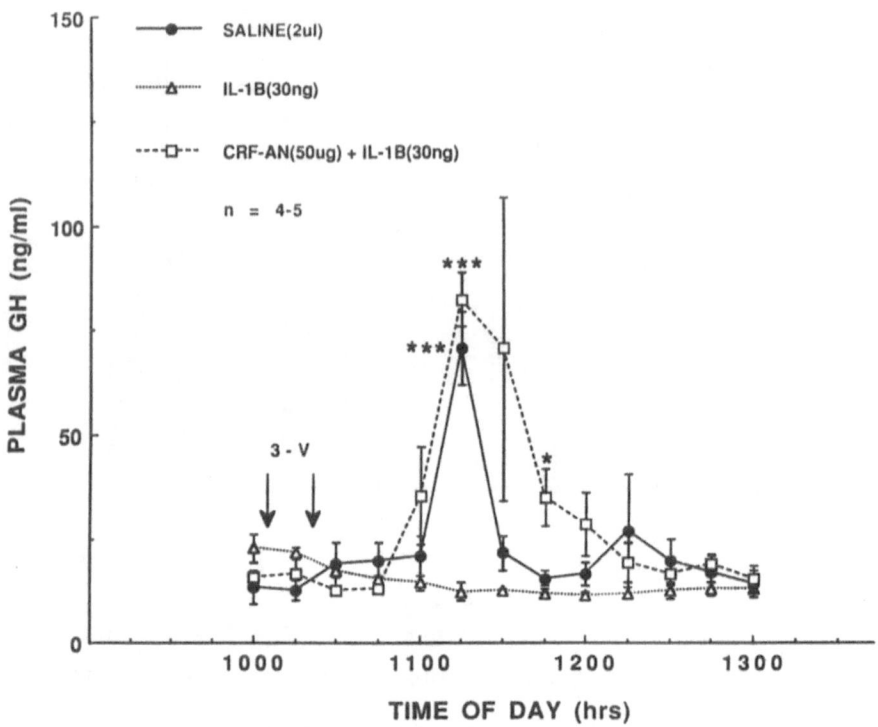

FIGURE 7.4. Immediately after withdrawing a baseline blood sample at 1000 hr, two groups of animals received sterile saline (2 μl) 3V and one group (*open boxes*) received 50 μg of the alpha-helical CRF_{9-41} antagonist (CRF-AN, Peninsula) into the 3V. After the next blood sample at 1015 hr, one group (*closed circles*) received 2 μl of saline (controls), another (*open triangles*) received 30 ng of IL-1β alone, and the CRF-AN–treated group (*open boxes*) also received 30 ng of IL-1β into the 3V. CRF-AN treatment prevented IL-1 from suppressing the spontaneous rise in plasma GH noted at 1115 hr in controls. Shown are mean ±SEM. ***vs. IL-1β (*open triangles*); *vs. IL-1β (*open triangles*). No significant differences were detected in the mean GH peaks between the saline controls and the CRF-AN plus IL-1 group.

IL-1 may regulate GH secretion via a modulation of CRF release. It has been shown by us[59] and others[60] that intraventricular injections of CRF will inhibit GH secretion, at least in part by stimulating SRIF release.[61,62] Therefore, we sought to determine whether IL-1 could exert its inhibition on GH secretion by a CRF-mediated mechanism. The experiment performed[54,63] was to inject the CRF peptide antagonist alpha-helical CRF_{9-41} (50 μg) into the third ventricle of the hypothalamus

15 min before the third ventricular microinjection of 30 ng of IL-1β (Fig. 7.4). The result was a complete blockade by the CRF antagonist of the IL-1–induced GH suppression. Since IL-1 acts acutely in the GRF-containing arcuate nucleus but not in the SRIF-containing periventricular region to decrease GH release,[45,54] the above result implies that CRF neurons mediate an IL-1 inhibition on GHRH neurosecretion and thus suppress GH release.

IL-1 and the Median Eminence

The ME is the most basal portion of the hypothalamus where the terminals of hypothalamic neurons containing releasing factor hormones are juxtaposed to the portal blood vessels that transport the releasing factors to the cells of the anterior pituitary gland. Dense innervation of the ME by immunoreactive IL-1β nerve terminals has been described in the human hypothalamus[32] and suggests the possibility of axo-axonal interactions among the intermingled terminals of IL-1, CRF, GHRH, SRIF, and other releasing factor neurons present in the ME. Furthermore, the juxtaposition of IL-1 terminals on the portal capillaries of the ME provides a possible route by which brain-secreted IL-1 could be transported by portal blood flow directly to the cells of the pituitary, thereby influencing hormone secretion. Similarly, astroglial and microglial cells of the brain, which are numerous in the ME region of the hypothalamus,[64] produce IL-1.[65,66] This places IL-1 in a favorable location to modify hypothalamic CRF, GHRH, and SRIF secretion, and hence, exert some regulatory influence over the patterns of ACTH and GH release.[67,68] Since glial cells of this area project their processes onto the portal capillaries of the ME,[64] it is also possible that glial-manufactured IL-1 could be delivered by portal blood flow directly to pituitary cells. The presence of IL-1 in portal blood, however, has not been demonstrated.

IL-1 and the Anterior Pituitary Gland

Recent work[69] has demonstrated the presence of both IL-1β peptide and mRNA in the cells of the anterior pituitary gland of the rat. The content of both the message and the final peptide product could be increased by injections of bacterial lipopolysaccharide. There are also preliminary reports of the characterization of IL-1 receptors in rat, mouse, and human pituitary membrane preparations.[70,71] However, the issue of whether IL-1 has a direct action on the cells of the anterior pituitary to modulate hormone release remains unsettled.[72] Uehara et al.[28] and others,[6,24] have found no direct effects of human recombinant IL-1α and IL-1β, on pituitary somatotrophs or corticotrophs or other cell types when examined in static cell incubations after acute exposure to this cytokine. Beach et al.,[73] in contrast, have found that human recombinant IL-1β and mouse IL-1α will stimulate GH, ACTH, LH, TSH, and PRL release when presented intermittently to pituitary cells maintained in a dynamic perifusion system. A study of the time course needed for IL-1β and IL-1α to stimulate hormone release from dispersed pituitary cells in vitro has provided some clarification to this controversy.[72] This study[27] revealed that up to 16 hr of direct exposure to IL-1 are needed to elicit ACTH release. No amount of time of exposure to IL-1 was sufficient to alter GH or luteinizing hormone (LH) secretion from pituitary cells. In any event, there is no current evidence to suggest that IL-1 acts directly on GH-secreting cells to inhibit GH release.

IL-1 and the Regulation of LH Secretion

Studies by Rivier and Vale[74] have shown that intracerebroventricular injections of recombinant human IL-1α reduces plasma LH but not follicle-stimulating hormone (FSH) concentrations in castrated male rats. Subsequent IV injection of gonadotropin-releasing hormone (GnRH) still produced normal LH release in all animals whether they had received prior IL-1 or control treatments. Furthermore, IV injection of IL-1α failed to alter gonadotropin secretion. Together, these results indicated that IL-1 must act centrally, rather than directly, at the pituitary to inhibit LH secretion. The explanation was offered[74] that because IL-1 stimulates CRF release, and since CRF suppresses LH secretion[59,75] and GnRH secretion,[76] the possibility existed that the suppressive action of IL-1 on LH secretion occurred via activation of CRF pathways in the brain.

In contrast, other investigators have reported that IL-1α and IL-1β act directly on pituitary cells in culture either to stimulate LH release[26,73] or to produce no changes in gonadotropin secretion.[27,28] Since stress,[77] infection, and inflammation can sometimes lead to reproductive failure as well as to increased IL-1 production, it is tempting to speculate that the predominant action of IL-1 in vivo may be to act on the hypothalamus to suppress GnRH secretion, with little or no direct effect on the pituitary gonadotropes.

IL-1 and the Regulation of TSH Secretion

Immune activation by infectious agents reduces thyroid function and endotoxin administration to rats[21,78] will result in the suppression of TSH release. A role for IL-1 is implicated in this process since injections of IL-1 into the third ventricle of the hypothalamus inhibit TSH release.[79] Since IL-1 stimulates CRF and SRIF secretion,[80] and since CRF can stimulate SRIF release,[61,62] it is possible that IL-1 suppresses plasma TSH levels via increased concentrations of SRIF flowing through the pituitary portal vasculature to the thyrotropes. It has also been found that IL-1 incubated with dispersed fetal rat hypothalamic neurons in culture decreases the content of TRH mRNA is these cells,[80] and thus, this could account for lowered TSH levels. Systemically administered IL-1β can also cause a rapid decline in serum TSH concentrations.[81] This result is consistent with the finding that IL-1β and its mRNA are colocalized with TSH in pituitary thyrotropes.[69] Thus, IL-1β may act at both the hypothalamic and pituitary levels to inhibit TSH secretion. However, some investigators report that IL-1 actually has a stimulatory effect on pituitary cells in vitro to release TSH.[16,73] However, the relevance of such in vitro results to the in vivo situation is not clear at this time.

IL-2 and Hormone Regulation

IL-1 can stimulate IL-2 production in certain immune cells.[82,83] IL-2 and IL-2 receptors have been found in the brain and in brain cells.[84,85] Thus, IL-2 may be able to function as a regulator of neuroendocrine function. When clinically tested in patients with cancer and acquired immunodeficiency syndrome, IL-2 produced an elevation in plasma ACTH levels.[86,87] However, when IL-2 was administered systemically in rats, it had no effect on ACTH or corticosterone levels.[5,88] Both GH and PRL secretion were stimulated by IL-2 therapy in cancer patients.[89]

IL-6 and Hormone Regulation

IL-6 is a cytokine produced by monocytes, T-cells, fibroblasts, and endothelial cells and is released in response to inflammation and injury of tissues.[90] IL-6 is more potent than IL-1 and TNF in stimulating acute-phase proteins from human hepatocytes.[91] IL-1 may stimulate the production of IL-6 in glial cell lines[92] and by viral infection in cultured astrocytes and microglia.[93] Recently, it was shown that in vitro incubations of rat medial basal hypothalamic fragments with lipopolysaccharide would stimulate the release of IL-6 into incubation media.[94] In addition, IL-6 has also been found in pituitary cells[95] and in folliculostellate cells[96] of the anterior pituitary gland. Such localizations of IL-6 place it in a position to modulate neuroendocrine function.

   Indeed, Fukata et al.[97] have observed that IL-6, as well as IL-1α and IL-1β, will stimulate the in vitro release of ACTH from AtT-20 tumor-derived pituitary cells after a latency of 24 to 72 hr. Additional work has shown that IL-6 can acutely stimulate the in vitro release of PRL, GH, and LH from normal dispersed pituitary cells.[98]

   IL-6 has also been implicated in the direct control of hypothalamic-releasing hormone neurons. Naitoh et al.[99] reported that the IV injection of recombinant IL-6 into rats significantly increased the secretion of ACTH. However, this action of IL-6 was blocked by the prior administration of an anti-CRH antiserum. These results indicated that IL-6 must stimulate the release of ACTH through activation of CRH neurosecretion. Other studies have shown that IL-6 can stimulate the release of GnRH[100] and SRIF[80] from hypothalamic neuronal cultures, findings that would be consistent with provoking LH but not GH secretion as described above.

TNF and Hormone Regulation

During infection and inflammation, tumor necrosis factor (cachectin) may be released from activated immune cell types along with IL-1 and Il-6. TNF alpha seems to be somewhat weaker than IL-6 in stimulating the synthesis of acute-phase proteins from human hepatocytes.[91] Not only is TNF produced by immune cells that orginate in the periphery, but TNF has also been identified in microglia and astrocytes in culture.[101] Lieberman et al.[102] reported that after astrocytes were exposed to lipopolysaccharide or Newcastle disease virus in vitro, they increased both the synthesis and secretion of TNF. Consistent with the glial cell production of TNF, Breder and Saper[103] have presented evidence of the presence of TNF in the mouse hypothalamus. Thus, TNF is positioned to regulate hypothalamic and pituitary hormone secretion.

Intravenous injections of TNF have been found to stimulate ACTH release in both the rat[104] and the human[105]; however, intraperitoneal injections have been found to be ineffective.[5] When TNF was given intravenously or intracerebroventricularly,[79,106] plasma levels of TSH declined, reminiscent of the same effect on TSH seen after IL-1 administration. Although IV administration of TNF to calves suppressed GH secretion,[107] it appears to stimulate PRL release in cancer patients,[108] but to have no effect on PRL secretion in rats.[104] Intraventricular administration of TNF, however, was effective in elevating plasma ACTH levels in the rat.[109] It has been reported that TNF has powerful direct effects on rat pituitary cells in vitro to stimulate the acute release of ACTH, TSH, and GH.[110] In contrast, Gaillard et al.[111] observed that exposure of pituitary cells in culture to TNF for a period of 8 to 24 hr will actually inhibit the stimulated levels of ACTH, GH, LH, and PRL.

At the hypothalamic level in the rat, TNF reduces the content of TRH and TRH mRNA.[80] Scarborough et al.[80] observed that TNFα was somewhat less active than IL-1 in stimulating SRIF release from fetal rat hypothalamic neurons in culture. Yet, IL-1β and TNF given together were synergistic and produced a much greater release of SRIF into incubation media than did the combination of IL-1α and IL-1β.[80] In addition to effects on SRIF, there are also data indicating that TNF can stimulate the release of GnRH from primary cultures of rat hypothalamic cells.[112] This finding is congruent with the recent report that TNF stimulates immunoreactive and bioactive LH secretion from rat pituitary cells in vitro.[113]

Gamma Interferon and Hormone-Regulation

Gamma interferon is known to be produced by T lymphocytes and natural killer cells and to posses macrophage-activating properties.[114] This cytokine is produced by immune cells when they mediate inflammatory reactions resulting from the injection of and invasion by tumor cells. It is generally believed that gamma-IFN is produced locally at the site of inflammation or infection and is then absorbed into the general circulation.[114] However, this view of local production may not apply to neural cells since the stimulation of astrocytes in culture with either lipopolysaccharide or the Newcastle disease virus failed to activate the gamma-IFN gene.[102]

Much like the effects of IL-1 when administered centrally, injections of subpicomole quantities of gamma-IFN into the third ventricle of the rat result in a lowering of plasma GH and TSH levels, an elevation of plasma ACTH concentrations, and no change in plasma PRL values.[115] However, in this same study, incubation of anterior hemipituitaries with levels of IFN found in the circulation failed to modify GH, TSH, and PRL release. Only pharmacologic doses of IFN were seen to exert a direct effect in vitro to stimulate ACTH release from hemipituitary glands. These findings led this group of investigators to conclude that IFN signals the neuroendocrine system via the the hypothalamus rather than the pituitary, but another group offers a somewhat different view. Vankelecom et al.[116] reported that gamma-IFN at physiologic concentrations inhibit the hypothalamic releasing factor–stimulated secretions of ACTH, PRL, and GH by anterior pituitary cells in culture, a model considered by the authors to be more representative of the in vivo situation where, of course, releasing factors may be continuously present. However, these results stand in contrast to clinical studies in which systemic injection of IFN has been to shown to elevate cortisol levels,[117,118] without altering plasma ACTH concentrations.[118] Whether such a finding indicates that IFN has a direct effect on the adrenal cortex is unknown.

Conclusions

Abundant evidence has now accumulated demonstrating that cytokines, whether of peripheral or central origin, modulate the secretion of hypothalamic-releasing factor and anterior pituitary hormone secretion. After stressful episodes of infection, inflammation, or trauma the levels of cytokines such as IL-1, IL-2, IL-6, TNF, and IFN rise in a number of a tissues and body fluids and are followed by appropriate changes in the stress-sensitive hormones. In general, the cytokine mediation of stressful events in the intact mammalian organism appears to result in activation of the CRF–ACTH–glucocorticoid axis, with attenuation of the secretory activity of the TRH–TSH, GHRH–GH (but increased SRIF production), and GnRH–LH axes. In most cases, the primary site of action for cytokine modulation of hormone secretion appears to be at level of the hypothalamus and brain; however, the anterior pituitary remains a potential direct target of cytokine regulation, especially when considering the chronic effects of elevated cytokine levels. Further support for the importance of the central actions of cytokines comes from the demonstration of the endogenous production of IL-1, IL-2, IL-6, and TNF by hypothalamic neurons and glial cells. Of all the cytokines, IL-1 currently appears to be the best situated to cordinate the approprite responses of the other cytokines and hypothalamic releasing factors at the onset of various stressors.

Acknowledgments. The author thanks Ms. Janet Bordeaux and Ms. Juanita Chipani for their expert assistance in preparing this chapter and greatefully acknowledges the research support of NIH grant RO1 NS 23036.

References

1. Dinarello CA. Interleukin-1. *Rev. Infect. Dis.* 1984;6:51–95.
2. March CJ, Mosley B, Larse A, et al. Cloning, sequence and expression of two distinct human interleukin-1 complementary DNAs. *Nature.* 1985; 315:641–647.
3. Besedovsky H, del Rey AE, Sorkin E. Immuneneuroendocrine interactions. *J Immunol.* 1985;135:750s–754s.
4. Daughaday WH. The anterior pituitary. In: Wilson JD, Foster DW, eds. *Williams' textbook of endocrinology.* Philadelphia: Saunders; 1985:568–613.
5. Besedovsky H, del Rey AE, Sorkin E, Dinarello CA. Immunoregulatory feedback between interleukin-1 and glucocorticoid hormones. *Science.* 1986;233:652–654.
6. Sapolsky R, Rivier C, Yamamoto G, Plotsky P, Vale W. Interleukin-1 stimulates the secretion of hypothalamic corticotropin-releasing factor. *Science.* 1987;238:522–524.
7. Snyder DS, Unanue, ER. Corticosteroids inhibit murine macrophage Ia expression and interleukin-1 production. *J Immunol.* 1982;129:1803–1809.
8. Gillis S, Crabtree GR, Smith K. Glucocorticoid induced inhibition of T cell growth factor production. I. The effect of mitogen induced lymphocyte proliferation. *J Immunol.* 1979;123:1624–1630.
9. Kelso A, Munck A. Glucocorticoid inhibition of lymphokine secretion by alloreactive T-lymphocyte clones. *J Immunol.* 1984;133:784–789.
10. Chatterton RTJr, Murray CL, Hellman L. Endocrine effects on leuko-cytopoiesis in the rat. I. Evidence for growth hormone secretion as the leukocytopoietic stimulus following acute cortisol-induced lymphopenia. *Endocrinology.* 1973;92:775–787.·
11. Comsa J, Schwarz JA, Neu H. Interaction between thymic hormone and hypophyseal growth hormone on production of precipitating antibodies in the rat. *Immunol Commun.* 1974;9:11–18.

12. Hayashida T, Li CH. The influence of adrenocorticotropic and growth hormones on antibody formation. *J Exp Med.* 1957;105:93–98.
13. Dantzer R, Kelley KW. Stress and immunity: an integrated view of relationship between the brain and the immune system. *Life Sci.* 1989;44:1995–2008.
14. Kelley KW. Growth hormone, lymphocytes and macrophages. *Biochem Pharmacol.* 1989;38:705–713.
15. Edwards III CK, Ghiasuddin SM, Schepper SM, Yanger LM, Kelly KW. A newly defined property of somatotropin: priming of macrophages for production of superoxide anion. *Science.* 1988;239:769–771.
16. Bernton EW, Meltzer MS, Holaday JW. Suppression of macrophage activation and T-lymphocyte function in hypoprolactinemic mice. *Science.* 1988;239:401–404.
17. Martin JB, Reichlin S. In: *Clinical neuroendocrinology.* 2nd ed. Philadelphia: Davis; 1987.
18. Terry LC, Willoughby JO, Brazeau P, Martin JB. Antiserum to sometostatin prevents stress-induced inhibition of growth hormone secretion in the rat. *Science.* 1976;192:565–567.
19. Fontana A, Weber E, Dayer JM. Synthesis of interleukin-1/endogenous pyrogen in the brain of endotoxinrelated mice: a step in fever induction? *J Immunol.* 1984;133:1696–1698.
20. Makara GB, Stark E, Meszaros T. Corticotrophin release induced by E. coli endotoxin after removal of the medial hypothalamus. *Endocrinology.* 1971;88:412–418.
21. Kasting NW, Martin JB. Altered release of growth hormone and thyrotropin induced by endotoxin in the rat. *Am J Physiol.* 1982;243:E332–338.
22. Smith EM, Meyer WJ, Blalock JE. Virus-induced corticosterone in hypophysectomized mice: a possible lymphoid adrenal axis. *Science.* 1982;218:1311–1313.
23. Dunn AJ, Powell ML, Moreshead WV, Gaskin JM, Hall NR. Effects of Newcastle disease virus administration to mice on the metabolism of cerebral biogenic amines, plasma corticosterone, and lymphocyte proliferation. *Brain Behav Immunol.* 1987;1:216–221.
24. Berkenbosch F, van Oers J, del Rey A, Tilders F, Besedovsky H. Corticotropin-releasing factor-producing neurons in the rat activated by interleukin-1. *Science.* 1987;238:524–526.
25. Uehara A, Gottschall PE, Dahl RR, Arimura A. Interleukin-1 stimulates ACTH release by an indirect action which requires endogenous corticotropin releasing factor. *Endocrinology.* 1987;121:1580–1582.
26. Bernton EW, Beach JE, Holaday JW, Smallridge RC, Fein HG. Release of multiple hormones by a direct action of interleukin-1 on pituitary cells. *Science.* 1987;238:519–521.
27. Kehrer P, Turnhill D, Dayer JM, Muller AF, Gaillard RC. Human recombinant interleukin-1 beta and-alpha, but not recombinant tumor necrosis factor alpha stimulate ACTH release from rat anterior pituitary cells in vitro in a prostaglandin E2 and cAMP independent manner. *Neuroendocrinology.* 1988;48:160–166.
28. Uehara A, Gillis S, Arimura A. Effects of interleukin-1 on hormone release from normal rat pituitary cells in primary culture. *Neuroendocrinology.* 1987;45:343–347.

29. Coceani F, Lees J, Dinarello CA. Occurrence of interleukin-1 in cerebrospinal fluid of the conscious cat. *Brain Res*. 1988;446:245–250.
30. Stitt JT. Evidence for the involvement of the organum vasculosum laminae terminalis in the febrile response of rabbits and rats. *J Physiol*. 1985;368:501–511.
31. Iriki M. Fever and fever syndrome—current problems. *Jap J Physiol*. 1988;38:233–250.
32. Breder CD, Dinarello CA, Saper CB. Interleukin-1 immunoreactive innervation of the human hypothalamus. *Science*. 1988;240:321–324.
33. Swanson LW, Sawchenko PE, Rivier J, Vale WW. Organization of ovine corticotropin-releasing factor immunoreactive cells and fibers in the rat brain: an immunohistochemical study. *Neuroendocrinology*. 1983;36:165–170.
34. Johansson O, Hokfelt T, Elde RP. Immunohistochemical distribution of somatostatin-like immunoreactivity in the central nervous system of the rat. *Neuroscience*. 1984;13:265–339.
35. Ibata Y, Obata HL, Kubo S, Fukui K, Okamura H, Ishigami T, Imagawa K, Sin S. Some cellular characteristics of somatostatin neurons and terminals in the periventricular nucleus of the rat hypothalamus and median eminance. Electron microscopic immunohistochemistry. *Brain Res*. 1983;258:291–295.
36. Bloch B, Brazeau P, Ling N, Bohlen P, Esch F, Wehrenberg WB, Benoit R, Bloom F, Guillemin R. Immunohistochemical detection of growth hormone-releasing factor in brain. *Nature*. 1983;301:607–608.
37. Merchenthaler I, Vigh S, Schally AV, Petrusz P. Immunocytochemical localization of growth hormone-releasing factor in the rat hypothalamus. *Endocrinology*. 1984;114:1082–1085.
38. Lechan RM, Toni R, Clark BD. Immunoreactive interleukin-1 beta localization in the rat forebrain. *Brain Res*. 1990;514:135–140.
39. Minami M, Yasushi K, Yamaguchi T, Nakai S, Hirai Y, Satoh M. Convulsants induce interleukin-1 beta messenger RNA in rat brain. *Biochem Biophys Res Commun*. 1990;171:832–837.
40. Farrar WL, Kilian PL, Ruff MR, Hill JM, Pert CB. Visualization and characterization of interleukin-1 receptors in brain. *J Immunol*. 1987;139:459–463.
41. Katsuura G, Gottschall PE, Arimura A. Identification of a high-affinity receptor for interleukin-1 beta in rat brain. *Biochem Biophys Res Commun*. 1988;156:61–67.
42. Suda T, Tozawa F, Ushiyama T, Sumitomo T, Nakagami Y, Demura H. Recombinant human interleukin-1 increases CRF mRNA levels in the rat hypothalamus. 71st Annual Meeting of the Endocrine Society #484. 1989. Abstract.
43. Tsagarakis S, Gilles G, Rees LH, Besser M, Grossman A. Interleukin-1 directly stimulates the release of corticotrophin releasing factor from rat hypothalamus. *Neuroendocrinology*. 1989;49:98–101.
44. Calogero AE, Bernardini R, Gold PW, Chrousos GP. Regulation of rat hypothalamic corticotropin-releasing hormone secretion in vitro: potential clinical implications. *Adv Exp Med Biol*. 1988;245:167–181.
45. Lumpkin MD, Hartmann DP. Recombinant human interleukin-1 beta acts within the hypothalamic arcuate nucleus to inhibit pulsatile growth hormone

secretion. 71st Annual Meeting of the Endocrine Society #789. 1989. Abstract.

46. Lumpkin MD, McDonald JK. Blockade of growth hormone-releasing factor (GRF) activity in the pituitary and hypothalamus of the conscious rat with a peptidic GRF antagonist. *Endocrinology*. 1989;124:1522–1531.

47. Lumpkin MD, Mulroney SE, Haramati A. Inhibition of pulsatile growth hormone (GH) secretion and somatic growth in immature rats with a synthetic GH-releasing factor antagonist. *Endocrinology*. 1989;124:1154–1159.

48. Wehrenberg WB, Brazeau P, Lubea R, Bohlen P, Gullemin R. Inhibition of the pulsatile secretion of growth hormone by monoclonal antibodies to the hypothalamic growth hormone releasing factor (GRF). *Endocrinology*. 1982;111:2147–2148.

49. Scarborough DE, Lee SL, Dinarello CA, Reichlin S. Interleukin-1 beta stimulates somatastatin biosynthesis in primary cultures of fetal rat brain. *Endocrinology*. 1989;124:549–551.

50. Samson WK, Lipton JM, Zimmer A, Glyn JR. The effect of fever on central alpha-MSH concentrations in the rabbit. *Peptides*. 1981;2:419–424.

51. Eskay RL, Giraud P, Oliver C, Brownstein MJ. Distribution of alpha-melanocyte-stimulating hormone in the rat brain: evidence that alpha-MSH-containing cells in the arcuate region send projections to extrahypothalamic areas. *Brain Res*. 1979;178:55–63.

52. Murphy MT, Richards DB. Lipton JM. Antipyretic potency of centrally administered alpha-melanocyte stimulating hormone. *Science*. 1983;221:192–193.

53. Rivier C, Chizzonite R, Vale W. In the mouse, the activation of the hypo-thalamic-pituitary-adrenal axis by a lipopolysaccharide (endotoxin) is mediated through interleukin-1. *Endocrinology*. 1989;125:2800–2805.

54. Lumpkin MD. Regulation of growth hormone and prolactin secretion by interleukin-1. In: Genazzani AR, Nappi G, Petraglia F, Martignoni E, eds. *Stress and related disorders*. Casterton Hall, England: Parthenon Publishing; 1991, pp. 17–30.

55. Cannon JG, Tatro JB, Reichlin S, Dinarello CA. Alpha melanocyte stimu-lating hormone inhibits immunostimulatory and inflammatory actions of interleukin-1. *J Immunol*. 1986;137:2232–2236.

56. Brown SL, Smith LR, Blalock JE. Interleukin-1 and interleukin-2 enhance proopiomelanocortin gene expression in pituitary cells. *J Immunol*. 1987;139:3181–3183.

57. Eipper BA, Mains RE. Structure and biosynthesis of pro-adrenocortico-tropin/endorpin and related peptides. *Endocr Rev*. 1980;1:1–26.

58. Smith EM, Morrill AV, Meyer III WJ, Blalock JE. Corticotropin releasing factor induction of leukocyte derived immunoreactive ACTH and en-dorphins. *Nature*. 1986;321:881–882.

59. Ono N, Lumpkin MD, Samson WK, McDonald JK, McCann SM. Possible intrahypothalamic action of corticotrophin-releasing factor (CRF) to inhibit growth and LH release in the rat. *Life Sci*. 1984;13:117–123.

60. Rivier C, Vale W. Corticotropin-releasing factor (CRF) acts centrally to inhibit growth hormone secretion in the rat. *Endocrinology*. 1984;114:2409–2411.

61. Peterfreund A, Vale WW. Ovine corticotropin-releasing factor stimulates somatostatin secretion from cultured brain cells. *Endocrinology*. 1983; 112:1275–1278.
62. Katakami H, Arimura A, Frohman LA. Involvement of hypothalamic somatostatin in the suppression of growth hormone secretion by central corticotropin-releasing factor in conscious male rats. *Neuroendocrinology*. 1985;41:390–393.
63. Lumpkin MD, Koenig JI, Tracey DE, Scott JW. Hypothalamic interleukin-1 (IL-1) and CRF receptors mediate IL-1-induced suppression of growth hormone secretion. 73rd Annual Meeting of the Endocrine Society. 1991. In press. Abstract.
64. Schiebler TH, Leranth C, Zarborsky L, Bitsch H. In: Scott DE, Kozlowski GP, Weindl A, eds. On the glia of the median eminence. *Brain-endocrine interaction III. Neural hormones and reproduction*. Basel: Karger; 1978: 46–56.
65. Giulian D, Baker TJ, Shih LN, Lachman LB. Interleukin-1 of the central nervous system is produced by ameboid microglia. *J Exp Med*. 1986;164: 594–604.
66. Giulian D, Young DG, Woodward J, Brown DC, Lachman LB. Interleukin-1 is an astroglial growth factor in the developing brain. *J Neurosci*. 1988;8:709–714.
67. Tannenbaum GS, Martin JB. Evidence for an endogenous ultradian rhythm governing growth hormone secretion in the rat. *Endocrinology*. 1976; 98:562–570.
68. Tannenbaum GS, Ling N. The interrelationship of growth hormone (GH)-releasing factor and somatostatin in generation of the ultradian rhythm of GH secretion. *Endocrinology*. 1984;115:1952–1957.
69. Koenig JI, Snow K, Clark BD, Toni R, Cannon JG, Shaw AR, Dinarello CA, Reichlin S, Lee SL, Lechan RM. Intrinsic interleukin-1 beta is induced by bacterial lipopolysaccharide. *Endocrinology*. 1990;126:3053–3058.
70. Marguette C, Ban E, Fillon G, Havor F. Receptors for interleukin-1,2, and 6 (IL-1 alpha and beta, IL-2, IL-6) in mouse, rat and human pituitary. *Neuroendocrinology*. 1990;52(Suppl 1):48.
71. Tracey DE, De Souza EB. Identification of interleukin-1 receptors in mouse pituitary cell membranes and AtT20 pituitary tumor cells. Society for Neuroscience, 18th Annual Meeting, Toronto, Ontario, Canada, 1988. Abstract 422.11.
72. Lumpkin MD. The regulation of ACTH secretion by IL-1. *Science*. 1987; 238:452–454.
73. Beach JE, Smallridge RC, Kinzer CA, Bernton EW, Holaday JW, Fein HG. Rapid release of multiple hormones from rat pituitaries perifused with recombinant interleukin-1. *Life Sci*. 1989;44:1–7.
74. Rivier C, Vale W. In the rat, interleukin-1 alpha acts at the level of the brain and gonads to interfere with gonadotropin and sex steroid secretion. *Endocrinology*. 1989;124:2105–2111.
75. Rivier C, Vale W. Influence of corticotropin-releasing factor (CRF) on reproductive functions in the rat. *Endocrinology*. 1984;114:914–919.
76. Petraglia F, Sutton S, Vale W, Plotsky P. Corticotropin-releasing factor decreases plasma LH levels in female rats by inhibiting gonadotropin-

releasing hormone release into hypophysial-portal circulation. *Endocrinology*. 1987;120:1083–1088.

77. Rivier C, Rivier J, Vale W. Stress-induced inhibition of reproduction functions: role endogenous corticotropin-releasing factor. *Science*. 1986; 231:607–610.

78. Reichlin S, Glaser RJ. Thyroid function in experimental streptococcal pneumonia in the rat. *J Exp Med*. 1958;107:219–235.

79. McCann SM, Rettori V, Milenkovic L, Jurcovicova J, Snyder G, Beutler B. Role of interleukin-1 and cachectin in control of anterior pituitary hormone release. *Neurol Neurobiol*. 1989;50:333–349.

80. Scarborough DE. Cytokine modulation of pituitary hormone secretion. In: O'Dorisio MS, Panerai A, eds. *Neuropeptides and immunopeptides. Messengers in a neuroimmune axis. Ann NY Acad Sci*. 1990;154:169–187.

81. Dubois J-M, Dayer J-M, Siegrist-Kaiser CA, Burger AG. Human recombinant interleukin-1 beta decreases plasma thyroid hormone and thyroid stimulating hormone levels in rats. *Endocrinology*. 1988;123:2175.

82. Dinarello CA. Interleukin-1 and its biologically related cytokines. *Adv Immunol*. 1989;44:153–205.

83. Le J, Vilcek J. Tumor necrosis factor and interleukin-1: cytokines with multiple overlapping biological activities. *Lab Invest*. 1987;56:234–248.

84. Nieto-Sampedro M, Chandy KG. Interleukin-2-like activity in injured rat brain. *Neurochem Res*. 1987;12:723–727.

85. Hofman FM, Vonhanwehr RI, Dinarello CA, Mizel SB, Hinton D, Merrill JE. Immunoregulatory molecules and IL 2 receptors identified in multiple sclerosis brain. *J Immunol*. 1986;136:3239–3245.

86. Lotze MT, Frana LW, Sharrow SO, Robb RJ, Rosenberg SA. In vivo administration of purified human interleukin-2. I. Half life and immunological effects of the Jurkat cell line derived interleukin-2. *J Immunol*. 1985;134:157–162.

87. Bindon C, Czerniecki M, Ruell P, Edwards A, McCarthy WH, Harris R, Hersey P. Clearance rates and systemic effects of intravenously administered interleukin-2 (IL-2) containing preparations in human subjects. *Br J Cancer*. 1983;46:123–128.

88. Del Rey A, Besedovsky H, Sorkin E, Dinarello CA. Interleukin-1 and glucocorticoid hormones integrate an immunoregulatory feedback circuit. *Ann NY Acad Sci*. 1987;496:85–95.

89. Atkins MB, Gould JA, Allegretta M, Li JJ, Dempsey RA, Rudders RA, Parkinson DR, Reichlin S, Mier JW. Phase I evaluation of recombinant interleukin-2 in patients with advanced malignant disease. *J Clin Oncol*. 1986;4:1380–1391–273.

90. Van Snick J. Interleukin-6: an overview. *Annu Rev Immunol*. 1990;8:253.

91. Castell JV, Gomez-Lechon MJ, David M, Andus T, Geiger T, Trullenque R, Fabra R, Heinrich PC. Interleukin-6 is the major regulatory of acute phase protein synthesis in adult human hepatocytes. *FEBS Lett*. 1989; 242:237–242.

92. Yasukawa K, Hirano T, Watanabe Y, Muratani K, Matsuda T, Nakai S, Kishimoto T. Structure and expression of human B cell stimulatory factor-2 (BSF-2/IL-6) gene. *EMBO J*. 1987;6:2939–2945.

93. Frei K, Malipiedro UV, Leist TP, Zinkernagel RM, Schwab ME, Fontana A. On the cellular source and function of interleukin-6 produced in the central nervous system in viral diseases. *Eur J Immunol.* 1989;19:689–694.

94. Spangelo BL, Judd AM, MacLeod RM, Goodman DW, Isakson PC. Endotoxin-induced release of interleukin-6 from rat medial basal hypothalami. *Endocrinology.* 1990;127(4):1779–1785.

95. Spangelo BL, MacLeod RM, Isakson PC. Production of interleukin-6 by anterior pituitary cells in vitro. *Endocrinology.* 1990;126:582–587.

96. Vankelecom H, Carmeliet P, Van Damme J, Biliau A, Denef C. Production of interleukin-6 by folliculo-stellate cells of the anterior pituitary gland in a histiotypic cell aggregate culture system. *Neuroendocrinology.* 1989;49: 102–106.

97. Fukata J, Usui T, Naitoh Y, Nakai Y, Imura H. Effects of recombinant human interleukin-1 alpha, 1 beta, 2 and 6 on ACTH synthesis and release in the mouse pituitary tumor cell line AtT-20. *J Endocrinol.* 1989;122:33–38.

98. Spangelo BL, Judd AM, Isakson PC, MacLeod RM. Interleukin-6 stimulates anterior pituitary hormone release in vitro. *Endocrinology.* 1989; 125:575–577.

99. Naitoh Y, Fukata J, Tominaga T, et al. Interleukin-6 stimulates the secretion of adrenocorticotropic hormone in conscious, freely moving rats. *Biochem Biophys Res Commun.* 1988;155:1459–1463.

100. Yamaguchi M, Yoshimoto Y, Komura H, et al. Interleukin 1 beta and tumor necrosis factor alpha stimulate the release of gonadotropin-releasing hormone and interleukin 6 by primary cultured cat hypothalamic cells. *Acta Endocrinol.* 1990;123:476–480.

101. Sawada M, Kondo N, Suzumura A, Marunouchi T. Production of tumor necrosis factor-alpha by microglia and astrocytes in culture. *Brain Res.* 1989;491:394–397.

102. Lieberman AP, Pitha PM, Shin HS, Shin ML. Production of tumor necrosis factor and other cytokines by astrocytes stimulated with lipopolysaccharide or a neurotropic virus. *Proc Natl Acad Sci USA.* 1989;86:6348.

103. Breder CD, Saper CB. Tumor necrosis factor immunoreactive innervation in the mouse brain. *Soc Neurosci.* 1988;14:1280. Abstract.

104. Sharp BM, Matta SG, Peterson PK, Newton R, Chao C, Mcallen K. Tumor necrosis factor-alpha is a potent ACTH secretogogue: comparison to interleukin-1 beta. *Endocrinology.* 1989;124:3131–3133.

105. Nichie HR, Spriggs DR, Manogue KR, Sherman ML, Revhaug A, O'Dwyer ST, Arthur K, Dinarello CA, Cerami A, Wolff SM, Kuge DW, Wilmore DW. Tumor necrosis factor and endotoxin induce similar metabolic responses in human beings. *Surgery.* 1988;104:280–286.

106. Pang X, Hershman JM, Mirell CJ, Pekary AE. Impairment of hypothalamic-pituitary-thyroid function in rats treated with human recombinant tumor necrosis factor-alpha (cachectin). *Endocrinology.* 1989;125:76–84.

107. Elsasser TH, Caperna TJ, Kenison DC, Fayer R. Recombinant bovine tumor necrosis factor-alpha can affect growth hormone secretion by a direct pituitary interaction. 71st Annual Meeting of the Endocrine Society #788. 1989. Abstract.

108. Nolten WE, Goldstein EN, Ekrlich IH, Carlson MV, McKenna PA, Rueckert EC, Trump DL. Endocrine changes associated with tumor necrosis factor administration in cancer patients. 71st Annual Meeting of the Endocrine Society #491. 1989. Abstract.
109. Rettori V, Milenkovie L, Beutler BA, McCann SM. Hypothalamic action of cachectin to alter pituitary hormone release. *Brain Res Bull*. 1989;23:471–475.
110. Milenkovic L, Rettori V, Snyder GD, Beutler E, McCann SM. Cachectin alters anterior pituitary hormone release by a direct action in vitro. *Proc Natl Acad Sci USA*. 1989;86:2418–2423.
111. Gaillard RC, Turnill D, Sappino P, Muller AF. Tumor necrosis factor alpha inhibits the hormonal response of the pituitary gland to hypothalamic releasing factors. *Endocrinology*. 1990;127:101–106.
112. Yamaguchi M, Yoshimoto Y, Komura H, Koike K, Matzuzaki N, Hirota K, Miyake A, Tanizawa O. Interleukin 1 beta and tumor necrosis factor alpha stimulate the release of gonadotropin-releasing hormone and interleukin 6 by primary cultured rat hypothalamic cells. *Acta Endocrinol*. 1990;123:476–480.
113. Yamaguchi M, Sakata M, Koike K, Matsuzaki N, Miyake A, Tanizawa O. Induction by tumor necrosis factor-alpha stimulates rapid release of immunoreactive and bioactive luteinizing hormone secretion from rat pituitary cells in vitro. *Neuroendocrinology*. 1990;52:468–472.
114. Dijkmans R, Ballian A. Interferon gamma: a master key in the immune septem. *Curr Opin Immunol*. 1988;1:269–280.
115. Gonzales MC, Riedel M, Rettori V, Yu WH, McCann SM. Effect of recombinant human gamma-interferon on the release of anterior pituitary hormones. *Prog Neuroendocrinimmunol*. 1990;3:49–54.
116. Vankelecom H, Carmeliet P, Heremans H, Van Damme J, Dijkmans R, Billian A, Denef C. Interferon-gamma inhibits stimulated adrenocorticotropin, prolactin, and growth hormone secretion in normal rat anterior pituitary cell cultures. *Endocrinology*. 1990;126:2919–2926.
117. Spath-Schwalbe E, Porzolt F, Digel W, Born J, Kloss B, Fehm HL. Elevated plasma cortisol levels during IFN-gamma treatment. *Immunopharmacol*. 1989;17:141–146.
118. Holsboer F, Stalla GK, Bardeleben U, Hamman K, Muller H, Muller OA. Acute adrenocortical stimulation by recombinant gamma interferon in human controls. *Life Sci*. 1988;42:1–7.

8
Bone Growth, Remodeling, and Repair: Interactions of Parathyroid Hormone, Calcitonin, Vitamin D, Growth Factors, and the Prostaglandins

JOSEPH E. ZERWEKH

One of the more remarkable events in the course of evolution was the development of the bony endoskeleton of higher animals. This endoskeleton was designed to fulfill two specific roles. These are (1) to provide structural support and protection to soft tissue organs and (2) to serve as an ion reservoir for calcium and phosphorus homeostasis. Bone also acts as a third line of defense in maintaining acid–base balance, after respiratory and renal responses, by providing additional buffers. Bone is also unique in that it is a dynamic tissue, constantly undergoing breakdown and then rebuilding to maintain its biomechanical competence. This latter process, termed remodeling, is the only activity that persists into adult life and is of concern because of its role in producing metabolic bone disease. In the following discussion, I examine bone growth, remodeling, and repair not only from a structure–function relationship but also from a consideration of events at the cellular level and to what extent the interactions of various hormones might exert on the remodeling process. It therefore seems appropriate to begin with a consideration of the macroscopic features of bone and skeleltal development.

Macroscopic Structure of Bone

On a gross level, all bones are composed of two basic architectural structures. They are termed cortical or compact and trabecular or cancellous bone. Cortical bone comprises approximately 80% of the skeleton and is located predominantly in the appendicular diaphysis. Cortical bone is characterized by a low surface-to-volume ratio and is largely responsible for the support and protective functions of the skeleton (Fig. 8.1). Trabecular bone is characterized by numerous bony spicules that traverse the marrow cavities of flat bones and the metaphyses of long bones. It has a high surface-to-volume ratio and often reflects calcium homeostasis.

With few exceptions, the outer surface of most bones is covered by a sheath of fibrous connective tissue and an inner cellular, or cambian,

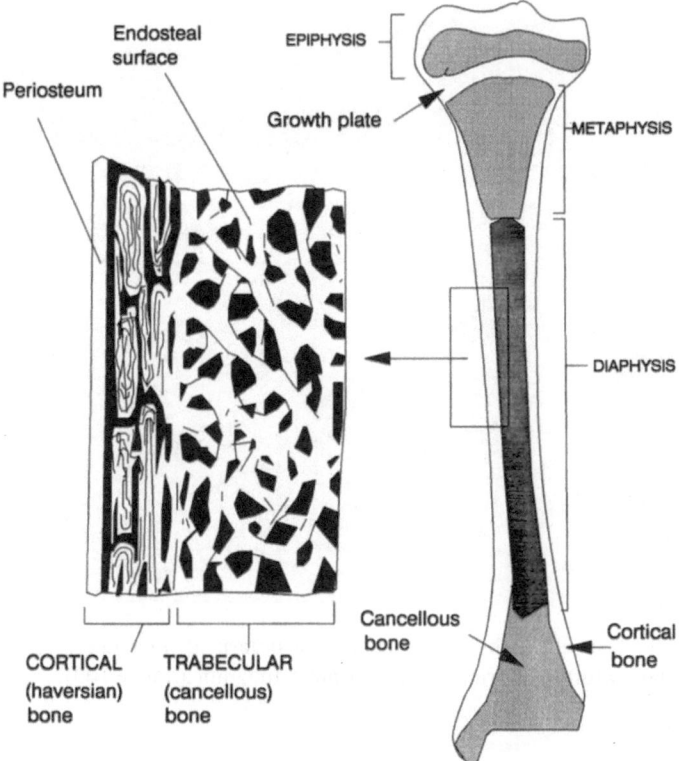

FIGURE 8.1. Macroscopic structure of bone demonstrating the areas of cortical and cancellous bone found in a typical skeletal long bone. The boxed area is enlarged on the left to demonstrate the compact nature of cortical bone and the latticelike structure of cancellous bone.

layer of undifferentiated cells, the periosteum. The periosteum has the potential to form bones during growth and fracture healing. It is not present in areas where tendons and ligaments insert on bones, on bone ends that are lined with articular cartilage, on surfaces of the sesamoid bones, or in subcapsular areas and the neck of the femur. A similar membrane of bone cells lines the marrow cavity of the diaphysis and the cavities of cancellous bone and is called the endosteum.

Microscopic Structure of Bone

Adult mammalian bone, whether compact or cancellous, is built in layers or sheets on preexisting bone. These layers or lamellae appear as stacks of parallel or concentrically curved sheets. Each lamella is approximately

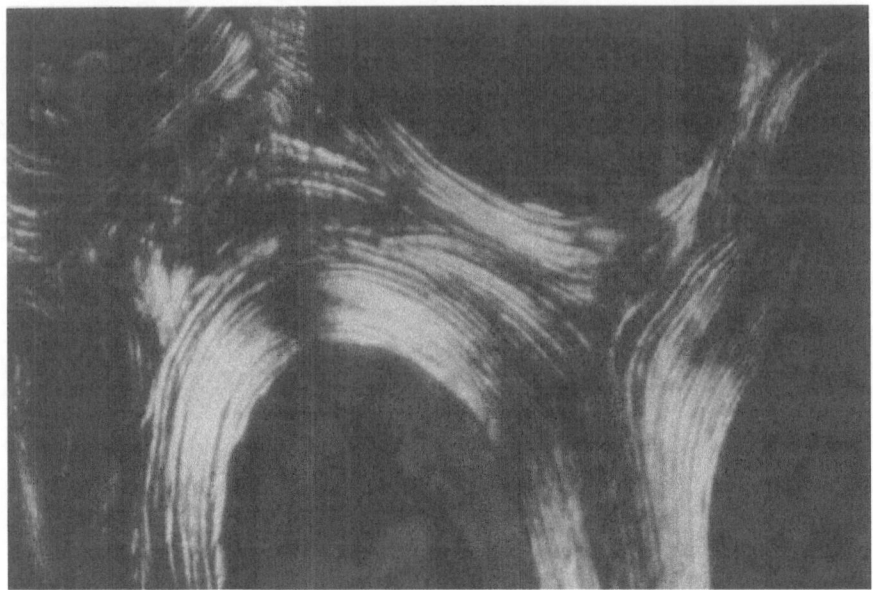

FIGURE 8.2. Cancellous bone as viewed by polarized microscopy demonstrating the lamellar nature of normal bone (original magnification ×100).

3 to 7 μm thick and its collagen fibers are oriented parallel to each other. In histological preparations, the lamellations are best seen by polarized light and appear as alternating bright and dark layers. This is a result of the differing orientation of collagen fibers within adjacent lamellae (Fig. 8.2).

Spaced throughout lamellar compact and cancellous bone are small cavities, or lacunae, connected by thin tubular channels called canaliculi. Entrapped bone cells, or osteocytes, and their long cytoplasmic processes occupy the lacunae and canaliculi, respectively. These cell processes within canaliculi communicate by gap junctions with processes of osteocytes lying in adjacent lacunae. Since canaliculi open to the extracellular fluid at bone surfaces, a joining network is formed for the nutrition and metabolic activities of the osteocytes as well as for a means of delivering calcium and phosphorus ions to the blood during conditions of need.

Most human compact bone is in the form of an osteon. The osteon has therefore been classified as the main bone structural unit of cortical bone. A typical osteon consists of a central Haversian canal surrounded by concentric lamellae. Each canal is 30 to 70 μm in diameter and contains nutrient vessels, nerves, and connective tissue. The Haversian canals communicate with periosteum, bone marrow, and each other through transverse or oblique channels called Volkmann's canals.

On the other hand, adult cancellous bone consists of a network of joining trabeculae with intertrabecular spaces containing bone marrow. Most trabeculae are less than 0.2 mm thick and contain no blood vessels. The trabecular osteocytes are nourished by diffusion from the trabecular surface by the canaliculi extending to the surface. Each trabecula is composed of a group of angular segments, each segment formed by parallel sheets of lamellae. These segments of lamellar bone are called trabecular packets and are functionally analogous to the osteon, the bone structural unit of cortical bone. Thus, the trabecular packet is the bone structural unit of cancellous bone. As with cortical bone, cement lines hold the trabecular packets together.

Composition of Bone Matrix

The two major components of mineralized bone are organic matrix and inorganic salts. Inorganic material makes up about 75% of cortical bone and organic matrix, 25%. The organic matrix is a general term for the extracellular organic phase, composed primarily of protein, glycoprotein, and polysaccharide, which is secreted by and surrounds the osteogenic cell. It is composed predominantly of collagenous fibers (90%) embedded in an amorphous ground substance (10%). Bone collagen is composed exclusively of type I collagen. However, unlike the type I collagen of skin and tendon, bone collagen calcifies. This difference is probably not due to specific properties of the collagen alone, but to the interaction of the collagen fibrils with macromolecules within the extracellular matrix. These agents have recently been proposed as possible regulators of mineral deposition and include osteocalcin, phosphoproteins, and osteonectin.[1-3] In addition, posttranslational modification of the collagen and the distribution of intermolecular cross-links may also contribute to type I collagen of bone undergoing mineralization.[4]

The amorphous ground substance is a noncollagenous cementing substance known to contain phosphoproteins, glycoproteins, osteocalcin, and small amounts of proteoglycans, lipids, and peptides. Collagen fibers and crystals become embedded in this substance and under the influence of these highly acidic macromolecules become mineralized.

Although mineral constitutes 75% of bone weight, by volume the mineral content of bone is 50%. The main mineral phase is similar to hydroxyapatite $[Ca_{10}(PO_4)_6(OH)_2]$. In mature bone, the hydroxyapatite is present as needles, thin plates, or leaves 15 to 30 Å thick and 100 Å long. Furthermore, there are substantial quantities of carbonate, citrate, sodium, and magnesium in bone mineral. Trace amounts of iron, zinc, copper, lead, manganese, tin, aluminum, strontium boron, and silicone have also been reported. The ground substance also surrounds and stabilizes the hydroxyapatite crystals. This interaction of hydroxyapatite with collagen fibers and noncollagenous proteins brings about the hardness and rigidity of bone.

During growth, the amount of organic material per unit volume of bone remains relatively constant, but the amount of water decreases and the proportion of bone mineral increases, attaining a maximum of about 65% of the fat-free weight of the tissue in adults by the third decade of life.

Bone Cells

Five kinds of bone cells can usually be recognized in the growing and adult skeleton. They are represented by osteoprogenitor cells, osteoblasts, osteocytes, osteoclasts, and bone-lining cells.

Osteoprogenitor Cells

Osteoprogenitor cells can be functionally defined as any cell that has the capacity for mitosis and further differentiation and specialization into mature bone cells. In general, two types of osteoprogenitor cells are found. One type (preosteoblasts) gives rise to bone-forming osteoblasts, and the other type (preosteoclasts) leads to the formation of bone-resorbing osteoclasts. Under light microscopy osteoprogenitor cells are spindle-shaped with oval or elongated nuclei and nonremarkable cytoplasm. They are most commonly found near bone surfaces and other bone cells. The two types of osteoprogenitor cells can be distinguished between one another by electron microscopy. One type, the osteoblast precursor, has some endoplasmic reticulum and a poorly developed Golgi region. The other type, the osteoclast precursor, has more mitochondria and free ribosomes. These latter osteoprogenitor cells (precursors of osteoclasts) are now believed to be mononuclear phagocytes of hematopoietic stem cell origin.[5] The osteoblast osteoprogenitor cell is believed to be derived from the marrow stromal cell lineage.

Osteoblasts

Osteoblasts are bone-forming cells. The principal function of the osteoblast is to synthesize an organic extracellular matrix (osteoid) and to regulate its subsequent mineralization. Phenotypically, active osteoblasts share the following characteristics: morphologically, active osteoblasts are cuboidal cells (that rarely undergo mitosis) with cellular processes, gap junctions, abundant endoplasmic reticulum, Golgi, and collagen-containing secretory vesicles. They contain abundant alkaline phosphatase, a marker of bone formation, in the plasma membrane. They also synthesize and secrete the following macromolecules: type I collagen, the noncollagenous matrix proteins, osteocalcin, osteonectin, and other matrix proteins; a putative bone-specific proteoglycan, specific growth factors, prostaglandins E_1, E_2, and I_2, collagenase, and tissue plasminogen activator.

Osteoblasts have been the focus of active research on the mechanisms of bone formation and remodeling. This is not only because both transformed and nontransformed osteoblastlike cells have been successfully maintained in vitro, but also because such studies have disclosed the osteoblast to be a target cell, with receptors for parathyroid hormone (PTH) and 1,25,-dihydroxyvitamin D_3 [$1,25(OH)_2D_3$]. Both hormones appear to have anabolic effects: $1,25(OH)_2D_3$ stimulates the synthesis of osteocalcin and PTH inhibits alkaline phosphatase activity. These hormonal effects on osteoblast activity will be considered more fully in the section on Hormonal Regulation of Bone Growth and Remodeling.

The life cycle of an osteoblast involves (1) the birth from a progenitor cell, (2) differentiation and participation in matrix elaboration and its subsequent calcification, and (3) either a return to the preosteoblast pool or transformation to a bone-lining cell and ultimately to an osteocyte or demise.

Osteocytes

The real end cell of the osteoblastic differentiation line is the osteocyte. During bone formation a number of osteoblasts become incorporated in the organic bone matrix. Subsequently the bone matrix around the osteocytes is calcified. The osteocytes in this osteoid layer share many ultrastructural features with osteoblasts but are sufficiently differentiated to display an osteocyte-specific surface antigen.[6] As one moves deeper into the bone, the osteocytes become smaller and lose many of their cytoplasmic organelles. The cells demonstrate many slender cell processes that can extend for considerable distances in canaliculi. These cell processes contain extensive microfilaments and often contact cell processes of other osteocytes and bone surface cells. At points of contact between osteocytes or between osteocytes and bone surface cells, gap junctions are present. This finding explains how the cells can survive in such an isolated environment. Nutrients may pass into and waste products out of the osteocytes by several possible pathways. Ion and small molecule passage through the gap junctions of adjoining cell processes or percolation of fluids through the space between the cells and their processes and the canalicular and lacunar walls are two possible mechanisms.

At one time osteocytic osteolysis, a concept of local osteocytic resorption of lacunar walls, was advanced to explain how rapid fluctuations in serum calcium might come about in response to bone-resorbing agents. However, this concept is not widely accepted today by investigators in this area in light of the paucity of experimental evidence in its support.

Osteoclasts

Osteoclasts are the cells responsible for the resorption of bone. The "textbook" osteoclast is a large (20 to >100 μm in diameter), multinucleated giant cell containing abundant mitochondria and lysosomes. Actively resorbing osteoclasts are usually found in or near cavities on bone surfaces called resorption pits or Howship's lacunae. At such sites, the osteoclast surface adjacent to the bone often has a striated appearance corresponding to an area of extensive membrane infoldings termed the ruffled border. In addition, the plasma membrane of the ruffled border appears to be coated with small bristlelike structures that may contribute to the transport of materials across the membrane.

The ruffled border is surrounded by an ectoplasmiclike zone devoid of cellular organelles yet containing many actin filaments. Besides being actin rich, these clear zones depend on matrix contact and enlarge promptly after resorptive stimulation. They are believed to bind osteoclasts to bone surfaces and to act as a permeability seal to maintain a microenvironment conducive to bone resorption. Osteoclasts that lack ruffled borders are not capable of resorbing bone. This is the situation in osteopetrosis, a genetic disease characterized by defective bone resorption.[7]

The osteoclast has been shown to possess a tartrate-resistant acid phosphatase, specific groups of dehydrogenases, carbonic anhydrase activities, and receptors for calcitonin. Interestingly, no receptors have been found for PTH or $1,25(OH)_2D_3$ on osteoclasts despite the fact that both hormones significantly stimulate bone resorption. One explanation is that the osteoblast may mediate the resorptive stimulating action of these agents on the osteoclast (see Remodeling below). At present, there is little information concerning the life cycle of osteoclast cells in vivo. The estimated osteoclast half-life is around 6 to 10 days. Cessation of bone resorption is associated with migration of osteoclasts from endosteal surfaces into adjacent marrow space where they are believed to degenerate and disintegrate.

Bone-Lining Cells

The remaining cell type, which is found on most bone surfaces of the adult skeleton, are flat and elongated with spindle-shaped nuclei. These cells are commonly referred to as bone-lining cells although they are also known by other names. Bone-lining cells are believed to be derived from osteoblasts and/or osteoblast precursors that have ceased their bone-forming activity or differentiation and flattened out on bone surfaces. Their function is still not understood. These cells may serve as an ion barrier separating fluids moving through the osteocyte and lacunar canalicular system from the interstitial fluids. This membrane barrier around bone may have a role in mineral homeostasis by regulating the

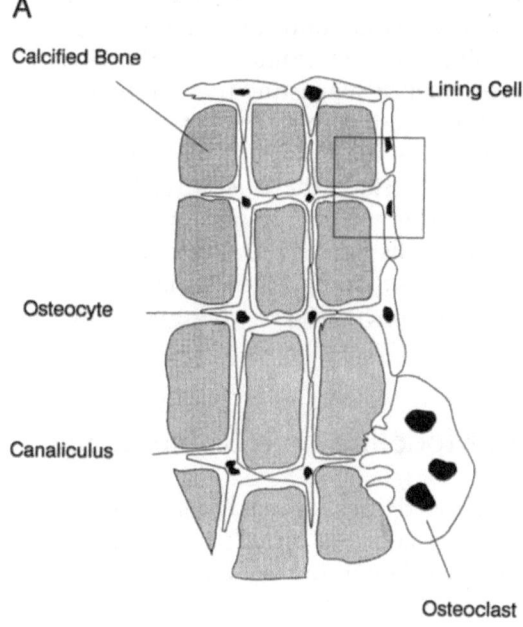

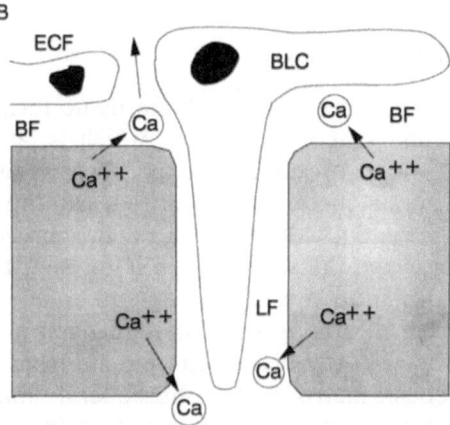

FIGURE 8.3. **A**: Model of cancellous bone surface demonstrating the close association of osteocytes with one another and with the surface bone-lining cells via cytoplasmic extension in the canaliculus. **B**: An enlarged view of the boxed area in **A**. Calcium ions (Ca^{2+}) released during resorption of mineralized bone pass into the lacunar fluid (*LF*) and ultimately to the bone fluid compartment (*BF*) where they enter the extracellular fluid (*ECF*) via gap junctions between the bone-lining cells (*BLC*).

fluxes of calcium and phosphate in and out of bone fluids (Fig. 8.3). Another potential action is in controlling the growth of bone crystals by maintaining a suitable microenvironment at the site of mineralization. There is also some evidence suggesting that the lining cell is a transducer for the effect of bone resorption stimulators, and the elaboration of collagenase by these cells may prepare the surface for osteoclastic resorption.[5,8] Furthermore, since there is an elaborate network of bone-lining cells and osteocytes making contact with each other, it has been postulated that these cells sense the shape of bone and its reaction to stress and strain. In so doing, these cells transmit these triggering events as signals to the bone surface, where new bone formation and/or resorption is possible.

Bone Growth, Modeling, and Remodeling

These three skeletal activities differ fundamentally from each other, operate under different controls, in different locations, and at different ages in the human even though the same types of bone cells are involved in all three instances. As might be anticipated, a derangement in each results in different clinical syndromes.

Bone Growth or Histogenesis

In the histogenesis of bone, the first bone tissue formed is an immature type known as primary or woven bone, which is later replaced by secondary (lamellar) bone or bone marrow. This sequence of formation occurs via two major modes of osteogenesis: intramembranous and endochondral ossification. Although these two processes differ by the initial material on which the woven bone is formed, the actual processes of bone tissue formation are essentially the same in both modes of ossification. The bone forming process is recognized to proceed by the following sequence: (1) osteoblasts differentiate from mesenchymal cells, (2) osteoblasts deposit matrix that is subsequently mineralized, (3) bone is initially laid down as a network of immature (woven) trabeculae, and (4) this immature bone is replaced by secondary bone, removed to form bone marrow, or converted into primary cortical bone by the filling of spaces between the trabeculae ("compaction").

Intramembranous ossification occurs predominantly in the flat bones of the skull, the facial skeleton, the clavicles, and terminal phalanges. This process is also responsible for the deposition of bone matrix in the subperiosteal regions of long bones. The process of intramembranous ossification occurs within an area of well vascularized primitive connective tissue generally comprised of mature mesenchymal and epithelial tissue. Direct bone formation occurs without prior formation of cartilage.

However, the epithelial–mesenchymal interaction appears to be necessary. During the eighth week of human fetal life, a cluster of mesenchymal cells differentiates into osteoid-secreting osteoblasts.[9] At this stage of ossification, the trabeculae are thin and needlelike. These fine trabeculae of woven bone are lengthened and thickened by continual osteoblastic apposition. Together, they form a network of trabecular bone known as the primary spongiosa.

The primary spongiosa may then undergo one of two further developments. In the tissue destined to become compact bone, bone is continually deposited on the existing trabecular surfaces until the intratrabecular spaces are filled. During this process, layers of woven bone are formed onto the existing trabecular surface as irregular concentric layers leading to the development of primary Haversian systems. Primary spongiosa destined to remain cancellous bone undergoes a cessation of trabecular thickening and the connective tissue between the trabeculae differentiates into the hematopoetic tissue of the bone marrow. Simultaneous with this process is the transformation of quiescent osteoblasts and mesenchymal cells located near the endosteal surface into an endosteum. The remaining envelope of connective tissue at the periosteal surface condenses, forming a periosteum.

Endochondral ossification is the bone forming process that results in formation of the bones of the base of the skull, the vertebral column, the pelvis, and the extremities. Unlike intramembranous ossification, these skeletal structures are initially formed within hyaline cartilage. This cartilage is synthesized by chondrocytes that result from the differentiation of local mesenchymal cells. As the cartilage templates of the bones enlarge, they undergo focal matrix mineralization, forming ossification centers. These changes are accompanied by the intramembranous deposition of bone on the superficial surface of the diaphyses resulting in periosteal bone formation. Ingrowth of blood vessels and osteoprogenitor cells into the tubular shaft area results in secretion of bone matrix onto the existing mineralized cartilage scaffold to create trabecular structures of mixed tissue composition (primary spongiosa). Most of the primary spongiosa is converted into secondary (lamellar) spongiosa by the simultaneous removal of regions of calcified cartilage cores and woven bone and addition of newly formed secondary trabeculae. Concurrently, a bone marrow forms within the perivascular space from mesenchymal connective tissue, which replaces the primary spongiosa.

This process of endochondral bone formation is responsible for the increases in length that occur in developing long bones. This continued, highly organized endochondral bone formation occurs in the tapered, metaphyseal region of the bone. Here the growth structure (collectively termed the physis or growth plate) is organized in such a way that the chondrocytes appear in vertical columns that can be identified as three distinct horizontal zones based on the degree of cell maturation. The first of these zones, located immediately subjacent to the bony epiphysis, is characterized by many mitotic figures, flattened chondrocytes, and synthesis and secretion of matrix. This "proliferative" zone is subtended by a second region in which the cells cease dividing and undergo hypertrophy. These hypertrophic chondrocytes are large vacuolated cells with swollen nuclei. The matrix of this area becomes reduced to thin longitudinal and transverse septae because of the enlargement of the lacunar spaces. This hypertrophic zone adjoins the third (provisional calcification), where the hyaline matrix becomes increasingly calcified. This results in the restricted diffusion of nutrients and eventually in the death of the hypertrophic chondrocytes. Between the columns of cells, the vertically oriented matrix becomes fully calcified, whereas the septae between lacunae fail to calcify. The uncalcified transverse septae are removed at the junction of the growth plate and the metaphysis by the lytic action of the dying chondrocytes and by chondroclasts. In contrast, approximately one third of the calcified longitudinal septae persist and eventually become the scaffolding for the formation of primary trabeculae.

As the epiphyseal growth plate advances and the bone elongates, primary spongiosa slowly replaces the regions previously occupied by the growth plate. At the front of the advancing metaphysis there is an invasion of blood vessels and perivascular connective tissue that carries osteoprogenitor cells into the newly formed osteogenic area. The pre-osteoblastic progenitor cells differentiate into osteoblasts that quickly lay down woven bone onto the surfaces of the longitudinal septae, thus creating the trabeculae of the primary spongiosa. Osteoclasts also differentiate from specific osteoprogenitors (preosteoclasts) and begin to resorb some of the calcificed septae to make room for the forming trabeculae and marrow cavity. In these regions of primary bone formation, osteoblasts outnumber osteoclasts 10 to 1. Eventually, most of the primary spongiosa is converted into secondary (lamelluar) spongiosa by the simultaneous removal of woven bone and calcified cartilage cores and addition of secondary (lamellar) bone.

Hormones such as growth hormone, thyroxine, and vitamin D can accelerate various stages of the endochondral ossification process. On the other hand estrogens, vitamins D and C deficiency, adrenocortical steroids, and excessive androgens can decelerate various stages of the endochondral ossification process.[10]

Modeling

The second major process to occur in the long skeleton is modeling. Growth and modeling occur before physeal closure and are responsible for the major, genetically determined architectural features of the skeleton. Growth refers to a net increase in skeletal mass, whereas

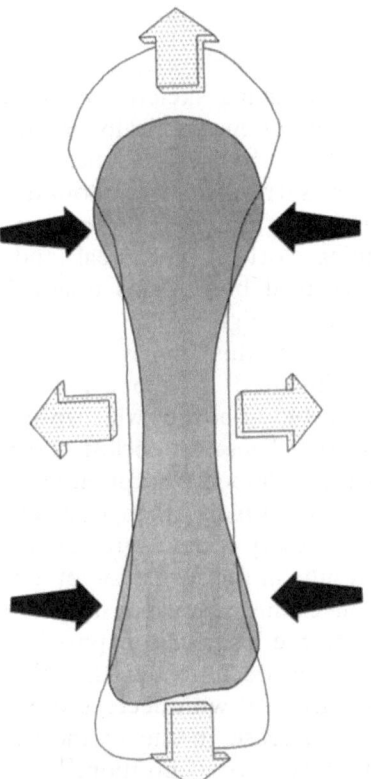

FIGURE 8.4. Growth and modeling of bone. Growth refers to the overall increase in size of a young (*shaded*) bone to its fully mature counterpart (*open*). Modeling refers to the site-specific sculpting of bone required to maintain the normal overall architecture during growth. Note that both bone resorption (*dark arrows*) and bone formation (*stippled arrows*) must occur to maintain the genetically predetermined structure of the bone.

modeling is the process responsible for the alteration of the size and shape of bones. This process is carried out by bone formation and resorption at different surfaces and rates during the growth process. Thus, for example, in immature long bones, net matrix synthesis occurs on the lateral surface and resorption on the medial aspect, resulting in the maintenance of the overall shape of such bones and their net displacement in a lateral direction (Fig. 8.4). Examples of the modeling process are (1) the shaping of the ends (flaring) of long bones during growth, (2) the drifting of the midshaft during growth, and (3) the enlargement of the cranial vault and the modification of the cranial curvature.

Remodeling

Remodeling is an ever-occurring skeletal activity that is responsible for producing and maintaining biomechanically and metabolically competent bone. It is best characterized by an anatomical coupling consisting of the appearance of osteoclasts (resorption) followed by the appearance of osteoblasts (formation). The remodeling occurs at scattered locations on bone surfaces including cortical–endosteal and trabecular–endosteal surfaces, and within cortical bone. This remodeling sequence for trabecular bone is illustrated in Figure 8.5.

The essential elements of this process can be described as follows. Remodeling sites are initiated by the activation of progenitor cells, probably both local and blood-borne, which proliferate into a group of newly formed osteoclasts. These osteoclasts resorb a packet of bone, thereby forming either a scalloped resorption bay (Howship's lacuna) on trabecular surfaces or, alternatively, a longitudinal cutting cone in cortical bone. This degradative activity is much greater for cortical bone than for trabecular bone and results in the formation of a new Haversian canal. It is still unknown what mechanism is responsible for arresting resorption or the precise duration of the osteoclastic phase of remodeling. Recent experimental evidence has documented that the osteoclast undergoes morphologic changes consistent with decreased bone degradative activity in the presence of an increase in the extracellular (and subsequently intracellular) ionized calcium concentration.[11,12] Such increases in free calcium concentration probably occur during active osteoclastic bone resorption. Thus, the rise in extracellular calcium concentration may serve as an initiating signal to arrest continued osteoclastic activity.

Upon completion of osteoclastic activity, an inactive reversal phase of varying duration ensues. This phase is characterized by the appearance of a variety of mononuclear cells within the resorption bays. These cells may originate from degradation of the multinucleated osteoclast to its mononuclear precursors.[5] These cells are generally found adjacent to a densely staining, metachromatic band marking the limits of osteoclastic resorption in a remodeling focus. This band is termed the cement line. This cement line may be deposited by the mononuclear cells of the reversal phase.[13]

Upon completion of the reversal phase, osteoblasts appear at the previously completed resorption foci. As previously discussed, these cells deposit bone matrix in amounts equal to the bone previously removed by the osteoclasts. In so doing, net skeletal balance is maintained. This balance of skeletal mass is maintained until the third decade of life whereafter the quantity of bone resorbed from a remodeling site exceeds that subsequently deposited and, consequently, leads to the well documented decline in skeletal mass associated with aging.

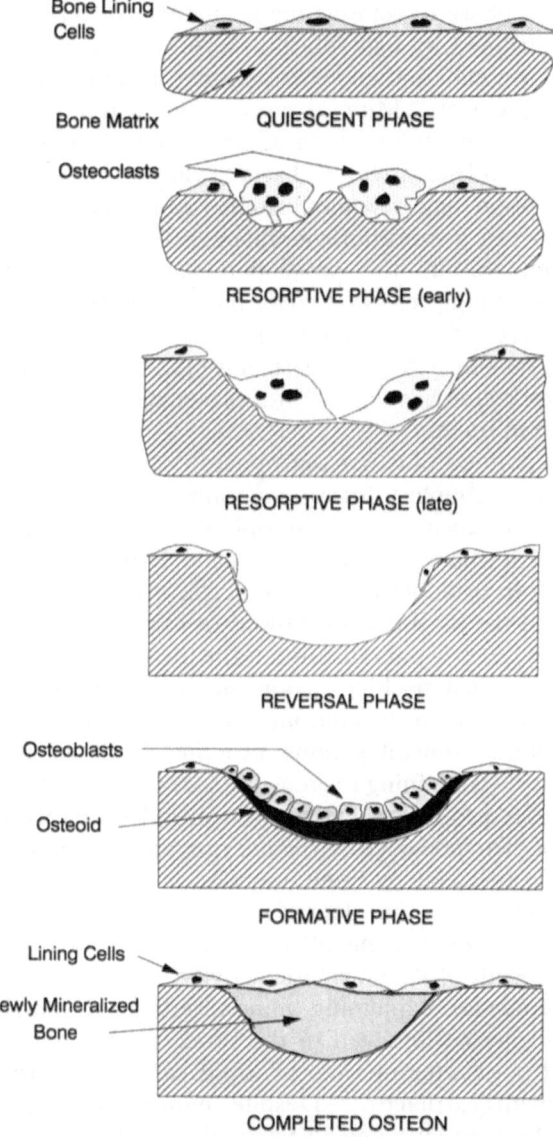

FIGURE 8.5. Pictorial representation of the remodeling sequence in normal cancellous bone. The process is initiated by the activation and appearance of osteoclasts at a remodeling site. After the osteoclasts have resorbed a discrete amount of bone, a reversal phase ensues whereby osteoclasts disappear from the resorption site. Subsequently, osteoblasts appear within the resorptive bay and synthesize matrix (formative phase) until a new packet of bone (the osteon) is produced.

The formative phase of remodeling is characterized by the presence of an osteoid seam lined by osteoblasts. During active osteoid deposition, the osteoblasts are cuboidal and the quantity of organic matrix deposited per day is at its greatest.[14] Mineralization of the osteoid is also greatest at this time but does not keep up with osteoid deposition and contributes to the increased osteoid seam width observed during early bone formation.

As the resorption bay becomes progressively filled with osteoid, the osteoblasts undergo a morphologic change. They become less cuboidal and more attenuated to the point that they resemble the flattened, relatively inactive lining cells that cover most bone surfaces. At this stage, the rate of matrix synthesis rapidly diminishes, as does mineralization. However, mineralizing activity persists until the width of the osteoid seam is decreased and eventually results in replacement of the osteoid seam by newly mineralized bone.

It should be pointed out that the mineralization process is still poorly understood. In its simplest form, mineralization requires (1) adequate concentrations of calcium and phosphorus, (2) nucleation sites for initiating crystal formation, and (3) a calcifiable matrix free of inhibitory substances.[15] Both osteoblasts and osteocytes are most likely involved in the regulation of the local concentrations of calcium and phosphate, thus promoting a phase of mineralization. The presence of nucleation sites created by the distinct packing arrangement of collagen in the osteoid may regulate the site of formation, type of mineral, and amount of mineral deposited. Noncollagenous proteins such as the gamma-carboxyglutamic acid–containing proteins (osteocalcin and osteonectin) have also been suggested to have roles in the controlled deposition of mineral onto the matrix. The presence of the enzyme alkaline phosphatase within osteoblasts and other cells in bone may also help regulate mineralization not only by catalyzing the degradation of pyrophosphate (a potent inhibitor of mineralization) but also by raising the local phosphate concentration as a result of its enzymatic activity.

Another theory for explaining mineralization of matrix holds that matrix vesicles may be involved in the process. This is based on observations of matrix vesicles in sites of rapid calcification (e.g., calcifying epiphyseal growth cartilage).[16] Despite numerous observations made about vesicles both in vivo and in vitro, the mechanisms by which they promote mineralization remain unclear. Furthermore, the lack of evidence for the presence of matrix vesicles in adult bone casts further doubt about the importance of this mechanism to the mineralization mechanism of normal remodeling bone.

It is known that the remodeling sequence is normally tightly controlled. Each bone remodeling unit can be activated (A) by one of many stimuli (see Hormonal Regulation of Bone Growth and Remodeling below), which promotes differentiation of precursor cells of osteoclasts into active osteoclasts followed by osteoclastic resorption (R) of preexisting bones and osteoblastic formation (F) to replace the bone resorbed in an un-varying succession first proposed by Frost as A → R → F.[17] In a normal adult, the time scale for one bone remodeling cycle has been proposed to be between 3 and 4 months.

As previously mentioned, the orderly remodeling of bone must involve the coupling of formation to resorption. Although this phenomenon has been proposed and supported in theory, the existence of a single mole-cular species that can perform this task in vivo is lacking. For coupling to proceed from resorption to formation, some factor must be elaborated from the resorbing bone cells or from the resorbed bone that suppresses further osteoclastic resorption and acts as a stimulator of osteoblastic differentiation and activity. Several growth factors have been proposed as potential coupling agents. Two factors that continue to receive attention as potential coupling agents are bone-derived growth factor (BDGF) and skeletal growth factor (SGF). BDGF has been purified to homogeneity and its amino acid sequence determined. BDGF has been shown to be identical to B_2-microglobulin.[18] SGF, on the other hand, has recently been identified to be insulinlike growth factor II (IGF-II).[19] In addition, IGF-I is also synthesized by bone cells.[20] Both IGF-I and IGF-II are known to have similar regulating effects on bone collagen synthesis in rat calvarial cultures.[21] Although none of these factors are produced by the bone-resorbing osteoclasts, they are found in sufficient concentrations in bone and are probably released from resorbed bone. These factors have been shown to have potent osteoblastic stimulating activities.

Another potential candidate is transforming growth factor B (TGF-B). This factor is produced by osteoblasts and subsequently sequestered in the newly formed osteoid. Because this cytokine has potent actions on osteoblastic maturation and activity,[22] it has been proposed to be a potential coupling agent. This notion is further supported by two addi-tional findings. The first is that TGF-B is released from bone with one or more binding proteins that regulate its activity.[23] When bound to the binding protein, TGF-B is inert. However, an acidic environment controls TGF-B activity by release of active 25-kDa TGF-B from its binding protein. Such an acid environement exists as part of the local microenvironment of active osteoclasts. The second recent finding is that TGF-B inhibits formation of osteoclastlike cells in long-term human marrow cultures.[24] Taken together, these findings suggest that TGF-B could serve as a bone-coupling agent since it would become active in an acid environment, decrease osteoclast formation, and promote osteo-blastic differentiation and collagen synthesis.

In addition to these bone-derived factors, various cytokines and lymphokines have been demonstrated to be present in bone and may also have a potential role in the regulation of bone remodeling.[25] Table 8.1 lists a number of these growth factors that are believed to be involved in the local regulation of bone remodeling.

TABLE 8.1. Growth factors involved in the local regulation of bone remodeling.

Growth factors synthesized by skeletal cells	
Transforming growth factor β	(TGF-β)
β₂-microglobulin	(β₂M or BDGF)
Platelet-derived growth factor	(PDGF)
Insulin-like growth factor I and II	(IGFI and II)
Granulocyte macrophage colony stimulating factor	(GM-CSF)
Growth factors found in bone matrix	
All of the above	
Acidic fibroblast growth factor	(aFGF)
Basic fibroblast growth factor	(bFGF)
Monokines and lymphokines from blood	
Interleukin-1	(IL-l)
Tumor necrosis factor α	(TNFα)
Tumor necrosis factor β	(TNFβ)
Gamma-interferon	(INF-γ)

An alternative theory to the idea of osteoclastic resorption elaborating a putative coupling factor that in turn stimulates osteoblastic bone formation is the notion of reverse coupling. In this schema, systemic hormones that influence the remodeling process interact with the osteoblast first. This interaction then leads to the production of a factor(s) that in turn activates osteoclastic bone resorption and subsequent bone formation. This idea is supported by several observations. As previously mentioned, the osteoblast appears to be the target cell for most of these hormones since it is on this cell that the specific hormonal receptors are found. Only calcitonin receptors have been reported to be present on the osteoclast. Second, if osteoclasts are disaggregated mechanically from neonatal mammalian long bones and sedimented onto slices of cortical bone, excavation of the bone surface by osteoclasts commences within a few hours, and by 18 hr each osteoclast has usually produced one to three resorption pits.[26] Agents that are known to stimulate osteoclastic bone resorption, such as PTH and $1,25(OH)_2D_3$, are without influence on the spontaneous resorption of cortical bone slices by isolated osteoclasts. This deficiency can be restored by incubating osteoclasts with either osteoblastic cell lines or primary cultures of normal osteoblastic cells. Such cocultures respond to concentrations of PTH as low as 10^{-4} IU/ml, with increases in bone resorption of two- to fourfold that seen without the hormone.[27] Thus, the enhanced bone resorption in osteoblast–osteoclast cocultures has been attributed to the secretion into the culture medium of an osteoclast resorption stimulating activity that in the presence of PTH directly stimulates osteoclasts: addition of supernatants from osteoblastic cells previously incubated with PTH or $1,25(OH)_2D_3$ to osteoclast cultures increases osteoclastic resorption to an extent similar to that seen in cocultures.[28,29] Although the nature of this osteoclast stimulating factor is unknown, it does not appear to be a prostaglandin.[28,30] A recent report

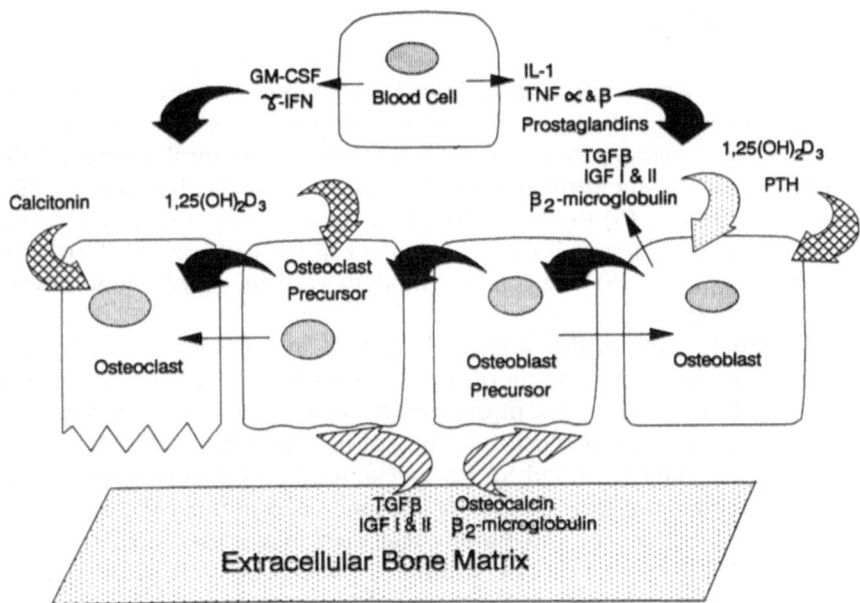

FIGURE 8.6. Model of the local regulation of bone cell activity. In addition to circulating hormones that act on the osteoblast or osteoclast in an endocrine manner (➡), the elaboration of local factors by the osteoblast or other cells found in bone marrow (—►) can function in both autocrine (⟜) and paracrine fashion (➡) to promote bone cell differentiation and proliferation (—►). The release of bound growth factors from bone matrix during osteoclastic resorption may provide an additional "matricrine" pathway (⟜) for bone cell regulation.

has suggested the factor to be a protein with an apparent molecular size of about 110,000 daltons. A second protein of about 70,000 daltons may also be involved.[30]

In summary, bone remodeling is a complex process wherein the two major processes of bone resorption and formation are closely regulated. Many of the local factors play a critical role in the control of bone remodeling. It is probable that these factors may also mediate the effects of systemic hormones (see section on Hormonal Regulation of Bone Growth and Remodeling, below), which could modify the synthesis or effects of the local factors. Since bone is a heterogeneous tissue comprised of a mixed cell population and is in direct contact with cartilage and marrow cells, the local regulators of bone remodeling may originate from a variety of cells. The release of matrix-associated growth factors during osteoclastic resorption suggests.the existence of a matricine control of bone remodeling in addition to both autocrine and paracrine mechanisms (Fig. 8.6).

Bone Healing

Repair of bone serves to heal injuries from traumatic causes and wall off harmful processes such as infection, neoplasm, or physicochemical irritants. The healing process includes the following natural stages: (1) the fracture itself, (2) the production of a temporary soft healing tissue called granulation tissue, (3) the replacement of this tissue by a temporary hard tissue called the callus, (4) the replacement of the callus by well-oriented lamellar bone, and (5) a reshaping (modeling) of the whole bone toward its normal shape.[31]

The initial insult to the skeleton (fracture in this instance) results in injury to the local marrow, periosteum, and adjacent soft tissues as well as to the bone itself. This injury results in two processes that initiate normal healing. First, it sensitizes some of the surviving local cells so that they may ultimately respond to the special local and systemic messengers such as the skeletal growth factors. Second, it promotes the release of these local biochemical and biophysical messengers that then prescribe how those cells respond. For example, platelet-derived growth factor (PDGF) is another polypeptide found in bone matrix.[32] PDGF stimulates bone DNA and protein synthesis and may be a systemic or a local regulator of skeletal growth. As a systemic growth factor, it could be released during platelet aggregation and have important effects in the early stages of fracture healing; as a local factor, it may interact with other hormones and growth factors and thereby make bone cells competent to respond to other factors present in skeletal tissue. Under the action of such systemic and local growth factors the first stage of bone healing is probably completed within 7 days after injury.

In the second stage, the sensitized and stimulated precursor cells begin to differentiate and produce new cells and new blood vessels. Collectively they form a soft granulation tissue in the space between the fracture fragments. This stage lasts about 2 weeks and may also be characterized by the appearance of osteoclasts.

Further cell proliferation, differentiation, and organization begin to promote the appearance of osteoblasts in the granulation tissue. The synthesis of the extracellular organic matrix of collagen and woven bone heralds the beginning of the appearance of the callus. This stage of fracture healing may also be influenced by other growth stimulating polypeptides, the fibroblast growth factors (FGFs), a family of polypeptides that have potent effects on endothelial cell replication and neovascularization.[33] In bone cultures, FGFs have been reported to stimulate DNA synthesis and cell replication; this results in an increased bone cell population capable of synthesizing collagen and noncollagen protein.[34,35] The FGF effect is not specific for collagen, but the collagen synthesized is type I, indicating that these growth factors are affecting cells of the osteoblastic lineage. Although FGFs increase the number of osteoblastic cells, they have no direct stimulatory effect on the differentiated function of these cells.

The stimulatory effects of the FGFs on neovascularization in association with those on bone cell replication may be important for wound healing and bone repair after fractures. This is further supported by the observation that the mitogenic effect of FGF on endothelial and bone cells is enhanced by heparin, and the fracture callus is rich in heparin-containing mast cells.[34] Heparin is known to increase the binding of FGF to its receptor, and this could have physiologic implications in the process of bone repair, particularly if mast cell degranulation and heparin release occur at the fracture site. In addition, FGFs are not secreted cell products but probably become available only after cell injury or death.[36]

After callus formation, the previously described remodeling stage begins to replace the callus with packets of new bone. Complete replacement of the callus with functionally competent lamellar bone requires 1 to 4 years. It proceeds quickly during the first one third of the replacement (probably a result of regional acceleratory phenomena[37] and slows progressively in the last two thirds. This regional acceleratory phenomena (RAP) somehow accelerates the normal healing process. By increasing the rapidity of the aforementioned healing stages, they make healing occur 2 to 10 times more quickly than it would otherwise occur. For example, published histomorphometric data show that, if only the pre-existing osteoblasts were available, healing of a femoral shaft fracture would reguire between 200 and 1000 years. As previously mentioned the osteoblastic life span is only 2 to 3 months. Thus, the RAP must be responsible not only for increasing the number of osteoblasts at the fracture site but also for the replacement of exhausted osteoblasts with new ones during the many months needed to heal a fracture. Although the mechanism by which RAP exerts its effects at the various stages of bone healing are unknown, it is probable that many of the locally produced growth factors previously discussed (e.g., TGF-B, FGFs, PDGF, etc.) are responsible, in part, for this response.

Hormonal Regulation of Bone Growth and Remodeling

The growth and development of the skeleton, the attainment of peak bone mass and strength during the third decade of life, and the repair of the skeleton after trauma or infection depend on the concerted action of a number of systemic and local growth-regulating hormones and factors. Some of these factors have been discussed above (see section on Remodeling and Bone Healing). In addition to local factors, systemic calcium-regulating hormones have powerful effects on bone that are essential to the skeleton's function as a mineral reservoir and on the differentiation of osteoprogenitor cells to their respective fully differentiated mature state. The action of these systemic hormones on bone cells is probably mediated, in part, by the elaboration of many of these local growth regulators. The discussion will therefore be limited to the three major systemic calcium-regulating hormones, namely PTH, vitamin D (specifically $1,25(OH)_2D_3$), and calcitonin (CT). In addition, the prostaglandins, particularly PGE_2, will be considered since it was one of the first local regulators to be defined and may act as a mediator of the responses observed with PTH.

Parathyroid Hormone

Although it has been known for nearly 30 years that PTH can stimulate bone resorption directly, a full understanding of its interaction with bone cells is still lacking. Early studies using organ cultures have shown that PTH can increase both the activity and number of osteoclasts in bone culture.[38] However, as previously discussed, PTH appears to have little effect on isolated osteoclasts[29] even though these cells are still capable of resorbing bone. It appears that the major determinant of the magnitude of the resorptive response to PTH appears to be an increase in osteoclast number rather than changes in activity of individual osteoclasts, but both processes do occur. Since the osteoblast has been shown to be the only bone cell with receptors for PTH,[39,40] the hypothesis that these cells mediate the PTH increase in bone resorption has been widely accepted. Evidence to date indicates that osteoblasts can produce substances that activate isolated osteoclasts, but there is no evidence that the increase in osteoclast number is mediated by an osteoblast product. In addition, it has been demonstrated in organ culture that this PTH-stimulated increase in the number of active osteoclasts does not depend on the multiplication of progenitor cells because it can occur in the presence of inhibitors of DNA synthesis.[41] Thus, it has been proposed that PTH appears to stimulate the fusion of existing osteoclast precursors into active osteoclasts.

It has been demostrated that PTH stimulates cyclic adenosine mono-phosphate (cAMP) production in osteoblasts and that agents that increase osteoblastic cAMP production can stimulate bone resorption.[42] However, when analogues that block PTH stimulation of adenylate cyclase are preincubated with bone, PTH stimulation of bone resorption is not in-hibited. Thus, it has been difficult to prove that cAMP is the major second messenger for the PTH response. One interesting possibility may be a rapid and transient increase in intracellular ionic calcium after exposure of osteoblastlike cells in culture to physiologic concentrations of PTH,[43-45] suggesting that the PTH-mediated stimulation of bone re-sorption may be mediated by the osteoblast. Since PTH has also been demonstrated to elevate inositol phosphates and diacylglycerol in a rat osteoblastlike cell, it is possible that this rapid change in intracellular calcium may signal the production of a factor(s) that promotes increases in osteoclastic bone resorption as discussed above (see section on Coupling). In addition to the previously mentioned growth factors, PTH has been reported to stimulate osteoblastic secretion of granulocyte macrophage–colony stimulating factor (GM-CSF).[46] GM-CSF is one of several cytokines that can stimulate the proliferation of bone marrow–derived osteoclast precursors[47] and increase fusion of the mononuclear osteoclast precursor cells into multinucleated osteoclasts.

The effects of PTH on bone formation are variable and highly dependent on the time, dose, and mode of administration. It has been fairly well established that in high concentrations PTH inhibits osteoblastic collagen synthesis and can reduce alkaline phosphatase activity.[48] Under these conditions, the osteoblasts retract somewhat and secrete collagenase and plasminogen activator, which may be involved in initiating the bone resorptive process by removing a thin layer of unmineralized organic material and thus preparing the surface for osteoclasts.[49] If the cells are continuously exposed to these high concentrations of PTH, this state of increased bone resorption and decreased formation persists. However, intermittent exposure to low concentrations of PTH results in an anabolic response both in vivo[50] and in vitro.[51] This action of PTH appears to represent an increase in the number as well as the activity of osteoblasts as supported by the presence of increased bone-forming surfaces and increased mineral apposition rates.[52] It has been proposed that this anabolic effect of PTH may be mediated by systemic hormones. However, this seems unlikely in view of the evidence for an in vitro effect. A second hypothesis is that PTH could increase bone formation by releasing a bone-derived growth factor from the matrix during resorption. Again, this seems unlikely since the resorptive and formative effects of PTH on bone can be dissociated in terms of dose and time. Furthermore, pretreatment of the bone with resorption blocking agents (e.g., diphosphonate, calcitonin) during the early phase of intermittent low-dose PTH administration does not decrease the anabolic response.[53] The best explanation of the PTH anabolic effect appears to be the stimulation of the production of a local bone growth factor. Indeed, PTH has been shown to stimulate the release of TGF-B from bone,[54] promote the osteoblastic synthesis of IGF-1,[55] and stimulate the production of PGE_2 in bone.[56] As will be discussed below, PGE_2 may be an important local stimulator of bone formation.

Vitamin D

The physiologic activation of vitamin D to $1,25(OH)_2D_3$ and its calcemic action on the skeleton have been known for nearly two decades. However, it is only in the last 5 years that an understanding of how $1,25(OH)_2D_3$ exerts its effects on the skeleton has begun to emerge.

Like PTH, $1,25(OH)_2D_3$ is without effect on the spontaneous resorption of cortical bone slices by isolated osteoclasts.[26] When the osteoclasts are coincubated with either osteoblastic cell lines or primary cultures of osteoblastic cells, this deficiency can be restored. Thus, it appears that cells of the osteoblastic lineage are also required to initiate $1,25(OH)_2D_3$-mediated bone resorption. This contention is further supported by the observation of specific $1,25(OH)_2D_3$ receptors on osteoblasts, but not on osteoclasts.[57,58] In this model, $1,25(OH)_2D_3$ would act on the osteoblast to stimulate the elaboration of factors that in turn would stimulate osteoclastic bone resorption. Although several factors have been proposed as possible osteoclastic resorption stimulating agents, there is no general agreement on which of these factor(s) is ultimately responsible for the stimulation of osteoclastic activity.

An alternative hypothesis holds that osteoblasts may respond to $1,25(OH)_2D_3$ by synthesizing procollagenase[59,60] and plasminogen activator.[49] Plasminogen activator may generate plasmin, a protease that can efficiently activate procollagenase by conversion to collagenase. Collagenase would then act on the thin nonmineralized matrix covering bone to remove it and thereby expose native mineral surfaces to the activated osteoclast. Although, attractive, the applicability of this mechanism may well be limited to bone tissue of the fetal, woven type, where a lining layer of unmineralized osteoid or collagen on all bone surfaces is believed to exist, and not to adult lamellar bone,[61] where its presence is disputed.

A third mechanism by which osteoclastic resorption may be influenced by $1,25(OH)_2D_3$ is through the regulation of osteoclastic differentiation. As discussed above, several lines of experimental evidence have established that osteoclasts originate from hematopoietic tissue.[62] According to this view, the osteoclast arises from the granulocyte-macrophage progenitor cell that itself is derived from the multipotential hematopoietic stem cell. At some stage along their differentiation pathway, a noncommitted progenitor can generate a committed osteoclast progenitor, which then follows an osteoclast-specific diverging differentiation pathway. Alternatively, the same noncommitted progenitor may differentiate irreversibly toward the monocyte, eventually giving rise to the tissue macrophage. Recent experimental evidence has suggested that $1,25(OH)_2D_3$, which is known to promote the differentiation of monocytes into macrophages[63] as well as the fusion of macrophages into multinucleated giant cells,[64,65] has the capacity to promote the differentiation of the preosteoclast to the osteoclast.[66,67]

This proposed role of $1,25(OH)_2D_3$ in the regulation of osteoclastic differentiation would be a direct effect and not require previous binding to the osteoblastic $1,25(OH)_2D_3$ receptor. Since mature osteoclasts lack $1,25(OH)_2D_3$ receptors, it may be that the osteoclastic precursors possess receptors but lose them during maturation. An alternative and more likely, possibility is that there is an accessory cell, among the many that respond to $1,25(OH)_2D_3$, that is present not only in bone but also in spleen (and is therefore unlikely to be osteoblastic) that mediates osteoclastic induction.

The presence of specific $1,25(OH)_2D_3$ receptors on osteoblasts suggests that this active vitamin D metabolite may also have an important role in bone mineralization and/or bone formation. $1,25(OH)_2D_3$ has been shown to stimulate several osteoblastic activities including production of alkaline phosphatase activity and osteocalcin.[68-70] However, many of the observed in vitro responses have demonstrated that these osteogenic effects of $1,25(OH)_2D_3$ are biphasic[71,72]: at low doses (10^{-11}–10^{-10} M) $1,25(OH)_2D_3$ stimulates bone cell proliferation, whereas at higher concentrations ($>10^{-10}$ M) the hormone inhibits proliferation and enhances bone cell differentiation as reflected by the stimulation of osteoblastic alkaline phosphatase activity and synthesis of osteocalcin and type I collagen. Others have found that $1,25(OH)_2D_3$ produces a consistent inhibition of type I collagen synthesis in osteoblastlike cells.[73] In addition, $1,25(OH)_2D_3$ was also observed to reduce procollagen mRNA levels.[74] Many of the differences reported for $1,25(OH)_2D_3$ action on osteoblasts can be explained, in part, by the various osteoblasticlike cell lines used and by the proliferation state of the cells.[75] Thus, at face value, it seems paradoxic that whereas calcitriol inhibits collagen synthesis in vitro, it is necessary for adequate bone formation in vivo. A teleogic explanation for this observation[76] is that the catabolic effects of $1,25(OH)_2D_3$ are useful when the exogenous supply of calcium or phosphorus is diminished. When the mineral supply is reduced, the synthesis of $1,25(OH)_2D_3$ increases. If dietary calcium and phosphorus is low, $1,25(OH)_2D_3$ will have little effect on the intestinal transport of these minerals. In this instance, $1,25(OH)_2D_3$ will promote the mobilization of calcium and phosphate from the skeleton by direct stimulation of osteoclastic bone resorption. Since this mineral is removed from the skeleton to maintain serum calcium, it could not be used for any net growth of bone mass. Hence, a decrease in overall bone-matrix formation would be appropriate and useful.

Calcitonin

Calcitonin is well recognized as a general inhibitor of bone resorption elicited by diverse stimuli. This antiresorptive action of calcitonin is believed to be characterized by ultrastructural changes in the osteoclast such that there is a diminution in the extent of the ruffled border and loss of contact between the osteoclast and the bone surface.[77-79] The inhibitory effect of calcitonin on bone resorption is most likely mediated by direct action of calcitonin on the osteoclast. This is supported by the demonstration of specific calcitonin receptors on rat osteoclasts.[80] In addition, when the hormone is added to cultures rich in osteoclasts, there is an inhibition of their intrinsic or stimulated mobility.[26,81] The sharp increases in skeletal cAMP seen during calcitonin addition suggest that this cyclic nucleotide may well be the mediator for this inhibitory effect. Recently, two reports[11,12] have demonstrated that osteoclasts are acutely sensitive to the extracellular ionized calcium concentration. When extracellular ionized calcium was increased there was a transient increase in intracellular ionized calcium followed by an appreciable cell retraction and significant alteration of the cell morphology. Calcitonin by itself caused only a small increase in intracellular ionic calcium but in the presence of high extracellular calcium the respective effects were synergistic. These observations suggest a possible role of intracellular ionized calcium in osteoclast physiology. During bone resorption the extracellular ionized calcium concentration probably becomes locally elevated, causing a change in intracellular ionic calcium and a contraction of the cytoskeleton. This would ultimately arrest bone resorption by loss of osteoclast–bone contact. It is proposed that calcitonin augments this ultrastructural change in osteoclast morphology by acting synergistically with the locally elevated extracellular calcium concentration in the promotion of increased intracellular ionized calcium.

It is less clear whether calcitonin has any effect on osteoblast and bone formation. Although it has been reported that this hormone increases bone formation in developing bone in vivo[82] and can stimulate bone formation and bone cell proliferation in vitro,[83] supportive studies are lacking. Previous reports that calcitonin (acutely) increased bone formation indices in humans[82,84] led to the hypothesis that calcitonin treatment might be clinically applicable for increasing bone volume in osteoporosis. This hypothesis was supported by subsequent studies indicating calcitonin-dependent increases in (1) total body calcium and iliac crest trabecular bone volume,[85] and (2) bone mineral content at the distant epiphysis[86] in osteoporotics. These effects were observed only acutely and may represent an action of calcitonin on osteoblast precursor cells. Others have failed to observe any cAMP response to calcitonin in osteoblast-enriched calvarial-derived cells[87,88] or in osteosarcoma-derived cells.[89,90] Thus, it is generally believed that differentiated osteoblasts are calcitonin-unresponsive, at least in terms of their cAMP content.

Prostaglandins

Since the initial report that prostaglandins, particularly of the E series, could stimulate the resorption of fetal rat long bones in organ culture, there have been many experiments confirming and extending this observation.[91–95] The extensive literature on the effects of prostaglandins on bone and mineral metabolism has been reviewed recently.[96] PGE_2 and PGE_1 appear to be the most potent stimulators and produce a dose-related increase in calcium release from bone over the concentration range of 10^{-9} to 10^{-5} M.

Prostaglandin-mediated bone resorption appears to be different from that mediated by parathyroid hormone or $1,25(OH)_2D_3$. In fetal rat long bone, the dose-response curve to the prostaglandins is not as steep as that for parathyroid hormone or $1,25(OH)_2D_3$. Moreover, the time course of increased ^{45}Ca release during exposure to prostaglandin is slower and prostaglandins are less able to elicit prolonged resorption after relatively brief exposure. Despite these differences, the morphologic response of osteoclasts to prostaglandin stimulation does not differ qualitatively from that produced by other stimulators.[38] Prostaglandins appear to increase the number and activity of osteoclasts in treated bone cultures by a mechanism involving an increase in cell membrane adenylate cyclase activity.[97] Increased entry of calcium into bone cells has also been implicated as a mediator of hormone action, particularly for PTH (see above), but evidence concerning the effect of prostaglandin on cell calcium entry is limited and conflicting. It now appears that this prostaglandin-responsive adenylate cyclase activity is enriched in osteoblast-like cells, suggesting that the osteoblast may be the prostaglandin target cell[75,96] and stimulate osteoclastic bone resorption by mechanisms previously discussed above (see section on Remodeling).

Although the effects of prostaglandins on bone resorption have been studied to a much greater extent than those on bone formation, it is well accepted that prostaglandins can increase collagen synthesis.[96,98] This response appears to be biphasic since low concentrations of PGE_2 (10^{-9}–10^{-7}M) stimulate collagen synthesis but concentrations greater than 10^{-6}M inhibit this response. PGE_2 also prevents the decrease in bone cell DNA content produced by prolonged exposure to cortisol.[96] This observation suggests that prostaglandins may stimulate the replication and differentiation of osteoblast precursors, at least in organ culture systems. In addition to these in vitro findings, a number of in vivo observations support the possible stimulatory effect of prostaglandins on bone formation. The first of these is that fetal bone growth is inhibited by indomethacin, a potent inhibitor of prostaglandin synthesis. The second is that prolonged infusion of PGE_1, used to maintain potency of the ductus arteriosus in children with congenital heart disease, stimulates periosteal new bone formation.[99-101] Furthermore, indirect evidence has accumulated that suggests that prostaglandins are involved in fracture healing since indomethacin delays this process[102] and increased prostaglandin production has been demonstrated at fracture sites.[103]

Endogenous skeletal prostaglandin production may participate in bone resorption. This is of special importance with regard to the stimulation of resorption by immune mechanisms and other local factors. Although organ culture experiments have clearly demonstrated significant prostaglandin production by bone, the responsible cell types have not been

TABLE 8.2. The effect of various agents on bone cell function.

Agent	Type of function[a]	Osteoclasts		Osteoblasts	
		Action[b]	Mechanism	Action	Mechanism
Parathyroid hormone	Endocrine	Indirect	↑ fusion of osteoclast precursors	Direct	↓ collagen synthesis; intermittent administration promotes ↑ collagen synthesis and ↑ osteoblast number activity
1,25(OH)$_2$D$_3$	Endocrine	Indirect	↑ Differentiation of osteoclast precursors ↑ fusion of osteoclast precursors	Direct	↑ Proliferation at high concentrations and ↑ proliferation at low concentrations; ↓ collagen synthesis
Calcitonin	Endocrine	Direct	↓ osteoclast adhesive to bone surface	No known action	
Prostaglandin	Local	Indirect	↑ osteoclast number and activity	Direct	↑ Proliferation and differentiation of osteoblast precursors; high concentrations ↑ collagen synthesis; low concentrations ↓ collagen synthesis

[a] Endocrine function implies synthesis and release from a nonskeletal tissue; local implies synthesis and action at the bone cell.
[b] Direct action implies specific action on the bone cell. Indirect represents an action mediated in a paracrine manner.
↑, increase; ↓, decrease.

identified. However, studies with clonal osteoblastlike cells and with osteoblast-rich calvarial cell populations suggest that cells of the osteo-blastic lineage may be the major prostaglandin-producing cell type in bone.[75,96,104]

The physiologic role of the prostaglandins in normal bone growth and remodeling is unknown. Prostaglandins are probably important local regulators of bone cellular activity and are unique in their ability to stimulate not only bone resorption, but cell replication and collagen synthesis and thereby increase bone formation. Such findings have led to the speculation that prostaglandins may be involved in regulating bone turnover in response to such stimuli as stress, growth, and repair. Pertubations such as mechanical stress, injury, inflammation, and changes in vascularization increase local prostaglandin production. This local synthesis of prostaglandins may then serve to (1) stimulate osteoblastic collagen synthesis, (2) promote osteoclastic bone resorption, and (3) stimulate the replication and maturation of preosteoblasts into fully differentiated osteoblasts.

Table 8.2 summarizes the actions of PTH, $1,25(OH)_2D_3$, calcitonin, and prostaglandin on osteoclast and osteoblast formation and function. In addition to these hormones, a number of classic hormones with primary effects on other tissues have distinct actions on bone as well. Estrogens, androgens, glucocorticoids, and glucagon are but a few of the hormones that may also play a significant role in the hormonal regulation of bone growth and remodeling.

References

1. Delmas PD, Malaval L, Arlot ME, Meunier PJ. Serum bone Gla-protein compared to histomorphometry in endocrine diseases. *Bone.* 1985;6:339–341.
2. Termine JD, Kleinman HK, Whitson SW, Conn KM, McGarvey ML, Martin GR. Osteonectin, a bone-specific protein linking mineral to collagen. *Cell.* 1981;26:99–105.
3. Termine JD, Belcourt AB, Conn KM, Kleinman HK. Mineral and collagen-binding proteins of fetal calf bone. *J Biol Chem.* 1981;256:10403–10408.
4. Prockop DJ, Kadler KE, Hojima Y, Constantinou CD, Dombrowski KE, Kuivaniemi H, Tromp G, Vogel B. Expression of type I procollagen genes. In: Evered D, Harnett S, eds. *Cell and molecular biology of vertebrate hard tissues.* Chichester: John Wiley and Sons; 1988:142–160.
5. Baron R, Vignery A, Horowitz M. Lymphocytes, macrophages and the regulation of bone remodeling. In: Peck WA, ed. *Bone and mineral research; annual 2.* Amsterdam: Elsevier; 1984:175–243.
6. Nijweide PJ, Mulder PP. Identification of osteocytes in osteoblast-like cell cultures using a monoclonal antibody specifically directed against osteocytes. *Histochemistry.* 1986;84:342–347.
7. Marks JC Jr. Osteopetrosis-multiple pathways for the interception of osteoclast function. *Appl Pathol.* 1987;5:172–183.

8. Malone JD, Teitelbaum SL, Griffin GL, Senior RM, Kahn AJ. Recruitment of osteoclast precursors by purified bone matrix constituents. *J Cell Biol.* 1982;92:227–230.

9. Caplan AI. Bone development. In: Evered D, Harnett S, ed. *Cell and molecular biology of vertebrate hard tissues.* Chichester: John Wiley and Sons; 1988:3–21.

10. Frost HM. *The physiology of cartilaginous, fibrous and bony tissue. Orthopaedic Lectures, vol. II.* Springfield, Ill: Charles C Thomas; 1972.

11. Malgaroli A, Meldolesi J, Zambonin Zallone A, Teti A. Control of cytosolic free calcium in rat and chicken osteoclasts: the role of extracellular calcium and calcitonin. *J Biol Chem.* 1989;264:14342–14347.

12. Zaidi M, Datta HK, Patchell A, Moonga B, MacIntyre I. "Calcium-activated" intracellular calcium elevation: a novel mechanism of osteoclast regulation. *Biochem Biophys Res Commun.* 1989;163:1461–1465.

13. Baron R, Vignery A, Tran Van P. The significance of lacunar erosion without osteoclasts: studies on the reversal phase of the remodeling sequence. *Met Bone Dis Rel Res* 1980;2S:35–40.

14. Parfitt AM. The physiologic and clinical significance of bone histomorphometric data. In: Recker R, ed. *Bone histomorphometry: techniques and interpretation.* Boca Raton, Fla: CRC Press; 1983:143–223.

15. Neuman W. Bone material and calcification mechanisms. In: Urist M, ed. *Fundamental and clinical bone physiology.* Philadelphia: Lippincott; 1980: 83–109.

16. Anderson HC. Vesicles associated with calcification in the matrix of epiphyseal cartilage. *J Cell Biol.* 1969;41:59–72.

17. Frost HM. *Mathematical elements of bone remodeling.* Springfield, Ill: Charles C Thomas; 1964.

18. Canalis E, McCarthy T, Centrella M. A bone-derived growth factor isolated from rat calvariae is beta 2 microglobulin. *Endocrinology.* 1987;121:1198–1200.

19. Mohan S, Jennings JC, Linkhart TA, Baylink DJ. Primary structure of human skeletal growth factor: homology with human insulin-like growth factor II. *Biochim Biophys Acta.* 1988;966:44–55.

20. Canalis E, McCarthy T, Centrella M. Isolation and characterization of insulin-like growth factor I (somatomedin-C) from culture of fetal rat calvariae. *Endocrinology.* 1988;122:22–27.

21. McCarthy T, Centrella M, Canalis E. Regulatory effects in insulin-like growth factors I and II on bone collagen synthesis in rat calvarial cultures. *Endocrinology.* 1989;124:301–309.

22. Centrella M, McCarthy T, Canalis E. Transforming growth factor B is a bifunctional regulator of replication and collagen synthesis in osteoblast enriched cell cultures from fetal rat bone. *J Biol Chem.* 1987;262:2869–2874.

23. Oreffo RO, Mundy GR, Seyedine SM. Activation of the bone-derived latent TGF beta complex by isolated osteoclasts. *Biochem Biophys Res Commun.* 1989;158:817–823.

24. Chenu C, Pfeilschifter J, Mundy GR, Roodman GD. Transforming growth factor B inhibits formation of osteoclast-like cells in long term human marrow cultures. *Proc Natl Acad Sci USA.* 1988;85:5683–5687.

25. Canalis E, McCarthy T, Centrella M. Growth factors and the regulation of bone remodeling. *J Clin Invest*. 1988;81:277–281.
26. Chambers TJ, McSheehy PMJ, Thomson BM, Fuller K. The effect of calcium-regulating hormones and prostaglandins on bone resorption by osteoclasts disaggregated from neonatal. *Endocrinology*. 1985;116:234–239.
27. McSheehy PMJ, Chambers TJ. Osteoblastic cells mediate osteoclastic responsiveness to PTH. *Endocrinology*. 1986;118:824–828.
28. McSheehy PMJ, Chambers TJ. Osteoblast-like cells in the presence of parathyroid hormone release soluble factor that stimulates osteoclastic bone resorption. *Endocrinology*. 1986;119:1654–1659.
29. McSheehy PMJ, Chambers TJ. 1,25-Dihydroxyvitamin D_3 stimulates rat osteoblastic cells to release a soluble factor that increases osteoclastic bone resorption. *J Clin Invest*. 1987;80:425–429.
30. Perry HM III, Skogen W, Chappel J, Kahn AJ, Wilner G, Teitelbaum SL. Partial characterization of a parathyroid hormone-stimulated resorption factor(s) from osteoblast-like cells. *Endocrinology*. 1989;125:2075–2082.
31. Frost HM. The biology of fracture healing. An overview for clinicians. Part I. *Clin Orthop Rel Res*. 1989;248:283–293.
32. Hauschka PV, Mavrakus AE, Iafrati MD, Doleman SE, Klagsbrun M. Growth factors in bone matrix. *J Biol Chem*. 1986;261:12665–12674.
33. Naufeld G, Gospoolarowicz D. Basic and acidic fibroblast growth factors interact with the same cell surface receptors. *J Biol Chem*. 1986;261:5631–5637.
34. Canalis E, Lorenza J, Burgess WH, Maciag T. Effects of endothelial cell growth factor on bone remodelling in vitro. *J Clin Invest*. 1987;79:52–58.
35. McCarthy T, Centrella M, Fox G, Arakawa T, Canalis E. Endothelial cell growth factor (ECGF) and basic fibroblast growth factor (bFGF) independently regulate bone cell replication and type I collagen transcription and translation. *J Bone Min Res* 1987;2(Suppl 1):252 (Abstr).
36. Abraham JA, Whang JL, Tumolo A, Mergia A, Friedman J, Gospoolarowicz D, Fiddes JC. Human bone fibroblast growth factor: nucleotide sequence and genomic organization. *Eur Mol Biol Organ J*. 1986;5:2523–2528.
37. Frost HM. The regional acceleratory phenomenon: a review. *Henry Ford Hosp Med J*. 1983;31:3–9.
38. Holtrop ME, Raisz LG. Comparison of the effects of 1,25-dihydroxycholecalciferol, prostaglandin E_2 and osteoclast activating factor with parathyroid hormone on the ultrastructure of osteoclasts in cultured long bones of fetal rats. *Calcif Tiss Intl*. 1979;29:201–205.
39. Komm BS, Terpening CM, Benz DJ, Graeme KA, Gallegos A, Korc M, Greene GL, O'Malley BW, Hassler MR. Estrogen binding, receptor mRNA and biologic response to osteoblast-like osteosarcoma cells. *Science*. 1988;241:81–83.
40. Eriksen EF, Colvard DS, Berg NJ, Graham ML, Mann KG, Spelsberg TC, Riggs BL. Evidence of estrogen receptors in normal human osteoblast-like cells. *Science*. 1988;241:84–86.
41. Lorenzo JA, Raisz LG, Hock JM. DNA synthesis is not necessary for osteoclastic response to parathyroid hormone in cultured fetal rat long bones. *J Clin Invest*. 1983;72:1924–1929.

42. Lorenzo JA, Sousa SL, Quinton J. Forskolin has both stimulatory and inhibitory effects on bone resorption in fetal rat long bone cultures. *J Bone Min Res.*.1986;1:313–318.
43. Löwik CWGM, Van Leeuween JPTM, Vander Meer JM, Van Zeeland JK, Scheven BA, Hermann-Erlee MP. A two-receptor model for the action of parathyroid hormone on osteoblasts: a role for intracellular free calcium and cAMP. *Cell Calcium.* 1985;6:311–326.
44. Reid IR, Civitelli R, Halstead LR, Krieger NS, Stern PH. Parathyroid hormone acutely elevates intracellular calcium in osteoblast-like cells. *Am J Physiol.* 1987;252:E45–E51.
45. Yamaguchi DT, Hahn TJ, Ildo-Klieman CR, Kleeman CR, Muallem S. Parathyroid hormone-activated calcium channels in an osteoblast-like clonal osteosarcoma cell line: cAMP dependent and cAMP-independent calcium channels. *J Biol Chem.* 1987;262:7711–7718.
46. Weir EC, Insogna KL, Horowitz MC. Osteoblast-like cells secrete granulocyte-macrophage colony-stimulating factor in response to parathyroid hormone and lipopolysaccharide. *Endocrinology.* 1989;124:899–904.
47. MacDonald BR, Mundy GR, Clark S, Wang EA, Kuehl TJ, Stanley ER, Roodman GD. Effects of human recombinant CSFGM and highly purified CSFl on the formation of multinucleated cells with osteoclast characteristics in long-term bone marrow cultures. *J Bone Min Res.* 1986;1:227–233.
48. Raisz LG, Kream BE. Regulation of bone formation. *N Engl J Med.* 1983; 309:29–35,83–89.
49. Hamilton JA, Lingelbach S, Partridge NC, Martin TJ. Regulation of plasminogen activator production by bone-resorbing hormones in normal and malignant osteoblasts. *Endocrinology.* 1985;116:2186–2191.
50. Gunnes-Hey M, Hock JM. Increased trabecular bone mass in rats treated with human synthetic parathyroid hormone. *Met Bone Dis Rel Res.* 1984; 5:177–181.
51. Howard GA, Bottemiller BL, Turnver RT, Rader JI, Baylink DJ. Parathyroid hormone stimulates bone formation and resorption in organ culture: evidence for a coupling mechanism. *Proc Natl Acad Sci USA.* 1981;78:3204–3208.
52. Tam C, Heersche J, Murray T, Parsons JA. Parathyroid hormone stimulates the bone apposition rate independently of its resorptive action: differential effects of intermittent and continuous administration. *Endocrinology.* 1982; 110:506–512.
53. Hummert J, Hock JM, Fonseca J. Resorption is not essential for the stimulation of bone growth by hPTH 1–34 in rats in vivo. *J Bone Min Res.* 1987;1:(Suppl 1):Abstr 83.
54. Pfeilschifter J, Mundy GR. Modulation of type B transforming growth factor activity in bone cultures by osteotropic hormones. *Proc Natl Acad Sci USA.* 1987:2024–2028.
55. McCarthy TL, Centrella M, Canalis E. Parathyroid hormone enhances the transcript and polypeptide levels of insulin-like growth factor I in osteoblast-enriched cultures from fetal rat bone. *Endocrinology.* 1989;124:1247–1253.
56. Raisz LG, Simmons HA. Effects of parathyroid hormone and cortisal on prostaglandin production by neonatal rat calvaria in vitro. *Endocr Res.* 1985;11:59–74.

57. Manolagas SC, Haussler MR, Deftos LJ. 1,25-Dihydroxyvitamin D_3 receptor-like macromolecule in rat osteogenic sarcoma cell lines. *J Biol Chem.* 1980; 255:4414–4417.
58. Narbaitz R, Stumpf WE, Sar M, Huang S, Deluca HF. Autoradiographic localization of target cells for $1\alpha,25$-dihydroxyvitamin D_3 in bones from fetal rats. *Calcif Tiss Intl.* 1983;35:177–182.
59. Heath JK, Atkinson SJ, Meikle MC, Reynolds JJ. Mouse osteoblasts synthesize collagenase in response to bone resorbing agents. *Biochim Biophys Acta.* 1984;802:151–154.
60. Partridge NC, Jeffrey JJ, Ehlich LS, Teitelbaum SL, Fliszar C, Welgus HG, Kahn AJ. Hormonal regulation of the production of collagenase and a collagenase inhibitor activity by rat osteogenic sarcoma cells. *Endocrinology.* 1987;120:1956–1962.
61. Jones SJ, Boyde A, Ali NN, Maconnachi E. A review of bone cell and substratum interactions. An illustration of the role of scanning electron microscopy. *Scann Elect Microsc.* 1985;7:5–24.
62. Mundy GR, Roodman D. Osteoclast ontogeny and function. In: Peck WA, ed. *Bone and mineral research; annual S.* Amsterdam: Elsevier; 1987:209–279.
63. Bar-Shavit Z, Teitelbaum SL, Reitsma P, Hall A, Pegg LE, Trial JA, Kahn AJ. Induction of monocytic differentiation and bone resorption by 1,25-dihydroxyvitamin D_3. *Proc Natl Acad Sci USA.* 1983;80:5907–5911.
64. Abe E, Miyaura C, Tanaka H, Shiina Y, Kuribayashi T, Suda S, Nishii Y, De Luca HF, Suda T. $1\alpha,25$-Dihydroxyvitamin D_3 promotes fusion of mouse alveolar macrophages both by a direct mechanism and by a spleen cell-mediated indirect mechanism. *Proc Natl Acad Sci USA.* 1983;80:5583–5587.
65. Abe E, Shiina Y, Miyaura C, Tanaka H, Hayashi T, Kanegasaki S, Saito M, Nishii Y, De Luca HF, Suda T. Activation and fusion induced by $1\alpha,25$-dihydroxyvitamin D_3 and their relation in alveolar macrophages. *Proc Natl Acad Sci USA.* 1984;81:7112–7116.
66. Fuller K, Chambers TJ. Generation of osteoclasts in cultures of rabbit bone marrow and spleen cells. *J Cell Physiol.* 1987;132:441–452.
67. Kurihara N, Civin C, Roodman GD. Identification of a pure populations of committed osteoclast precursors. *J Bone Min Res.* 1989;4:(Suppl) S199 (Abstr. #324).
68. Skjodt H, Gallagher JA, Beresford JN, Couch M, Poser JW, Russell RGG. Vitamin D metabolites regulate osteocalcin synthesis and proliferation of human bone cells in vitro. *J Endocrinol.* 1985;105:391–396.
69. Kyeyune-Nyombi E, Law W, Baylink DJ, Strong DD. Stimulation of cellular alkaline phosphatase activity and its messenger RNA level in a human osteosarcoma cell line by 1,25-dihydroxy vitamin D_3. *Arch Biochem Biophys.* 1989;275:363–370.
70. Price PA, Baukol SA. 1,25-Dihydroxyvitamin D_3 increases synthesis of the vitamin K-dependent bone protein by osteosarcoma cells. *J. Biol Chem.* 1980;255:11660–11663.
71. Majeska RJ, Rodan GA. The effect of 1,25-$(OH)_2D_3$ on alkaline phosphatase in osteoblastic osteosarcoma cells. *J Biol Chem.* 1982;256:3362–3365.

72. Beresford JN, Gallagher JA, Russell RGG. 1,25-Dihydroxyvitamin D₃ and human bone derived cells in vitro: effects on alkaline phosphatase, type I collagen and proliferation. *Endocrinology*. 1986;119:1776–1785.

73. Kream BE, Rowe D, Smith MD, Maher V, Majeska R. Hormonal regulation of collagen synthesis in a clonal rat osteosarcoma cell line. *Endocrinology*. 1986;119:1922–1928.

74. Genovese C, Rowe D, Kream BE. Construction of DNA sequences complementary to rat α, and α₂ collagen mRNA and their use in studying the regulation of type I collagen synthesis by 1,25-dihydroxyvitamin D. *Biochemistry*. 1984;23:6210–6216.

75. Rodan GA, Rodan SB. Expression of the osteoblastic phenotype. In: Peck WA, ed. *Bone and mineral research; annual 2*; Amsterdam: Elsevier; 1984:244–285.

76. Raisz LG, Kream BE. Regulation of bone formation. *N Engl J Med*. 1983;309:29–35.

77. Baron R, Vignery A. Behavior of osteoclasts during a rapid change in the number induced by high doses of parathyroid hormone or calcitonin in intact rats. *Met Bone Dis Rel Res*. 1981;2:339–346.

78. Holtrop WE, Raisz LG, Simmons HA. The effects of parathyroid hormone, colchicine, and calcitonin on the ultrastructure and the activities of osteoclasts in organ culture. *J Cell Biol*. 1974;60:346–355.

79. Kallio DM, Garant PR, Minkin C. Ultrastructural effects of calcitonin on osteoclasts in tissue culture. *J Ultrastruct Res*. 1972;39:205–216.

80. Nicholson GC, Moseley PM, Sexton PM, Mendelsohn FAO, Martin TJ. Abundant calcitonin receptors in isolated rat osteoclasts. *J Clin Invest*. 1986;78:355–360.

81. Chambers TJ. Osteoblasts release osteoclasts from calcitonin-induced quiescence. *J Cell Sci*. 1982;57:247–260.

82. Weiss RE, Singer FR, Gorn AH, Hofer DP, Nimni ME. Calcitonin stimulates bone formation when administered prior to initiation of osteogenosis. *J Clin Invest*. 1981;68:815–818.

83. Farley JR, Tarbaux HM, Hall SL, Linkhart TA, Baylink DJ. The anti-bone-resorptive agent calcitonin also acts in vitro to directly increase bone formation and bone cell proliferation. *Endocrinology*. 1988;123:159–167.

84. Krane SM, Harris ED, Singer FR, Potts JT Jr. Auto effects of calcitonin on bone formation in man. *Metabolism*. 1973;22:51–58.

85. Gruber HE, Ivey J, Baylink DJ, Matthews M, Nelp WB, Sisom K, Chesnut CH III. Long-term calcitonin therapy in postmenopausal osteoporosis. *Metab Clin Exp*. 1984;33:295–303.

86. Mazzuoli GF, Passeri M, Gennari C, Minisola S, Antonelli R, Valtorta C, Palummeri E, Cervellin GF, Gonnelli S, Francini G. Osteoporosis and calcitonin therapy: a new horizon. In: Pecile A, ed. *Calcitonin 1984*. Amsterdam: Elsevier Excerpta Medica; 1985:129–136.

87. Peck WA, Burks JK, Wilkins J, Rodan SB, Rodan GA. Evidence for preferential effects of parathyroid hormone, calcitonin and adenosine on bone and periosteum. *Endocrinology*. 1977;100:1357–1364.

88. Wong GL, Cohen DV. Target cells in bone for parathormone and calcitonin are different: enrichment for each cell type by sequential digestion of mouse

calvaria and selective adhesion to polymeric surfaces. *Proc Natl Acad Sci USA*. 1975;72:3167–3171.

89. Majeska RJ, Rodan SB, Rodan GA. Maintenance of parathyroid hormone response in clonal rat osteosarcoma lines. *Exp Cell Res*. 1978;111:465–468.

90. Partridge NC, Alcorn D, Michelangeli VP, Kemp BE, Ryan GB, Martin TJ. Functional properties of hormonally responsive cultured normal and malignant rat osteoblastic cells. *Endocrinology*. 1981;108:213–219.

91. Klein DC, Raisz LG. Prostaglandins: stimulation of bone resorption in tissue culture. *Endocrinology*. 1970;86:1436–1440.

92. Dietrich JW, Goodson JM, Raisz LG. Stimulation of bone resorption by various prostaglandins in organ culture. *Prostaglandins*. 1975;10:231–240.

93. Raisz LG, Dietrich JW, Simmons HA, Seyberth HW, Hubbard W, Oates JA. Effect of prostaglandin endoperoxides and metabolites on bone resorption in vitro. *Nature*. 1977;267:532–534.

94. Tashjian AH Jr., Tice JE, Sides K. Biological activities of prostaglandin analogues and metabolites on bone in organ culture. *Nature*. 1977;266:645–646.

95. Sakamoto SM, Sakamoto P, Goldhaber MJ, Glimcher MJ. Collagenase activity and morphological and chemical bone resorption induced by prostaglandin E_2 in tissue culture. *Proc Soc Exp Biol Med*. 1979;161:99.

96. Raisz LG, Martin TJ. Prostaglandins in bone and mineral metabolism. In: Peck WA, ed. *Bone and mineral research: annual 2*. Amsterdam: Elsevier; 1984:286–310.

97. Chase LR, Aurbach GD. The effect of parathyroid hormone on the concentration of adenosine 3',5'-monophosphate in skeletal tissue in vitro. *J Biol Chem*. 1970;245:1520–1526.

98. Blumenkrantz M, Sondergaard J. Effect of prostaglandins E_1 and $F_{1\alpha}$ on biosynthesis of collagen. *Nature New Biol*. 1972;239:246.

99. Ueda K, Saito A, Nakano H, Aoshima M, Yokota M, Muraoka R, Iwaya T. Cortical hyperostosis following long-term administration of prostaglandin E_1 in infants with cyanotic congenital heart disease. *J Pediatr*. 1980;97:834–836.

100. Ringel RE, Brenner JI, Haney PJ, Burns, JE, Moulton AL, Berman MA. Prostaglandin induced periostitis: a complication of long-term PGE_1, infusion in an infant with congenital heart disease. *Radiology*. 1982;142:657–658.

101. Drvaric DM, Parks WJ, Bradley J, Dooley KJ, Plauth Wm H Jr., Schmitt EWM. Prostaglandin-induced hyperostosis: a case report. *Clin Orthop Rel Res*. 1989;246:300–304.

102. Ro J, Sudmann E, Martin PF. Effect of indomethacin on fracture healing in rats. *Acta Orthop Scand*. 1976;47:588–599.

103. Dekel S, Lenthall G, Francis MJO. Release of prostaglandins from bone and muscle after tibial fracture. An experimental study in rabbits. *J Bone Joint Surg*. 1981;63:185–189.

104. Shuprik MA, Tashjian AH Jr. Epidermal growth factor and phorbol ester actions on human osteosarcoma cells: characterization of responsive and non-responsive lines. *J Biol Chem*. 1982;257:12161–12164.

9
Endocrine, Iatrogenic, and Nutritional Causes of Osteopenia

ADEL B. KORKOR

Osteopenia is a decrease in bone mineral content. This can be either a result of osteoporosis where the mineral-collagen ratio of bone is normal or of osteomalacia where demineralization of collagen prevails. The term osteoporosis is also used to describe osteopenia in patients who have sustained fractures. In this discussion I will review the different causes of osteoporosis with special emphasis on clinical and laboratory features, diagnosis, and management according to the following outline:

Endocrine Osteopenia:
1. estrogen deficiency
2. hyperparathyroidism
3. thyrotoxicosis
4. Cushing's disease
5. diabetes mellitus
6. hypogonadism and hyperprolactinemia

Iatrogenic Osteopenia:
1. thyroid supplementation
2. exogenous glucocorticosteroids
3. other drugs (heparin, Dilantin, chemotherapy, lithium)

Osteopenia due to other causes:
1. bone loss in athletes
2. anorexia nervosa and other eating disorders
3. alcohol and cigarette use
4. other dietary factors

Endocrine Osteopenia

Estrogen Deficiency

After reaching its peak during the mid- or late 20s, the bone mineral content starts to decline. This decline is gradual and similar in men and women. However, once women reach menopause, the decline accelerates. In Caucasian women, loss may amount for as much as 2% per year.[1,2] Indeed, a significant decline in bone mass may bacome apparent even before the complete cessation of menstruation.[3] Estrogen replacement can prevent this decline in bone mass; however, if discontinued the acceleration will resume.[4] Although other agents such as calcium, vitamin D, or exercise may have some beneficial effect, estrogen replacement remains the most effective means of prevention.

Bone biopsy studies in postmenopausal women have provided evidence of accelerated bone resorption,[5] which appears to be the most likely mechanism of bone loss in the postmenopausal years. This phenomenon may be related directly to lack of estrogen or may be the indirect result of decreased serum levels of 1,25-dihydroxyvitamin D_3 and hence of decreased intestinal calcium absorption, hypocalcemia, and secondary hyperparathyroidism.[6-8] In some patients with postmenopausal osteoporosis, the bone cells may be normally active or inactive[5] and the inactivity may represent "a terminal state" of the disease itself or may reflect the influence of other external or internal factors such as age, diet, physical factors, and level of activity. Estrogen deficiency leading to increased bone turnover and decrease in bone mass appears to be more prominent among Causasian and Asian women.[9] It may also be an important contributing factor to osteopenia present in women smokers.[10,11]

Loss of bone mass in the post- and perimenopausal periods is often asymptomatic. Once bone mass reaches the fracture threshold, a fracture (vertebral, wrist, or hip) may be the first manifestation of osteoporosis. Such fracture may actually occur at bone mass level above the fracture threshold, suggesting that there may be other risk factors, such as the magnitude of trauma or the quality of bone. In patients with low peak bone mass or in those in whom additional factors enhance the rate of bone loss (immobilization, osteomalacia, alcohol use, multiple myeloma, etc.), bone fractures may occur early in the course of the postmenopausal period.

Results of laboratory evaluation in postmenopausal osteoporosis often indicate subtle changes. Serum calcium, phosphorus, and alkaline phosphatase are often normal, although occasionally serum alkaline phosphatase may be at the upper limits. Urine calcium-to-creatinine ratio may be normal or slightly elevated. Urinary hydroxyproline may also be increased. Serum osteocalcin level may be increased in patients with high turnover osteoporosis. However, this level may be normal or low in patients with inactive osteoporosis.[12] The presence of other abnormalities such as a) hypocalcemia, hypophosphatemia or elevated serum alkaline phosphatase level, b) anemia, increase in serum lactic dehydrogenase (LDH), uric acid and globulins, suggests the possibility of additional underlying processes such as osteomalacia or multiple myeloma, respectively.

TABLE 9.1. Evaluation of patients with osteopenia or osteoporosis.

Medical history
— General, present, and past medical history: age, sex, race, ethnic background, fracture
  history, height loss, constitutional symptoms, malabsorption, diabetes mellitus,
  rheumatoid arthritis, renal failure, gastric or intestinal surgery
— Dietary history: anorexia, intake of protein, caffeine, calcium, and sodium
— Menstrual history and estrogen use
— Personal habits: alcohol and cigarette use
— Medication: anticonvulsant drugs, glucocorticosteroids, heparin, lithium, sedatives
  (increase tendency to fall)
Physical examination
— General appearance: weight, height, arm span, kyphosis, gait, balance and coordination,
  breast mass, lymphadenopathy, hepatosplenomegaly, testicular atrophy, gynecomastia
Laboratory studies
— Serum calcium, phosphorus, multichannel chemical analysis, 25-hydroxyvitamin D_3
— Studies to evalute secondary causes: thyroid function tests, PTH, serum and urine
  protein electropheresis, FSH, cortisol, prolactin, 1,25-dihydroxyvitamin D_3
Radiographic studies
— Radiograph of the involvement region (by fracture or pain)
Bone mineral content
— Essential in patients with four or more risk factors and to monitor the result of therapy
Bone biopsy and bone histomorphometric studies
— Useful in some patients to rule out the possibility of osteomalacia and to determine the
  rate of bone turnover

Axial and appendicular measurements of bone mass in the initial and subsequent evaluation of estrogen deficient women are useful for assessing the rate of decline in bone mass and the results of therapy.[13,14] Precision and accuracy may the achieved through the dual energy x-ray absorptiometry (DEXA) method, although not as well using single or dual photon absorptionmetry (SPA, DPA) or computed tomography (CT) methods.[15] Bone biopsy is seldom required for the diagnosis of osteoporosis. However, it is often useful in determining the level of bone cell activities. Such knowledge can be useful in determining the most desirable mode of therapy. For example, in patients with increased bone resorption, antiresorptive agents would be useful whereas in patients with decreased bone cell activities, agents such as sodium fluoride and possibly 1,25-dihydroxyvitamin D_3 may be more effective. Table 9.1 summarizes our approach to the evaluation of patients with osteopenia or osteoporosis. The table provides a general scheme but variation in this approach may be necessary in individual patients.

The management of estrogen deficiency syndrome (Table 9.2) is essentially simple.[16] Hormone replacement therapy (estrogen 0.3–1.25 mg/day) can prevent the decrease in bone mass associated with estrogen withdrawal. To minimize the likelihood of estrogen-induced carcinoma of the endometrium, progesterone should be added to the regimen. Several combinations are possible: continuous estrogen and progesterone admin-

TABLE 9.2. Treatment of osteoporosis.

Preservation of bone mass
    Estrogen (oral or transdermal)
    Calcium
    Vitamin D
    Exercise
    Stop smoking/alcohol and reduce caffeine intake
Restoration of bone mass
    Estrogen (oral or transdermal)
    Calcitonin (subcutaneous or nasal)[a]
    Etidronate (cyclic)[b]
    Sodium fluoride[c]

[a] Useful in situations of severe pain associated with new or recent fractures. Effective especially in active (high turnover) osteoporosis.
[b] Not approved by the FDA. Preliminary studies on combined use of calcitonin and etidronate appear promising.
[c] Use of moderate doses in cyclic manner is useful especially in patients with inactive osteoporosis.

istration (a method gaining popularity because it does not cause uterine bleeding), cyclic estrogen and progesterone (estrogen from day 1 through 25 of each month and progesterone from day 12 through 25), and cyclic estrogen for a similar number of days and Provera during the first 14 days of each 31-day cycle.[17] In patients who have undergone hysterectomy, continuous estrogen replacement (without progesterone) is given. Although the risk of breast cancer is minimal with estrogen replacement therapy,[18] mammograms before starting this therapy and at yearly intervals thereafter are recommended. Manual examinations must be encouraged.

An added advantage of estrogen replacement in postmenopausal women is its beneficial effect on reducing the risk of severe coronary artery disease.[19,20] Although estrogen replacement therapy has favorable effects on bone mass at any time in the postmenopausal period, it appears to be more effective during the early postmenopausal years.[21] Since most postmenopausal women are in negative calcium balance,[22] maintenance of adequate dietary calcium intake is essential. If dietary intake does not provide the amount necessary to reverse this negative balance, calcium supplementation (40 mmol of calcium or approximately 1600 mg/day) in the form of calcium salts will achieve a slightly positive calcium balance in most osteoporotic patients. Table 9.3 summarizes the available calcium salts and the percent elemental calcium in each. Calcium carbonate remains the richest in elemental calcium and the cheapest. Since constipation is a frequent occurrence with the use of calcium carbonate, other salts may be considered.

TABLE 9.3. Elemental calcium content of calcium salts.[89]

Calcium salt	% Calcium	mg Ca/g
Calcium glubionate	6.5	66
Calcium gluconate	9	90
Calcium lactate	13	130
Calcium citrate	21	200
Dibasic calcium Phosphate dihydrate	23	230
Tricalcium phosphate	39	390
Calcium carbonate	40	400

Patients who do not maintain adequate dietary intake of calcium by drinking vitamin D–fortified milk should use vitamin D supplements. Similarly, patients who live at latitudes where ultraviolet content of sun light (even on sunny winter days) is limited should receive oral vitamin D supplements in the form of ergocalciferol drops (300–400 IU/day) or tablets (50,000 units/month).

Weight-bearing exercise, although somewhat beneficial in preventing the loss of bone mass,[23–25] has never been proven to be an adequate form of therapy for the prevention of postmenopausal osteoporosis. Nevertheless, such exercise should also be encouraged not only to decrease the rate of bone loss but also for its desirable cardiovascular effects.

Antiresorptive substances, such as calcitonin or bisphosphonates, also increase bone mass and decrease the incidence of new vertebral fractures in osteoporotic women. Calcitonin administered subcutaneously (50–100 units 3–7 times per week) or intranasally (available only in Europe) can result in the preservation or even in an increase of bone mass, especially in women with high turnover osteoporosis.[12] Generally, this therapy is well tolerated. However, it is expensive and its long-term benefits (more than 2–3 years) have yet to be established. Nausea can be avoided by administering the drug at bedtime. A major advantage of calcitonin, especially in patients with recent fractures, is its analgesic effect, which allows the patient to move with less discomfort and to become active sooner. Cyclic etidronate (400 mg/day for 2 weeks followed by 10–13 weeks of oral calcium supplementation) also has been shown to increase bone mass and decrease the number of vertebral body fractures among postmenopausal osteoporotic women.[26,27] The cyclic form of therapy avoids the potential inhibitory effects of biphosphonate on bone formation. Although this therapy has not yet been approved by the FDA for the treatment of postmenopausal osteoporosis, it appears to be well tolerated and quite effective at least for a period of 2 to 3 years. Whether the combined use of low doses of calcitonin and etidronate offers an additional advantage remains to be determined. The results of preliminary studies in our center on severely osteoporotic patients with multiple vertebral body fractures and bone pain have given us reason for optimism.

Sodium fluoride can increase trabecular bone mass by promoting the formation of woven bone. The mechanical strength of this bone is inferior to that of normal bone. Thus, although the use of fluoride in patients with osteoporosis does result in an increase in vertebral body bone mass, it does not significantly reduce the occurrence of vertebral or hip fractures. Indeed, evidence of an increase in appendicular bone fractures after fluoride administration has been reported recently.[28] Whether these effects on fractures are related to the dose of sodium fluoride is not clear, since the administration of low doses of sodium fluoride (40–50 mg given in the form of enteric-coated tablets) can increase bone mass and decrease fractures as well.[29] It is possible that the cyclic administration (3 months on and 3 months off) of sodium fluoride at these low doses may prove to be safe and beneficial in patients with inactive osteoporosis. Such therapy has been tried effectively in a number of patients seen in our center.

The administration of anabolic steroids, oral phosphate, 1,25-dihydroxyvitamin D_3, and/or intravenous parathyroid hormone in the treatment of patients with osteoporosis, although theoretically sound (particularly in low turnover osteoporosis), has not undergone adequate clinical trials.

Hyperparathyroidism

Primary hyperparathyroidism increases bone turnover and thus could lead to a decrease in bone mass, at least theoretically. The frequency of hyperparathyroidism increases with age and is 3 times more common in women than in men. Patients affected by this disease are usually 50 years old or older. Osteopenia (cancellous and cortical) is often present,[30,31] but rarely is the skeletal disorder the sole manifestation of the disease. Asymptomatic hypercalcemia is the most common finding. Indeed, the combination of premature osteopenia, renal involvement (such as renal stones or failure), and hypercalcemia, especially in women with low normal or low serum phosphorus, is a strong indication of hyper-parathyroidism. Once the diagnosis is established by the measurement of PTH level and parathyroid scan, surgery is the therapy of choice to decrease the risk of worsening bone disease and/or renal stones. Of 132 osteoporotic women evaluated in our bone and mineral metabolism center, 4 had primary hyperparathyroidism and 2 had only minimal elevation in serum calcium (10.5 and 10.4 vs. a maximum normal of 10.3). Surgery resulted in either preservation or improvement in bone mass, as measured 1 year postoperatively.

Thyrotoxicosis

Excess thyroid hormone is associated with an increase in bone turnover.[32] Although osteopenia can be a manifestation of this disorder, seldom is it a major one. In patients with premature or severe osteoporosis and other symptoms suggestive of hyperthyroidism, such as weight loss, palpitations, cardiac arrhythmias, and so forth, thyroid function tests are essential. Treatment of hyperthyroidism with antithyroid drugs or with radioactive iodine offers substantial relief and could restore bone mass.

Cushing's Disease

Osteopenia is an important manifestation of excessive glucocorticosteroid activity. In fact, in some patients, osteopenia can be an early manifestation of Cushing's syndrome whether cuased by pituitary tumor (Cushing's disease), ectopic adrenocorticotropic hormone (ACTH) production, or adrenocortical tumors. Cushing's disease is by far the most common cause and occurs 4 times more frequently in women aged 20 to 40 years. Fractures involving trabecular bones are present. In fact, premature osteoporosis associated with hypertension with or without additional manifestations of the disease (such as weight gain, obesity, moon face, muscle weakness, depression, etc.) must always raise the possibility of Cushing's disease. Diagnosis is based on the measurement of serum cortisol levels, loss of the normal diurnal pattern of cortisol secretion, and lack of dexamethasone suppressibility. A prolonged suppression test may be necessary to rule out ectopic ACTH syndrome. Radiographic studies including CT scanning of the adrenals, magnetic resonance (MRI) scanning of the pituitary, and venous catheterization to confirm the pituitary origin of the circulating ACTH are essential to confirm the diagnosis.

Treatment of Cushing's syndrome and in particular Cushing's disease by transsphenoidal surgery will most likely correct the underlying osteoporosis, increase bone mass, and prevent the recurrence of fractures.

Diabetes Mellitus

Cortical and trabecular osteopenia, involving axial and appendicular bones, is present in diabetic patients (especially young patients with insulin-dependent diabetes).[33-35] Vertebral body and appendicular fractures may occur within 10 to 15 years of onset of the disease. The mechanism of osteoporosis associated with diabetes mellitus is uncertain. One possibility is the negative calcium balance triggered by a decrease in intestinal calcium absorption as a result of a decreased serum 1,25-dihydroxyvitamin D_3 level.[36] Such a decrease, noted in diabetic animals, has raised the possibility that the production and degradation of 1,25-dihydroxyvitamin D_3 and/or the effects of the vitamin may be altered as a result of insulin deficiency. Histologic studies in diabetic osteoporotic individuals have demonstrated mainly inactive osteoporosis.[37] Therapy with 1,25-dihydroxyvitamin D_3 has been suggested. It is generally recommended that patients with type I diabetes maintain adequate dietary calcium and vitamin D intake, be encouraged to use estrogen replacement therapy early during menopause, and maintain an active lifestyle along with adequate metabolic control. The evaluation of the serum 25-hydroxyvitamin D_3 level is essential to assure adequate vitamin D stores. Therapeutic doses of vitamin D are necessary in patients whose levels are low.

Hypogonadism and Hyperprolactinemia

This disorder, especially in men, can be accompanied by a decrease in bone mass.[38] Osteoporosis in men with galactorrhea, visual field abnormalities, headaches, impotence, and small testicles deserves careful evaluation of possible hyperprolactinemic hypogonadism by measurements of serum testosterone and prolactin levels as well as by the evaluation of anterior pituitary function. Radiographic studies of the pituitary are essential. Therapy may be either surgical or medical (bromocriptine).

Iatrogenic Causes of Osteopenia

Thyroid Supplementation

Thyroid supplementation is often associated with evidence of decreased bone mass.[39-41] Over the past year, nearly 40% of the patients referred to our bone and mineral metabolism center were receiving thyroid supplementation. Although frequently the serum T_4 levels were elevated, normal values did not always preclude the possibility of iatrogenic hyperthyroidism. Measurement of thyrotropin-stimulating hormone (TSH) by a highly sensitive radioimmunoassay is useful in the evaluation of these patients, as low or undetectable levels of TSH despite a normal T_4 level are an indication of hyperthyroidism. Thyroid supplementation, often used in middle-aged and older women in whom additional risk factors of osteoporosis are often present, may lead to increased bone resorption and accelerated decline of bone mass. Women receiving thyroid supplementation must have a semiannual evaluation of thyroid function, including measurements of TSH, and their medication must be adjusted to maintain the values in the median range of normal. In addition, it is essential that individuals receiving thyroid supplementation maintain adequate dietary intake of calcium and vitamin D, receive estrogen replacement therapy if indicated, and undergo annual measurements of bone mass.

Exogenous Glucocorticosteroids

The administration of glucocorticosteroids is associated with a decrease in bone mass[42-44] to a degree that depends on the dose and duration of therapy.[45,46] Patients receiving large single doses of glucocorticosteroids (such as bolus Solu-Medrol for the treatment of renal transplant rejection) may undergo a rapid decline in bone mass within a few weeks.[46] Smaller doses and more prolonged use of glucocorticosteroids are associated with an increased bone turnover and significant decrease in bone mass.[47] This decrease appears much more pronounced in the first year of glucocorticosteroid administration[45,48] and involves the vertebral body and the hip to a similar extent.[45]

The pathogenesis of glucocorticosteroid-induced osteopenia is complex. The glucocorticosteroids have a direct inhibitory effect on bone cell activities. Low bone turnover osteoporosis may be identified by bone biospy in patients receiving large doses of steroids for an extended period of time.[47] The glucocorticosteroids may have a negative effect on vitamin D metabolism leading to decreased serum 25-hydroxy- and/or 1,25-dihydroxyvitamin D_3 levels.[49-53] This will directly affect intestinal calcium absorption and bone formation. The administration of glucocorticosteroids enhances the urinary excretion of calcium and phosphorus, probably through a direct effect on the renal tubular cells. And, the steroids may have an effect on the parathyroids. Although the nature of this effect is not clear, some studies have demonstrated an increase in serum PTH levels[54,55] and action[56] that could lead to an increase in bone resorption. It is always important to keep in mind the underlying disease being treated with glucocorticosteroids (such as rheumatoid arthritis, Crohn's disease, etc.) may in itself have a significant effect on bone metabolism.

Patients receiving glucocorticosteroids may exhibit symptomatic or asymptomatic fractures. The fractures often involve trabecular bone (vertebral body, hip, wrist)—seldom the appendicular bone.

Laboratory studies are usually not helpful in the evaluation of glucocorticosteroid-induced osteopenia. Serum calcium is often normal, phosphorus may be slightly reduced, and alkaline phosphatase may be increased, normal, or decreased. Urinary calcium and phosphorus excretion may be increased and serum 25-hydroxyvitamin D_3 and 1,25-dihydroxyvitamin D_3 levels may be reduced or normal. Serum PTH may be increased, normal, or decreased. Serum osteocalcin level is low.[57] The diagnosis of glucocorticosteroid-induced osteopenia is made on the basis of clinical history, evaluation of radiographic studies, and bone mineral content. Bone biopsy may be useful in assessing the activity of bone cells. This information is often helpful in determining the form of therapy. The best treatment is prevention and includes using the smallest possible effective doses of glucocorticosteroids, the rapid tapering off of the steroids as soon as possible, the simultaneous administration of oral calcium and vitamin D (or 25-hydroxyvitamin D_3), the maintenance of weight-bearing activities, and an active lifestyle while receiving steroids. Once glucocorticosteroid-induced osteopenia has become established, the above-mentioned approach remains useful, although additional measures may be required, especially for the control of pain associated with compression fractures of the vertebral body. Recently, antiresorptive therapy (calcitonin or biphosphonates) has been shown to be effective in controlling both skeletal complaints as well as the rate of bone loss in glucocorticosteroid-induced osteopenia.[58-62] Calcitonin given in doses of 50 to 100 units 3 to 4 times a week is recommended. In a patient with inactive osteopenia (proven by bone biospy), the use of small doses of sodium fluoride (40–50 mg/day of the oral enteric-coated preparation) may be useful.[63-64] Such therapy was used with favorable results in the treatment of several of our patients.

Other Drugs

The prolonged administration of several drugs, such as heparin, Dilantin, chemotherapeutic agents, and lithium, can be associated with a decrease in bone mass. The exact mechanism of each drug effect is uncertain. In patients receiving chronic Dilantin therapy, osteopenia as well as osteomalacia may be present. Thus, the history of patients with premature osteopenia or accelerated decline in bone mass must include questions related to the use of these drugs and management must include consideration of other risk factors and the administration of calcium and vitamin D supplements (especially in the case of nutritional deficiency). Assessment of bone mass during therapy with these drugs is also useful to determine whether additional diagnostic studies or therapy such as estrogen or calcitonin are necessary.

Other Causes of Osteopenia

Bone Loss in Athletes

Low estrogen levels and amenorrhea are observed in up to 50% of competitive runners and 44% of ballet dancers.[65-69] Osteopenia in these amenorrheic women has been confirmed.[70-72] This premature "aging" of the skeleton can certainly have a significant impact on the development of premature osteoporosis and fractures. The exact mechanism of bone loss in amenorrheic athletes is not clear. Evidence of altered hypothalamic gonadotropin release was confirmed by the studies of Fisher et al.,[73] who found low serum luteinizing hormone (LH), follicle-stimulating hormone (FSH) and estrogen levels in amenorrheic runners. Studies by Nelson et al.[74] failed to demonstrate any significant difference in dietary calcium intake. However, amenorrheic runners had significantly lower daily energy and protein intake. Cook et al.[75] confirmed the presence of low bone mass in amenorrheic runners and demonstrated lower PTH levels in these patients. Since a reduction in the PTH level was found to occur in the low estrogen peroid of the menstrual cycle,[76] the lower serum PTH levels of amenorrheic women may be related to lower overall estrogen levels. The treatment of osteopenia in amenorrheic runners can be effective. This includes estrogen replacement therapy, modification of the level and the intensity of training, and maintenance of adequate dietary intake of calories, protein, calcium, and vitamin D.

Anorexia Nervosa and Other Eating Disorders

Amenorrhea and osteoporosis are often encountered in patients with eating disorders, especially anorexia nervosa.[77-79] Evidence of decreased gonadotropin activity (LH and FSH levels) are present in these patients. In addition to these hormonal abnormalities, a poor dietary intake of calcium, vitamin D, magnesium, protein, and total calories plays a role in the development of osteopenia and occasionally of osteomalacia. The therapy of bone loss in this disorder is difficult and requires attention to all the above-mentioned factors. Adjunctive psychologic therapy is essential in all cases.

Alcohol and Cigarette Use

Excessive use of alcohol is a significant risk factor for the development of osteopenia, especially in men,[80,81] although the daily use of alcohol is not uncommon among middle-aged or older women, especially those who live alone and have a limited social life. Of 32 men seen in our bone and mineral metabolism center over the past 3 years, 21 were heavy alcohol users (more than 20 drinks per week). The mechanism of alcohol-induced osteopenia is uncertain and may involve a reduced dietary calcium and vitamin D intake, an alteration in vitamin D metabolism, such as a decrease in 25-hydroxylation, leading to a low serum 25-hydroxyvitamin D_3 level,[82-84] a direct effect of alcohol on bone cell metabolism,[81-85] or an increased urinary calcium and phosphorus excretion.[86] Thus, questioning patients with osteopenia about alcohol consumption is important and a discontinuation of alcohol is essential to prevent a further decline in bone mass.

The use of cigarettes is also associated with a decrease in bone mass in women.[87] The exact mechanism is uncertain and is probably related to a decrease in basal serum estrogen level and in the estrogen response to estrogen replacement therapy.[88] In turn, this may be due to an effect of nicotine on the hepatic metabolism of estrogen.[89] Thus, women of every age must be discouraged from smoking and osteoporosis must be added to the list of the numerous other untoward effects of this distasteful habit.

Other Dietary Factors

Excessive dietary intake of protein and caffeine can result in negative calcium balance,[90] predominantly by an increase in urinary calcium excretion.[90,91] The effect of caffeine on urinary calcium excretion is acute and can be demonstrated after the use of relatively small amounts of the drug (150 mg equivalent to 3–4 oz of coffee).[91] In situations of extreme sodium intake,[92] urinary calcium excretion is also increased. Although this situation (dietary intake of 800–1500 mg sodium/day) is seldom encountered, reducing dietary sodium intake by osteoporotic patients (especially in high users) may have some added benefits. High dietary phosphorus can lead to an increase in serum PTH level.[93] Such an increase may be advantageous in adult patients with inactive osteoporosis. However, its long-term effect on the skeleton in young individuals consuming low dietary calcium is probably not favorable.

References

1. Khairi MRA, Johnston CC. What we know—and don't know—about bone loss in the elderly. *Geriatrics*. 1978;Nov:67–70.
2. Elders PJM, Netelenbos JC, Lips P, van Ginkel FC, van der Stelt PF. Accelerated vertebral bone loss in relation to the menopause: a cross-sectional study on lumbar bone density in 286 women of 46 to 55 years of age. *Bone Min*. 1988;5:11–20.
3. Johnston Jr CC, Hui SL, Witt RM, Appledorn R, Baker RS, Longcope C. Early menopausal changes in bone mass and sex steroids. *J Clin Endocrinol Metab*. 1985;61:905–911.
4. Lindsay R, MacLean A, Kraszewski A, Hart DM, Clark AC, Garwood J. Bone response termination of oestrogen treatment. *Lancet*. 1978;June 24:1325–1327.
5. Malluche HH, Faugere MC. Osteoporosis. In: Malluche HH, Fengere MC, eds. *Atlas of mineralized bone histology*. New York: S. Karger; 1986:50–59.
6. Gallagher JC, Riggs BL, Jerpbak M, Arnald CD. The effect of age on serum immunoreactive parathyroid hormone in normal and osteoporotic patients. *J Lab Clin Med*. 1980;94:373–385.
7. Arnaud SB. Intestinal calcium absorption and serum vitamin D metabolites in normal subjects and osteoporotic patients. *J Clin Invest*. 1979;64:729–736.
8. Eastell R, Yergey AL, Vieira NE, Cedel SL, Kumar R, Riggs BL. Interrelationship among vitamin D metabolism, true calcium absorption, parathyroid function, and age in women: calcium absorption, parathyroid function, and age in women: evidence of an age-related intestinal resistance to 1,25-dihydroxyvitamin D action. *J Bone Min Res*. 1991;6:125–132.
9. Nelson DA, Kleerekoper M, Parfitt AM. Bone mass, skin color and body size among black and white women. *Bone Min*. 1988;4:257–264.
10. Wilson PWF, Garrison RJ, Castelli WP. Postmenopausal estrogen use, cigarette smoking, and cardiovascular morbidity in women over 50. *N Engl J Med*. 1985;313:1038–1043.

11. Baron JA. Smoking and estrogen-related disease. *Am J Epidemiol.* 1984;119:9–22.
12. Civitelli R, Gonnelli S, Zacchei F, Bigazzi S, Vattimo A, Avioli LV, Gennari C. Bone turnover in postmenopausal osteoporosis. Effect of calcitonin treatment. *J Clin Invest.* 1988;82:1268–1274.
13. Ross PD, Davis JW, Wasnich RD, Vogel JM. The clinical application of serial bone mass measurements. *Bone Min.* 1991;12:189–200.
14. Johnston Jr CC, Melton LJ III, Lindsay R, Eddy DM. Clinical indications for bone mass measurements. A report from the Scientific Advisory Board of the National Osteoporosis Foundation. *J Bone Min Res.* 1989;4(Suppl 2):1–28.
15. Cohn SH. Noninvasive measurements of bone mass. In: Avioli LV, Krane SM, eds. *Metabolic bone disease and clinically related disorders.* Philadelphia: Saunders; 1990.
16. Ettinger B. Prevention of osteoporosis: treatment of estradiol deficiency. *Obstet Gynecol.* 1988;72:12S.
17. Ettinger B. Optimal use of postmenopausal hormone replacement. *Obstet Gynecol.* 1988;72:31S.
18. Wingo PA, et al. The risk of breast cancer in postmenopausal women who have used estrogen replacement therapy. *JAMA.* 1987;257:209.
19. Knopp RH. The effects of postmenopausal estrogen therapy on the incidence of arteriosclerotic vascular disease. *Obstet Gynecol.* 1988;72:23S.
20. Stampfer MJ, Willett WC, Colditz GA, Rosner B, Speizer FE, Hennekens CH. A prospective study of postmenopausal estrogen therapy and coronary heart disease. *N Engl J Med.* 1985;313:1044–1049.
21. Avioli LV, Lindsay R. The female osteoporotic syndrome(s). In: Avioli LV, Krane SM, eds. *Metabolic bone disease.* 2nd ed. Philadelphia; Saunders: 1990:397–451.
22. Hasling C, Charles P, Jensen FT, Mosekilde L. Calcium metabolism in postmenopausal osteoporosis: the influence of dietary calcium and net absorbed calcium. *J Bone Min Res.* 1990;5:939–946.
23. Krolner B, Toft B, Nielsen SP, Tondevold E. Physical exercise as prophylaxis against involutional vertebral bone loss: a controlled trial. *Clin Sci.* 1983;64:541–546.
24. Pocock NA, Eisman JA, Yeates MG, Sambrook PN, Ebert S. Physical fitness is a major determinant of femoral neck and lumbar spine mineral density. *J Clin Invest.* 1986;78:618–621.
25. Dalsky GP, et al. Weight-bearing exercise training and lumbar bone mineral content in postmenopausal women. *Ann Intern Med.* 1988;108:824.
26. Storm T, et al. Effect of intermittent cyclical etidronate therapy on bone mass and fracture rate in women with postmenopausal osteoporosis. *N Engl J Med.* 1990;322:1265.
27. Watts NB, et al. Intermittent cyclical etidronate treatment of postmenopausal osteoporosis. *N Engl J Med.* 1990;323:73.
28. Riggs BL, Hodgson SF, O'Fallon WM, Chao EYS, Wahner HW, Muhs JM, Cedel SL, Melton JL III. Effect of fluoride treatment on the fracture rate in postmenopausal women with osteoporosis. *N Engl J Med.* 1990;322:802–809.
29. Mamelle N, Meunier PJ, Dusan R, Guillaume M, Martin JL, Gaucher A, Prost A, Zeigler G, Netter P. Risk-benefit ratio of sodium fluoride treatment in primary vertebral osteoporosis. *Lancet.* 1988;II:361–364.

30. Kochersberger G, Buckley NJ, Leight GS, Martinez S, Studenski S, Vogler J, Lyles KW. What is the clinical significance of bone loss in primary hyperparathyroidism? *Arch Intern Med.* 1987;147:1951–1953.

31. Silverberg SJ, Shane E, De La Cruz L, Dempster DW, Feldman F, Seldin D, Jacobs TP, Siris ES, Cafferty M, Parisien MY, Lindsay R, Clemens TL, Bilezikian J. Skeletal disease in primary hyperparathyroidism. *J Bone Min Res.* 1989;4:283–291.

32. Meunier PJ, Bianchi GGS, Edouard CM, Bernard JC, Courpron P, Vignon GE. Bone manifestations of thyrotoxicosis. *Orthop Clin North Am.* 1972;3:745–774.

33. Levin ME, Boisseau VC, Avioli LV. Effects of diabetes mellitus on bone mass in juvenile and adult-onset diabetes. *N Engl J Med.* 1976;294:241.

34. Rosenbloom AL, Lezotte DC, Weber FT, et al. Diminution of bone mass in childhood diabetes. *Diabetes.* 1977;26:1052.

35. Santiago JV, McAlister WH, Rabzan SK, et al. Decreased cortical thickness and osteopenia in children with diabetes mellitus. *J Clin Endocrinol Metab.* 1977;45:845.

36. Schneider LE, Omdahl J, Schedl HP. Effects of vitamin D and its metabolites on calcium transport in the diabetic rat. *Endocrinology.* 1976;99:793–799.

37. Tamayo R, Goldman J, Villanueva A, Walczak N, Whitehouse F, Parfitt M. Bone mass and bone cell function in diabetes. *Diabetes.* 1981;30:31A.

38. Jackson JA, Kleerekoper M, Parfitt AM. Symptomatic osteoporosis in a man with hyperprolactinemic hypogonadism. *Ann Intern Med.* 1986;105:543–545.

39. Fallon MD, Perry HM III, Bergfeld M, Droke D, Teitelbaum SL, Avioli LM. Exogenous hyperthyroidism with osteoporosis. *Arch Intern Med.* 1983;143:442–444.

40. Coindre J-M, David J-P, Riviere L, Goussot J-F, Roger P, de Mascarel A, Meunier PJ. Bone loss in hypothyroidism with hormone replacement. A histomorphometric study. *Arch Intern Med.* 1986;146:48–53.

41. Perry HM III. Thyroid replacement and osteoporosis. *Arch Intern Med.* 1986;146:41.

42. Baylink DJ. Glucocorticoid-induced osteoporosis. *N Engl J Med.* 1983;309:306–308.

43. Avioli LV. Effects of chronic corticosteroid therapy on mineral metabolism and calcium absorption. *Adv Exp Med Biol.* 1984;171:81–89.

44. Hahn TJ, Boisseau VC, Avioli LV. Effect of chronic corticosteroid administration on diaphyseal and metaphyseal bone mass. *J Clin Endocrinol Metab.* 1974;39:274–282.

45. Sambrook P, Birmingham J, Kempler S, Kelly P, Eberl S, Pocock N, Yeates M, Eisman J. Corticosteroid effects on proximal femur bone loss. *J Bone Min Res.* 1990;5:1211–1216.

46. Rickers H, Deding A, Christiansen C, Rodbro P. Mineral loss in cortical and trabecular bone during high-dose prednisone treatment. *Calcif Tissue Int.* 1984;36:269–273.

47. Dempster DW. Bone histomorphometry in glucocorticoid-induced osteoporosis. *J Bone Min Res.* 1989;4:137—141.

48. Lo Cascio V, Bonucci E, Ballanti P, Adami S, Rossini M, Imbimbo B, Bertoldo F, Della Rocca C. Glucocorticoid osteoporosis: a longitudinal

study. In: Christiansen C, Johnson JS, Riis BJ, eds. *Osteoporosis 1987*. Copenhagen: Osteo Press, ApS; 1987:1044–1046.

49. Hahn TJ, Halstead LR, Teitelbaum SL, Hahn B. Altered mineral metabolism in glucocorticoid induced osteopenia. Effect of 25-OH-D administration. *J Clin Invest*, 1979;64:655.

50. Bressot C, Meunier PJ, Capuy MC, Lejeune E, Edouard C, Darby AJ. Histomorphometric profile, pathophysiology and reversibility of corticoid-induced osteoporosis. *Met Bone Dis Rel Res*. 1979;1:303.

51. Slovik DM, Neer RM, Ohman JL, Lowell FC, Potts Jr JT. Parathyroid hormone (PTH) and 25-hydroxyvitamin D (25-OH-D) levels in glucocorticoid (GC)-treated patients. *Clin Endocrinol*. 1980;12:243.

52. Seeman E, Kumar R, Hunder GG, Scott M, Heath H III, Riggs BL. Production, de-radiation, and circulating levels of 1,25-dihydroxyvitamin D in health and in chronic glucocorticoid excess. *J Clin Invest*. 1980;66:664.

53. Chesney RW, Hamstra AJ, Mazees RB, DeLuca HF. Reduction of serum 1,25-dihydroxyvitamin-D3 in children receiving glucocorticoids. *Lancet*. 1978; 2:1123.

54. Baylink DJ. Glucocorticoid-induced osteoporosis. *N Engl J Med*. 1983; 309:306–308.

55. Avioli LV. Effects of chronic corticosteroid therapy on mineral metabolism and calcium absorption. *Adv Exp Med Biol*. 1984;171:81–89.

56. Korkor A, Martin KJ, Olgaard K, Bergfeld M, Teitelbaum S, Klahr S, Slatopolsky E. Altered adenosine 3′,5′-monophosphate release in response to parathyroid hormone by isolated perfused bone from glucocorticoid-treated dogs. *Endocrinology*. 1983;113:625–631.

57. Reid IR, Chapman GE, Fraser TRC, Davies AD, Surus AS, Meyer L, Huq NL, Ibbertson HK. Low serum osteocalcin levels in glucocorticoid-treated asthmatics. *J Clin Endocrinol Metab*. 1986;62:379–383.

58. Reid IR, King AR, Alexander CJ, Ibbertson HK. Prevention of steroid-induced osteoporosis with (3-amino-1-hydroxypropylidene)-1,1-bisphosphonate (APD). *Lancet*, 1988;1:146–146.

59. Nishioka T, Kurayama H, Yasuda T, Udagawa J, Matsumura C, Niimi H. Nasal administration of salmon calcitonin for prevention of glucocorticoid-induced osteoporosis in children with nephrosis. *J Pediatr*. 1991;118:703–707.

60. Rizzaio G, Tosi G, Schiraldi G, Montemurro L, Zanni D, Sisti S. Bone protection with salmon calcitonin (sCT) in the long-term steroid therapy of chronic sarcoidosis. *Sarcoidosis*. 1988;5:99–103.

61. Palmieri GM, Dvorak J, Bottomley R. Calcitonin in steroid-induced osteoporosis. *N Engl J Med*. 1974;290:1490–1491.

62. Thompson JS, Palmieri GM, Eliel DP, Crawford RL. The effect of porcine calcitonin on osteoporosis induced by adrenal cortical steroids. *J Bone Joint Surg*. 1972;54:1490–1500.

63. Conrozier T, Meunier PJ. Traitement de l'osteroporose cortisonique par le fluorore de sodium: effects cliniques et donnees histomorphometriques. *Ann Endocrinol (Paris)*. 1985;46:369–370.

64. Meunier PJ, Briancon D, Chavassieux P, Edouard C, Boivin G, Conrozier T, Marcelli C, Pastoureau P, Delmas PD, Casez JP. Treatment with fluoride: bone histomorphometric findings. In: Christiansen C, Johansen JS, Riis BJ, eds. *Osteoporosis 1987*. Copenhagen: Osteopress ApS; 1987:824–828.

65. Nelson ME, Fisher EC, Catsos PD, Meredith CN, Turksoy RN, Evans WJ. Diet and bone status in amenorrheic runners. *Am J Clic Nutr.* 1986;43: 910–916.
66. Dale E, Gerlack DH, Wilhite AL. Menstrual dysfunction in distance runners. *Obstet Gynecol.* 1979;54:47.
67. Feicht CB, Johnson TS, Martin BJ, Sparkes KE, Wagner Jr WE. Secondary amenorrhoea in athletes. *Lancet.* 1981;2:1145.
68. Malina RM, Spirduso WW, Tate C, Baylor AM. Age at menarche and selected menstrual characteristics in athletes at different competitive levels and in different sports. *Med Sci Sports.* 1978;10:218.
69. Shangold MM, Levine HS. The effect of marathon training upon menstrual function. *Am J Obstet Gynecol.* 1982;143:882.
70. Cann CE, Martin MC, Genant HK, Jaffe RB. Decreased spinal mineral content in amenorrheic women. *JAMA.* 1984;251:626.
71. Drinkwater BL, Nilson K, Chestnut CH, Bremmer WJ, Shainholtz S, Southworth MB. Bone mineral content of amenorrheic and eumenorrheic athletes. *N Engl J Med.* 1984;311:277.
72. Marcus R, Cann C, Madvig P, Minkoff J, Goddard M, Bayer M, Martin M, Gaudiane L, Haskell W, Genant H. Menstrual function and bone mass in elite women distance runners. *Ann Intern Med.* 1985;102:158.
73. Fisher EC, Nelson ME, Frontera WR, Turksoy RN, Evans WJ. Bone mineral content and levels of gonadotropins and estrogens in amenorrheic running women. *J Clin Endocrinol Metab.* 1986;62:1232–1236.
74. Nelson ME, Fisher EC, Catsos PD, Meredith CN, Turksoy RN, Evans WJ. Diet and bone status in amenorrheic runners. *Am J Clin Nutr.* 1986;43:910–916.
75. Cook SD, Harding AF, Thomas KA, Morgan EL, Schnurpfeil KM, Haddad RJ. Trabecular bone density and menstrual function in women runners. *Am J Sports Med.* 1987;15:503–507.
76. Baran D, Whyte M, Haussler M, et al. Effect of menstrual cycle on calcium-regulating hormones in normal young women. *J Clin Endocrinol Metab.* 1980;50:377–379.
77. Rigotti NA, Nussbaum SR, Herzog DB, Neer RM. Osteoporosis in women with anorexia nervosa. *N Engl J Med.* 1984;311:1601–1606.
78. Rigotti NA, Neer RM, Jameson L. Osteopenia and bone fractures in a man with anorexia nervosa and hypogonadism. *JAMA.* 1986;256:385–388.
79. Riggs BL, Eastell R. Exercise, hypogonadism and osteopenia. *JAMA.* 1986;256:392–393.
80. Seeman E, Melton LJ III, O'Fallon WM, Riggs BL. Risk factors for spinal osteoporosis in men. *Am J Med.* 1983;75:977–983.
81. Bikle DD, Genant HK, Cann C, Recker RR, Halloran BP, Strewler GJ. Bone disease in alcohol abuse. *Ann Intern Med.* 1985;103:42–48.
82. Anderson RA, Willis BR, Oswald C, Reddy JM, Beyler SA, Zaneveld LJD. Hormonal imbalance and alterations in testicular morphology induced by chronic ingestion of ethanol. *Biochem Pharmacol.* 1980;29:1409–1419.
83. Baran DT, Teitelbaum SL, Bergfeld MA, Parker G, Cruvant EM, Avioli LV. Effect of alcohol ingestion on bone and mineral metabolism in rats. *Am J Physiol.* 1980;238:E507–E510.
84. Pitts TD, Van-Thiel DH. Disorders of divalent ions and vitamin D metabolism in chronic alcoholism. *Rec Dev Alcohol.* 1986;4:357–377.

85. DeVernejoul MC, Bielakoff J, Herve M, Gueris J, Hott M, Modrowski D, Kuntz D, Miravet L, Ryckewaert A. Evidence for defective osteoblastic function. A role for alcohol and tobacco consumption in osteoporosis in middle-aged man. *Clin Orthop Rel Res.* 1983;179:107–115.
86. Adler AJ, Fillipone EJ, Berlyne G. Effect of chronic alcohol intake on muscle composition and metabolic balance of calcium and phosphate in rats. *Am J Physiol.* 1985;249:E584–E588.
87. Jensen GF. Osteoporosis of the slender smoker revisited by epidemiologic approach. *Eur J Clin Invest.* 1986;16:239–242.
88. Jensen J, Christiansen C, Rodbro P. Cigarette smoking, serum estrogens, and bone loss during hormone-replacemenprintAp therapy early after menopause. *N Engl J Med.* 1985;313:973–975.
89. *Drug facts and comparisons.* Philadelphia: Lippincott. 1990.
90. Heaney RP, Recker RR. Effects of nitrogen, phosphorus, and caffeine on calcium balance in women. *J Lab Clin Med.* 1982;99:46–55.
91. Massey LK, Wise KJ. The effect of dietary caffeine on urinary excretion of calcium, magnesium, sodium and potassium in healthy young people. *Nutr Res.* 1984;4:43–50.
92. McCarron DA, Rankin LI, Bennett WM, Krutzik S, McClung MR, Luft FC. Urinary calcium excretion in extremes of sodium intake in normal man. *Am J Nephrol.* 1981;I:84–90.

10
Bone Mineral Metabolism at the Menopause: Determinants and Markers

IAN A. KATZ and SOL EPSTEIN

This chapter details the changes in circulating and urinary parameters of bone mineral metabolism during the perimenopausal period. Hormonal changes that could affect or be the result of alterations in bone metabolism, as well as circulating and urinary markers of bone metabolism, are discussed. The relevance of these changes to the pathophysiology and diagnosis of postmenopausal osteoporosis is emphasized.

Estrogen deficiency after the menopause probably has the greatest effect on bone mineral metabolism, but there is no doubt that other hormones, such as progesterone, also play a role. Although post-menopausal women as a group have reasonably consistent alterations in biochemical markers of mineral metabolism, there is no single marker that can successfully screen an individual perimenopausal patient for bone disease. Local growth factors may also have important roles in the pathogenesis of postmenopausal osteoporosis.

The chapter is divided into four main sections:

1. changes in calcitropic hormones including parathyroid hormone, vitamin D, and calcitonin
2. changes in "epicalcitropic" factors including growth hormone and the insulinlike growth factors, thyroid hormones, steroid hormones, gonadotropins, prolactin, insulin, and vitamin K
3. changes in serum biochemical markers of bone resorption and formation including calcium, phosphate, bone Gla-protein (BGP), alkaline phosphatase, acid phosphatase, procollagen peptides, and bicarbonate
4. changes in urine biochemical markers of bone resorption and formation including calcium, hydroxylysine and hydroxyproline, and Gla and BGP.

Two types of bone loss are postulated to occur in women: Type 1 or postmenopausal osteoporosis, and Type 2 or age-related osteoporosis.[1] Riggs and Melton speculate that Type 1 osteoporosis is due primarily to increased bone loss and excessive calcium release, which suppresses parathyroid hormone (PTH) secretion and reduces 1,25-dihydroxyvitamin

D $(1,25(OH)_2D)$ production; Type 2 osteoporosis is primarily due to an impairment of $1,25(OH)_2D$ production and resistance to vitamin D resulting in secondary hyperparathyroidism and subsequent bone loss.[1] However, these theoretical entities are not distinct in clinical practice, and not all studies have been able to show alterations in calcium, $1,25(OH)_2D$, and PTH values at the menopause compatible with the above classification.

Bone constantly undergoes remodeling with a cycle of resorption by osteoclasts, followed closely by, and coupled to, bone formation by osteoblasts. The coupling of osteoclast to osteoblast activity maintains bone mineral homeostasis under normal circumstances when the cells are in equilibrium.[2] However, both osteoclasts and osteoblasts are subject to local and humoral regulation, and alterations in this regulation, both "normal" as occurs with the menopause, and abnormal as occurs for example in hyperparathyroidism, can lead to disorders of bone mineral metabolism. Increased osteoblastic activity is postulated to be a response to the increased osteoclastic bone resorption, and if formation cannot keep up with resorption, net bone resorption occurs, even in the face of increased bone formation. This "high-turnover" osteoporosis is characteristic of the postmenopausal period.

Numerous physiologic and biochemical changes occur in the perimenopausal period of a woman's life. Alterations in bone mineral metabolism are important, as postmenopausal women have a well documented increased incidence of bone loss (see refs. 3–5 for reviews). Despite the widely held view of the menopause as a state of estrogen deficiency exclusively, numerous other hormonal changes occur that could influence bone mineral metabolism. Although Albright et al. originally postulated that postmenopausal osteoblasts were deficient in laying down osteoid tissue,[6] it is now clear that the predominant effect of estrogen is to decrease bone resorption, and that the main defect in postmenopausal osteoporosis is uninhibited bone resorption with uncoupled, inappropriately low bone formation.[7]

Although estrogen deficiency appears to have a central role in this regard, and although the loss of bone can be prevented by estrogen replacement,[8] the questions of why formation and resorption are uncoupled at the menopause and why only approximately 50% of fractures are prevented by estrogen therapy[8] remain to be answered. The rate of bone loss varies widely among postmenopausal women and only some develop osteoporosis.[9] The women in this subgroup are sometimes termed "fast bone losers." It can be speculated that the answers to these questions lie in the associated hormonal alterations that occur around the menopause. This chapter describes the alterations in the hormonal milieu and the effects of these hormones on bone mineral metabolism (the determinants), as well as the biochemical consequences of these alterations (the markers) as they relate to screening for abnormalities and detecting women at risk of and suffering from osteoporosis. Additionally, changes in the parameters of bone metabolism due to the menopause will be contrasted to those due to normal aging.

Calcitropic Hormones

The calcitropic hormones influence bone mineral metabolism in health and disease. Consequently, it should come as no surprise that there has been extensive study of these hormones in patients with postmenopausal osteoporosis.

The Vitamin D System

The role of $1,25(OH)_2D$ in the pathogenesis of postmenopausal osteoporosis is unclear. The serum level of $1,25(OH)_2D$, the physiologically active vitamin D metabolite, may decrease by about 50%[10-13] or remain constant[14,15] with aging. Similarly, 25-hydroxyvitamin D may be unchanged[11,12] or reduced[14,16] with aging. These conflicting results probably reflect the varied proportions of men and women studied, as well as seasonal, dietary, and methodologic differences.

There is also controversy regarding the levels of vitamin D metabolites in pre- and postmenopausal women. A recent study that was controlled for age reported reduced serum $1,25(OH)_2D$ and reduced vitamin D binding protein (DBP) in post- compared to premenopausal women.[17] Although the basal $1,25(OH)_2D$ levels were lower in postmenopausal women, the response to a low calcium diet was normal and restored serum $1,25(OH)_2D$ to control values. Thus, there did not seem to be a primary abnormality in the $1,25(OH)_2D$ secretory reserve due to the menopause. The lower $1,25(OH)_2D$ levels in this study are more likely due to either a low DBP reducing the bound fraction of $1,25(OH)_2D$ or a secondary suppression of $1,25(OH)_2D$ by an elevated serum calcium and lowered PTH, which some authors have found in postmenopausal women.[17,18] Other workers, however, have described both impaired secretion and unchanged basal levels of vitamin D metabolites.[19-21] In a recent longitudinal study, Hartwell et al. could not detect alterations in $1,25(OH)_2D$ metabolism during natural menopause.[21]

A recent longitudinal study found no correlation between bone loss and vitamin D levels in normal women.[20] In another study Scharla et al.[22] induced an artificial menopause with a gonadotropin hormone–releasing hormone (GnRH) agonist in patients with endometriosis and uterine leiomyoma and studied indices of bone metabolism before, during, and after treatment. A positive correlation between pretreatment values of $1,25(OH)_2D$ and its decrease and the reduction in bone mass induced by the artificial menopause was found,[22] and they postulated that a patient's metabolic condition predicts her response to estrogen deficiency.

25-Hydroxyvitamin D levels are reported to be elevated in postmenopausal women.[17,21] This would be consistent with impaired $1,25(OH)_2D$ synthesis during the menopause resulting in inadequate product inhibition of liver 25-hydroxylase[23]; although such a phenomenon has been demonstrated in vitro in kidney slices of aged rats,[24] the significance of this in postmenopausal bone loss in unclear. Postmenopausal osteoporotic women have been found to have significantly reduced serum levels of 25-hydroxyvitamin D when compared to age-matched normal women,[25] or unchanged 25-hydroxyvitamin D and significantly reduced 24,25-dihydroxyvitamin D when compared to non–age-matched premenopausal women.[26] The importance of the reduced 24,25-dihydroxyvitamin D and the validity of the controls in this latter study, however, are questionable.

Postmenopausal estrogen therapy increases total $1,25(OH)_2D$, but because DBP also increases, the elevated $1,25(OH)_2D$ is not necessarily accompanied by a rise in the biologically effective free level.[27]

In summary, based on these conflicting results, in part due to methodologic difficulties,[28] the current hypothesis holds that vitamin D deficiency by itself is a pathogenic factor only in a small group of patients with postmenopausal osteoporosis,[29] and that any changes in vitamin D metabolites are probably a consequence and not a cause of the altered bone mineral metabolism.

Parathyroid Hormone

The role of PTH in the pathogenesis of postmenopausal osteoporosis is also controversial. A major problem of the accurate determination of biologically active PTH in the presence of inactive fragments of the PTH molecule has been alleviated by recently developed two-site radiometric and chemiluminescent assays. Measurement of renal phosphate transport and of urinary cyclic adenosine monophosphate (cAMP) assess PTH biologic activity indirectly.[30] The value of a novel bioassay for PTH, reported by Goltzman et al.,[31] has not been fully assessed for determination of biologically active PTH around the menopause.

Serum levels of immunoreactive PTH increase with age.[32–34] One group of investigators reported that when renal function was controlled for in the analysis, the increase in PTH with age was no longer significant.[33] This study, however, used an insensitive PTH assay. In an age-controlled study,[17] PTH, measured using a sensitive two-site immunoradiometric assay, was found to be unchanged perimenopausally under basal conditions and in response to a calcium stress. The fact that the PTH was unchanged despite a higher plasma calcium in the postmenopausal patients is in accordance with the hypothesis that estrogen deficiency directly influences parathyroid function by altering the set-point for PTH secretion. The result of this is that PTH production remains constant despite the rise in serum calcium.[35,36] Gallagher et al.[34] compared osteoporotic postmenopausal women and age-matched controls, and found that the osteoporotic women generally had lower PTH levels than their controls. This is in accordance with the Riggs and Melton classification, which postulates decreased PTH in Type 1 osteoporosis.[1]

Although it has been reported that estrogen therapy increases serum PTH level in postmenopausal women,[37] a recent study demonstrated that serum PTH is reduced in the first 2 weeks after estrogen therapy and then rises.[30] The significance of this is doubtful in view of the fact that the reduced PTH was not associated with evidence of a change in the biologic action of PTH (urinary cAMP excretion and Tm_{p04}/GFR).[30]

In summary, the menopause appears to have little direct effect on PTH metabolism. It is possible that osteoporotic women may have lower serum PTH levels than age-matched controls, but this phenomenon may be secondary to alterations in mineral metabolism. Furthermore, the results need to be interpreted and compared with caution in view of the different assay systems for PTH used in these studies.

Calcitonin

The physiologic role of calcitonin (CT) remains uncertain, despite intensive investigation since its discovery in 1962.[38] In light of the current hypothesis that osteoclasts have CT receptors, that CT functions as a physiologic inhibitor of osteoclasts (see ref. 39 for review), and that its secretion is regulated by the ambient calcium concentration, it has been postulated that CT participates in the regulation of calcium homeostasis and protects the skeleton from excessive osteoclastic bone resorption.[38,39]

The role of CT in the pathogenesis of both Type 1 and Type 2 osteoporosis is controversial.[40–59] The principal reasons for this uncertainty are, that as outlined above, the physiologic role of CT is not fully understood, and that researchers have been hampered by the lack of sensitive and specific assays for different forms of serum CT, resulting in wide variations between the results reported by different laboratories. This problem should be resolved as new techniques for the extraction and concentration of CT, and more specific antibodies for radioimmunoassay of the monomeric form of CT, are used.[42,45,48] The monomeric form of CT is believed to be the active form of the hormone.[52]

Although most recent work indicates that the basal plasma CT does not vary with age,[44,45,50] decreasing levels with increasing age have been reported.[41] Levels are generally lower in women than in men,[40,44,45] and increase during pregnancy.[39]

It is believed that one mechanism whereby estrogens protect bone is by stimulating CT secretion.[42,43,58] After the menopause an estrogen deficiency results in reduced CT secretion, and this in turn produces osteoporosis. Although the consensus of opinion is that postmenopausal basal CT levels are lower than premenopausal levels,[50,58] there is a paucity of information regarding the levels of CT in postmenopausal women with osteoporosis compared with age-matched controls. Not all researchers agree with the postulate that basal CT levels should be, or are, low in postmenopausal osteoporotic patients. For example, Prince et al.[48] recently reported that they could find no difference in serum CT between postmenopausal osteoporotic women and age-matched controls. Two other studies that had similar results[49,59] were performed earlier with less specific antibodies, and are probably not as relevant.

CT production rate (PR) and metabolic clearance rate (MCR) in pre- and postmenopausal women with and without osteoporosis have been measured.[58] The PR was reported to be insignificantly reduced in normal postmenopausal women compared with premenopausal women, but significantly reduced in osteoporotic patients. Both basal CT levels and PR correlated strongly with circulating estrone levels in all groups, and the researchers speculated that CT metabolism is in fact influenced by estrogen levels. MCR does not appear to be altered by menopause.[58] Tiegs et al.[45] obtained similar results, but studied women of a much wider age range and did not investigate the effects of the menopause specifically.

With regard to CT reserve, Perez Cano et al.[47] reported recently that a CT reserve deficiency is present in the first years after the menopause and recovers later. There was no difference in CT reserve between normal and osteoporotic women with recent menopause, nor were basal levels different in the premenopausal and postmenopausal women.[47] Another group has reported a reduced secretory reserve of CT in postmenopausal osteoporosis compared with age-matched controls,[54] but the CT levels in this study were very high and their assay probably measured multiple forms of plasma CT.

According to earlier studies, the administration of the estrogen–progesterone contraceptive pill, or of estrogen alone, increases plasma CT, probably as a result of indirect stimulation by estrogen-evoked alterations in bone mineral metabolism.[46,51] A more recent study, however, could not demonstrate an effect of estrogen on CT levels.[42] This result is supported, indirectly, by two recent reports that the administration of estrogen to postmenopausal women had no effect on basal CT levels when serum calcium supplementation was given as well.[42,57] This suggests that variations in serum CT are due to alterations in bone metabolism, particularly calcium, induced by the estrogens. In the earlier reported work contending that estrogens did in fact stimulate CT, the serum calcium levels were not maintained with supplemental calcium.[51]

In summary, although the consensus of opinion is that postmenopausal women have lower CT levels than premenopausal women, it is not clear whether postmenopausal osteoporotic women have CT levels different

TABLE 10.1. "Epicalcitropic" factors.
Growth hormone and the insulinlike growth factors
Thyroid hormones
Steroid hormones
Gonadotropins
Prolactin
Insulin
Vitamin K

from those of matched controls. There does appear to be a reduced CT reserve in postmenopausal osteoporotic women. Alterations in CT levels may be the cause or the result of altered bone mineral metabolism. Thus, although the use of CT in the prevention and management of postmenopausal osteoporosis is assuming wider recognition,[60] the pathophysiologic basis for this treatment is still uncertain.

"Epicalcitropic" Factors

Although the calcitropic hormones are the better known regulators of bone mineral metabolism, other factors, including hormones, also have effects on bone and are important in the perimenopausal period. These we shall term the "epicalcitropic factors" (Table 10.1).

Growth Hormone and the Insulinlike Growth Factors

Growth hormone (GH) secretion is impaired with aging,[61] and administration of GH to individuals over 60 years of age has been reported to increase vertebral bone density.[62]. There have been no studies on the serum levels of growth hormone around the menopause. In women, serum insulinlike growth factor (IGF)-I declines markedly with age whereas serum IGF-II decreases only slightly with age.[63] Neither IGF-I nor IGF-II concentration correlated with bone mineral density when age was held constant, and in women with postmenopausal osteoporosis, serum IGF-I and IGF-II did not differ significantly from the concentrations in normal women of the same age.[63] Duursma et al. demonstrated a decrease in the concentration of somatomedin and an increase in GH in estrogen-treated postmenopausal women.[64]

Thyroid Hormones

Thyroid hormones exert profound effects on skeletal growth, maturation, and turnover, and thyrotoxicosis is associated with the development of osteopenia (see ref. 65 for review). Alterations in the hypothalamus–

pituitary–thyroid axis have been reported in some women after the menopause.[66–69] Custro et al.[66] reported that up to 10% of recently postmenopausal women have borderline high values of triiodothyronine (T$_3$) and thyroxine (T$_4$), associated with (and perhaps due to) high thyroid-stimulating hormone (TSH).[67] Bottiglioni et al.[68] measured TSH, T$_3$, T$_4$, free T$_3$, free T$_4$, and thyroxine-binding globulin (TBG) in pre- and postmenopausal women, and found a significant alteration in only free T$_3$, which was reduced in women who had been postmenopausal for more than 3 years. As there were no age-matched controls, this finding could have been age-related.[70] More recently, another group reported that although T$_4$ was constant around the time of menopause, free T$_3$ levels fell before the menopause, rose significantly immediately after the menopause, before being significantly reduced in late postmenopausal women.[69] The level of TSH was significantly higher in women who had just undergone menopause than in pre- and late postmenopausal women.[69] Rico et al.[71] found no differences in any thyroid hormone between postmenopausal women with or without osteoporosis. The administration of sex steroids to postmenopausal women has been reported to have a varying effect on thyroid hormone levels, depending on the amount of estrogens and progestogens given,[72] a phenomenon probably related to alterations in the amount of TBG.

In summary, it is conceivable that alterations in thyroid hormone levels in the perimenopausal period have a significant influence on bone loss. Thyroid medication, however, is not indicated, except to correct any definite thyroid dysfunction.

Other Hormones and Vitamin K

Prolactin levels appear to drop significantly after menopause,[69] but there have been no reports of any effects on bone. A relationship between circulating insulin and bone mass has been postulated.[73] However, Sambrook et al.[74] could find no association between a single measurement of insulin and bone density in women over a wide age range. Vitamin K has been postulated to play a role in the pathogenesis of postmenopausal osteoporosis. Although serum vitamin K levels were similar in pre- and potsmenopausal women, Knapen et al.[75] reported that the administration of vitamin K to postmenopausal women increased serum bone Gla protein (BGP) levels and reduced urinary calcium excretion in women who were fast losers of calcium. In this study, a commercial kit was used for the BGP assay and may not have been reliable.

Steroid Hormones

Steroid hormones (Table 10.2) are believed to play an important role in maintaining skeletal integrity and quantitative abnormalities can be

TABLE 10.2. Sex steroid hormones affecting bone
mineral metabolism.

Estradiol
Estrone
Progesterone
Testosterone
Androstenedione
Dehydroepiandrostenedione

Also:
Sex hormone binding globulin

associated with bone abnormalities. Since menopause is associated with major alterations in circulating levels of these hormones,[76,77] it is important to review these changes in relation to perimenopausal bone mineral metabolism (see ref. 78 for review).

ESTROGEN

Estrogen deficiency is believed to have the pivotal role in the pathogenesis of postmenopausal osteoporosis, because of the temporal relationship between rapid bone loss and surgical or natural menopause (see refs. 3, 5 for reviews), the loss of bone in amenorrheic young women,[79] and the response of bone to estrogen replacement therapy.[8] Although the identification of estrogen receptors on osteoblastlike cells[80,81] provides a mechanism for the direct action of estrogen on bone, numerous additional factors are thought to modulate the response of bone to estrogen.

In premenopausal women the principal circulating estrogen is 17β-estradiol, with estrone present in much lower concentrations. As women approach the menopause, levels of estrogen begin to drop mainly during the late follicular and luteal phase of the menstrual cycle,[82] and irregular cycles result. In most women, the time lapse between regular cycles and the cessation of menstrual periods, often called the perimenopausal transition, is 2 to 7 years.[83] The average estrogen concentration declines during this time until a low noncycling estrogen level, accompanied by elevations in plasma follicle-stimulating hormone (FSH) and luteinizing hormone (LH), the hallmark of menopause, is reached.[84] On the other hand, in individual women hormone levels fluctuate widely,[85,86] and some studies have not documented a consistent premenopausal fall in estrogen levels,[69,87] depending on the study population characteristics. Postmenopausally, both estradiol and estrone are reduced, but the concentration of estrone exceeds that of estradiol because of the conversion of adrenal androstenedione into estrone by fat cells.[88]

Two patterns of bone loss are believed to occur in women (see above): a slow "normal" loss of bone that may occur in both sexes from the age of about 35 years, after peak bone mass is reached, and a superimposed, transient, accelerated bone loss occurring in perimenopausal women. In some women (the "fast losers"), the latter results in postmenopausal osteoporosis. It is further believed that ovarian deficiency and associated alterations, and not aging, are the predominant causes of bone loss during the first two decades after menopause.[89] Whether the Type 1 bone loss begins before or after menopause is a matter of continuing controversy. Bone mineral density measurements have shown that bone loss begins to increase in normal women before the menopause.[86,90,91] Indeed, Riggs et al.,[87] in a well designed longitudinal study, reported that more than 50% of all vertebral bone loss occurred before the menopause. However, in this study there was no premenopausal decline in serum estrogen, and thus no correlation between bone loss and estrogen levels was found. The authors concluded that factors other than estrogen deficiency must have contributed to this osteoporosis.

Although there is a well known association between bone loss and estrogen deficiency, attempts to correlate bone density and estrogen levels have been surprisingly unsuccessful.[90] The levels of estrone and estradiol in postmenopausal women with osteoporosis and in age-matched controls without osteoporosis have been compared in numerous studies, but no significant differences have been found.[71,92,93] Marshall et al.[94] reported an absolute deficiency of estrogen associated with (and probably due to) low androstenedione levels in postmenopausal osteoporotic women compared to matched controls. In one study, women with osteoporosis were found to have lower concentrations of free estrone but this probably was due to higher concentrations of sex hormone binding globulin (SHBG) in thinner patients who coincidently (or not!) were osteoporotic.[93] In another study, Slemenda et al.[95] calculated that, in the entire population, the mean estrogen concentration by itself was a strong predictor of the rate of bone loss when plotted against bone mass over time.

Although it is well recognized that sex steroid hormones circulate in the serum as free hormone, or bound with high affinity and low capacity to SHBG, and with low affinity and high capacity to albumin,[96,97] there is some controversy as to which of these fractions is biologically active. Because of speculation that changes in SHBG before and during the menopause might result in altered serum levels of biologically active estrogen despite constant total serum levels, Steinberg et al.,[90] in a cross-sectional study, investigated SHBG in pre- and perimenopausal women and correlated the results to bone mineral density as measured by single photon absorptiometry. Although SHBG concentrations were similar in pre- and perimenopausal women, their data indicated that in pre-menopausal women only was there a negative association between bone density and SHBG. The authors speculate that in view of the inverse relationship believed to exist between the availability of estrogen and testosterone and the level of SHBG, the latter may influence bone density by regulating available steroid hormones. There is no evidence that SHBG acts directly on bone.[90] In accordance with these results, a negative relationship between SHBG and bone density in postmenopausal women[98] and higher values of SHBG in postmenopausal patients with fractured femur necks than in age-matched controls have been reported.[93] McPherson et al.[18] found unaltered serum globulin levels before and after the menopause, but did not examine SHBG specifically.

Cigarette consumption is reportedly associated with an increased incidence of osteoporosis in postmenopausal women,[99] perhaps through an effect on estrogen metabolism. It has been reported that smokers have an earlier menopause than nonsmokers[100] and both pre- and postmenopausal smokers have lower estrogen levels than age-matched controls.[101,102] However, in a different study, no effect of smoking on the metabolism and production of estrogens and androgens was found when the variables were adjusted for body weight.[103] Thus, the rates of bone loss in the perimenopausal period do not seem to be directly influenced by smoking.[104] This area is further clouded by the recent finding that smoking overall does not appear to increase the risk for hip fracture but may counteract the protective effect of estrogens in women taking them.[104a]

PROGESTERONE

Clearly, ovarian failure at menopause reduces not only the supply of estrogen but also that of progesterone.[105] Progesterone levels are high during the luteal phase of the normal menstrual cycle, and because menstrual abnormalities (irregular ovulation) usually precede the actual menopause, the progesterone levels may be abnormally low for a long time before menopause.[69] Additionally, steroidogenic output of the corpus luteum is generally lower in older woman.[88]

Recently experimental, epidemiologic, and clinical data have demonstrated that progesterone is active in bone metabolism, possibly playing a role in the coupling of bone resorption and bone formation (see ref. 106 for review). It is believed that progesterone may exert its action directly on bone by engaging an osteoblast receptor, or indirectly through competition for a glucocorticoid receptor.[107,108] Progesterone–glucocorticoid competition for an osteoblast receptor may be relevant to glucocorticoid-induced osteoporosis as progesterone may inhibit the negative corticoid actions on osteoblasts. Thus, menopausal bone loss may be the result of a combination of decreased osteoclast inhibition by reduced estrogen and increased cortisol inhibition of osteoblasts, in turn due to reduced competition with progesterone. This theory is supported by the observations of Johnston et al.[109] that postmenopausal women with higher progesterone levels were slow losers of bone, whereas those with lower progesterone levels lost bone at a much faster rate. Progesterone deficiency in ovari-ectomized women might contribute to bone loss.[110] Progesterone alone is not effective in preventing postmenopausal bone loss,[111] but with estrogen is reported to uncouple formation and resorption so that the diminished bone resorption induced by estrogen is accompanied by increased bone formation.[112]

GONADOTROPINS

Yen has provided data on the mean plasma level of a variety of pituitary hormones in pre- and postmenopausal women.[77] LH and especially FSH increase five- to tenfold, starting some time before the menopause.[69,85] There have been no reports of either hormone having any direct effect on bone mineral metabolism, although Steinberg et al.[90] found that in perimenopausal women there was an inverse correlation between bone density and the plasma level of FSH, probably reflecting the level of estrogen. Johansen et al.[113] reported that the administration of a GnRH analog to premenopausal women stimulated bone turnover with net bone loss. Although these effects of GnRH on bone were attributed to the induction of an artificial menopause with diminished circulating estrogen, a direct effect of GnRH on bone was not ruled out. Similar results were reported by Scharla et al.,[22] who investigated the indices of bone metabolism before, during, and after artificial menopause induced by a GnRH agonist. Hartwell et al.[21] found that women receiving a GnRH agonist continued to show significant alterations in $1,25(OH)_2D$ while on estrogen therapy, indicating that GnRH agonists (and GnRH itself) may induce changes in other unknown pituitary hormones involved in vitamin D regulation. These observations suggest that alterations in estrogen levels are only one of the factors affecting bone mineral metabolism after GnRH therapy.

ANDROGENS

Before menopause, 50% of the circulating androstenedione is derived from the ovaries and 50% from the adrenals. Therefore, a reduction of approximately 50% in serum androstenedione after menopause reflects ovarian failure.[97] According to one group of investigators, testosterone levels generally increase slightly in many women with aging, although they are significantly higher in late premenopausal than early post-menopausal women.[69] Similarly, another group found that the average levels of serum testosterone in postmenopausal women were approxi-mately 33% lower than in premenopausal women.[97]

A relationship between adrenal function after menopause and rate of loss of bone has been reported,[94,114-116] and it has been suggested that the peripheral conversion of androgens to estrogens may be the main factor protecting bone. Additionally, free testosterone was reported to be lower in osteoporotic patients than nonosteoporotic controls.[93] Steinberg et al.[90] reported that free testosterone and the percentage of free testosterone correlated positively and significantly with bone density. This positive correlation, coupled with the discovery that testosterone competes with estrogen for estrogen binding sites in rat osteoblastlike osteosarcoma cells,[80] suggests that testosterone deficiency may play an important role in the development of postmenopausal osteoporosis. The reports that androgens can increase bone density in osteoporotic women,[117] and that hypogonadal men and men with delayed puberty[118a] develop osteoporosis,[118] support this suggestion.

This notion, however, is controversial because no differences in adrenal function have been found between postmenopausal women who are slow or fast losers of bone,[114] and because postmenopausal patients with osteoporosis have been reported to have levels of total androgens similar to those of age-matched controls without osteoporosis.[92,93] As discussed above, this probably reflected the higher SHBG levels observed in thinner patients who are prone to osteoporosis.[93]

CORTISOL

There appear to be no significant perimenopausal changes in plasma cortisol levels,[69,71,77] although significantly reduced levels have been reported in late postmenopausal women.[69] Increased urinary free cortisol has been found in postmenopausal women who are fast losers of bone.[114] A better correlation has been found between the dehydroepiandrosterone: cortisol (DHA:C) ratio and vertebral mineral density than between DHA and vertebral mineral density,[116] but the significance of this observation is unclear. Duke has proposed that chronic alterations in the levels of circulating physiologically active cortisol due to changes in the serum levels of its carrier protein could play a role in the pathogenesis of postmenopausal osteoporsis.[119] Reduced competition by progesterone for cortisol receptors in osteoblasts could also be a factor in the pathogenesis of postmenopausal osteoporosis by contributing an element of glucocorticoid-induced osteopenia.

In summary, it is possible that all the steroid hormones may contribute to pathogenesis of osteoporosis in an individual perimenopausal woman. A conceivable hypothesis is that reduced estrogen levels, perhaps associated with reduced androgen levels, would allow excessive osteoclastic activity, whereas reduced progesterone levels, in association with some degree of cortisol activity, would allow underactivity of the osteoblasts, with resultant net bone loss. Examination of the whole hormonal milieu in an individual patient is thus an essential part of the management of postmenopausal osteoporosis.

TABLE 10.3. Perimenopausal markers of bone mineral metabolism.

Serum	Urine
Formation	**Formation**
Noncollagenous bone proteins	? Urinary Gla/BGP
Bone Gla-protein	**Resorption**
Other	Calcium
Alkaline phosphatase	Hydroxyproline
Total	Hydroxylysine
Bone-specific	Pyridinium cross-links
Procollagen extension peptides	
Resorption	
Acid phosphatase	
Tartrate resistant	
Calcium	
Phosphate	
Other	
Bicarbonate	

Biochemical Markers

In recent years efforts have been made to discover specific markers of bone turnover that would assist in the screening and management of various diseases, including postmenopausal osteoporosis (see refs. 120, 121 for reviews). Consequently, it is of utmost importance to review the changes in these markers during the perimenopausal period, and examine their significance in the pathophysiology and diagnosis of perimenopausal bone loss. These markers are listed in Table 10.3.

Serum Markers

Bone Gla-Protein

Serum bone Gla-protein (BGP) (see refs. 120–125 for reviews), also termed osteocalcin, is considered to be the only noncollagenous protein that is virtually specific for bone tissue and dentin. It is the most abundant noncollagenous protein in mature bone constituting 1% to 2% of the total protein,[126,127] and has a molecular weight of 5669 daltons in the human.[128] It is synthesized mainly by osteoblasts[129,130] and is incorporated into the extracellular matrix of bone, where it may have a role in the recruitment of osteoclasts.[131,132] A fraction of newly synthesized BGP is released into the circulation where it can be measured by radioimmunoassay.[133] Serum BGP is derived exclusively from this newly synthesized protein and thus reflects largely, if not exclusively, osteoblastic activity.[134] It correlates well with bone histomorphometric parameters of bone formation in normal adults.[135] For these reasons, the use of serum BGP as an indicator of osteoporosis in both symptomatic and asymptomatic individuals aroused high expectations, especially as a promising substitute for bone biopsy. Unfortunately, these initial high expectations have not been totally fulfilled.

Although most authors believe that in women BGP increases gradually from the fourth to the tenth decade,[136–140] reflecting an age-related increase in bone turnover, it has also been reported that BGP remains constant[141] or decreases with age.[142] Superimposed on any age-related effect, menopause induces a marked and transient acceleration of bone turnover,[136,143,144] associated with a twofold increase of serum BGP that can be reduced to premenopausal levels by estrogen therapy.[143] There is histomorphometric evidence of increased bone formation in these women.[145] Serum BGP is considered to be the best single biochemical marker of bone turnover in postmenopausal women and its levels accurately reflect bone remodeling.[145] Reports of normal BGP levels at menopause probably reflect measurements from a subgroup of women with low bone turnover.[146,147] The main problem is to distinguish the normal postmenopausal elevations in BGP due to normal increases in bone turnover, from the pathologic elevations in BGP due to excessive remodeling characteristic of osteoporosis.[148]

Yasumura and colleagues,[146] in a comprehensive study, compared BGP and total body calcium in normal pre- and postmenopausal women and postmenopausal osteoporotic patients. They reported that many of the normal perimenopausal women had increased serum BGP values, but that 15 years after the menopause most values had returned to normal. In normal women, as a whole, the mean BGP levels before and after the menopause were not significantly different despite the significantly postmenopausal reduction in total body calcium. However, when normal postmenopausal women were classified according to their serum BGP, women with the highest BGP levels had the lowest skeletal mass, as measured by total body calcium and phosphorus, and forearm linear bone density. Postmenopausal women with osteoporosis had mean serum BGP concentrations significantly higher than age-matched postmenopausal women without osteoporosis.[146] It remains to be seen whether normal women with high BGP values, reflecting a high bone-remodeling rate at the time of menopause, are predisposed to rapid skeletal loss, or vice versa.

Combined estrogen–progresterone therapy is believed to increase serum BGP, by virtue of the coupling effect by progesterone of bone formation to resorption.[112] with resulting increased bone formation. Indeed, in postmenopausal women estrogen alone depresses both bone formation and resorption and reduces serum BGP.[149]

Alkaline Phosphatase

Serum total alkaline phosphatase (AP) is the most commonly used marker of bone formation, but because it reflects the activity of different isoenzymes of AP derived from bone, liver, intestine, and placenta,[150] it lacks sensitivity and specificity for bone disease. Total AP increases with age, especially in women after the menopause, indicating increased osteoblastic activity,[91,151–153] but it correlates poorly with bone histo-morphometry.[154,155] Bone-specific AP assays using monoclonal antibodies should provide more specific data and might correlate better with the bone loss associated with the menopause. Recently it has been reported that wheat germ lectin–precipitated AP correlates well with bone mineralization rates.[156] Estrogen therapy in normal postmenopausal women reduces the elevated AP level to within the premenopausal range.[152]

Procollagen 1 Extension Peptides

Cleavage of the amino- and of the carboxyl-terminal sequence of type 1 procollagen results in the production of circulating peptides that might be useful markers of bone formation.[157,158] However, data on the sources, clearance, and degradation of these peptides are virtually nonexistent and there have been as yet no published studies of serum levels around the menopause.

Tartrate-Resistant Acid Phosphatase

The bone isoenzyme of acid phosphatase (isoenzyme 5), a lysosomal enzyme found in many tissues, can be distinguished from other isoenzymes by its resistance to tartrate.[159] It is believed to be a product of osteoclasts and, thus, a marker of bone resorption,[160] but sufficient data for comparison and evaluations are not yet available. Recently, Kraenzlin et al.[162] reported the development of an enzyme-linked immunoassay for osteoclast-specific tartrate-resistant acid phosphatase (TR ACP), which should be helpful in the evaluation of bone resorption, but the clinical value of TR ACP for the assessment of bone resorption at the natural menopause has not been tested. Stepan et al.[110] reported that changes in plasma TR ACP in response to artificial menopause and its treatment with estrogens and norethisterone were comparable in direction, magnitude, and time course to those of urinary hydroxyproline excretion, and, therefore, should be a useful marker of osteoclastic function in peri-menopausal women.[110]

Calcium

Although most studies have documented that the total fasting serum calcium increases significantly after the menopause, there is controversy regarding the origins of this elevation. Marshall et al.,[162] in a study that did not include age-matched controls, concluded that most of it could be accounted for by a rise in ionized calcium and that any contribution by alterations in proteins or ligands had to be small, since in this study there were no significant differences in mean albumin, globulin, and total protein concentrations between the pre- and postmenopausal women. In contrast to these results, Sokoll and Dawson-Hughes, in a study limited to women in the fifth decade of life, reported that total serum calcium but not ionized calcium increased significantly at menopause,[163] and that this increase in the protein-bound component of total calcium was associated with (and could be attributed to) small increases in mean serum globulin concentration and in serum pH. In agreement with these results, Nordin et al.[164] reported that total serum calcium declined with age in both the pre- and postmenopausal women. Previous workers had studied only total plasma calcium and had attributed the increase in total calcium to an increase in bone resorption, perhaps resulting from increased PTH secretion.[165,166] Estrogen therapy in normal post-menopausal women is associated with significant decreases in both total and ionized serum calcium.[35]

It has been reported that the malabsorption of dietary calcium may be a risk factor for psotmenopausal osteoporosis,[11,116] and that the rate of calcium absorption is positively related to vertebral density in post-menopausal osteoporotic women.[116] Nordin and Morris have proposed a "calcium deficiency model for osteoporosis,"[167] whereby a negative calcium balance in postmenopausal women results in osteoporosis, yet therapeutic replacement with calcium has never been proved to be consistently beneficial in restoring or attenuating bone loss in post-menopausal osteoporosis.

Phosphate

With age, the levels of inorganic phosphate may decrease slightly in both sexes,[18] or they may remain constant in women and decrease in men.[15,124] In women, at the time of menopause the level of inorganic phosphate begins to rise steeply, finishing 17% higher than premenopausal women and men.[18] This finding was confirmed by a recent cross-sectional and longitudinal study in a large number of postmenopausal women.[153]

Bicarbonate

A rise in serum bicarbonate has been described in postmenopausal women.[163,164] This rise may be due to alterations in the levels of serum

PTH at the menopause and consequent changes in acid excretion.[168] Any resultant change in serum pH may alter the protein-bound calcium fraction and this could have consequences for calcium turnover.[163]

Urine Markers

Calcium

Urinary calcium, which is a general reflection of bone resorption,[169] is usually measured in relation to the urinary excretion of creatinine. The 24-hr urinary calcium decreases with age, whereas the fasting calcium: creatinine ratio increases with age as renal function deteriorates.[170] Urinary calcium has been found to be higher in postmenopausal women than in premenopausal nonmatched controls,[153,171] an elevation that could be corrected by estrogen replacement.[112,152]

Hydroxyproline

Hydroxyproline (HP) is found almost exclusively in collagen,[170] and since free HP cannot be reincorporated into collagen, its urinary excretion reflects collagen breakdown.[171] Because bone, which contains about half of the collagen in the body, is the largest reservoir of collagen and the one with the most rapid turnover, urinary HP is commonly regarded as a reflection of bone resorption.[121] In normal women urinary HP increases with age,[137,140] and after the menopause indicates accelerated bone resorption.[152] Estrogen therapy decreases HP excretion of postmenopausal normal women to premenopausal levels.[112,152]

Recently it has been claimed that the combined measurements of urinary HP, body fat index, alkaline phosphatase, and urinary calcium can identify 79% of postmenopausal women with accelerated bone loss.[9] There have been no studies on the predictive value of combined urinary HP, serum BGP, and estrogen measurements and the risk of fracture.

Hydroxylysine

Hydroxylysine is also found uniquely in collagen and collagenlike proteins. Thus, like hydroxyproline, it is a potential marker of collagen breakdown.[121] Different forms of hydroxylysine, glucosylgalactosylhydroxylysine and galactosylhydroxylysine (GHL), are found in different proportions in different tissues. Recently, Moro et al.[172] reported that high urinary GHL excretion was associated with increased bone mineral loss. In another report, the same group reported that urinary GHL, corrected for creatinine, was as sensitive a marker for predicting the risk of osteoporosis as other more expensive and invasive tests.[173]

Pyridinium Crosslinks

Extracellular matrix is composed of collagen molecules stabilized by the formation of covalent cross-links between adjacent collagen chains. These cross-links consist of pyridinoline molecules, a reduced derivative of the amino acid lysine. Recently, methods have been developed to measure the urinary excretion of these cross-linking molecules as an index of mature collagen degradation.[174,175] Although there have been no studies on the effect of menopause, Black et al.[176] reported that measurements of urinary free pyridinoline (and its derivative deoxypyridinoline) appear to be a good index of bone resorption induced by estrogen deficiency in ovariectomized rats. A recent report concludes that "urinary hydroxylysylpyridinoline and lysylpyridinoline represents the first sensitive and specific marker of bone resorption,"[177] but this statement requires further substantiation. A review[177a] on the usefulness as well as the problems of measuring pyridinoline crosslinking amino acids in urine puts these biomarkers of bone resorption in perspective.

BGP and Gla

According to a recent report, the determination of 24-hr urinary BGP can distinguish normal from osteoporotic individuals better than either skeletal alkaline phosphatase or serum BGP,[178] because the urinary excretion of BGP fragments is less subject to circadian variations. The influence of renal function on this measurement is unclear.

Gla arises from numerous protein sources, including BGP. Under conditions when the amount derived from nonbone sources remain constant, it may be useful as a general marker of osteoporosis.[179]

Local Factors

Local growth factors such as interleukins, prostaglandins, transforming growth factors, and tumor necrosis factor are gaining recognition as important players in the pathogenesis of postmenopausal osteoporosis as their roles in bone mineral metabolism become more clear (see refs. 180, 181, 181A for reviews). Although little is known about changes in their local, let alone in their systemic, concentrations and excretion rates during the perimenopausal period (probably for lack of easily available or reliable assay techniques), a report by Pacifici et al.[182] states that blood monocytes from postmenopausal patients with fast bone turnover secrete significantly more interleukin-1 (IL-1) than age-matched women with low turnover. In light of the observation that IL-1 is a potent stimulator of bone resorption in vitro,[183] this might provide some insight into the association between menopause and osteoporosis. More recently, Pacifici et al.[184] have confirmed their initial report,[182] and added that monocytes from those patients who had been treated with ovarian steroids secreted less IL-1, suggesting that ovarian steroids may inhibit the increased IL-1 production by postmenopausal fast losers. In a prelimenary report, Hustmyer et al.[185] have disputed the above finding, and reported that monocytes from osteoporotics do not secrete more IL-1 than controls. Another interleukin which is receiving prominence as a regulater of bone resorption is interleukin-6,[181a] especially as its behaviour appears to be influenced by estrogens in experimental studies. Rosen et al.[186] described a difference in T-lymphocyte subsets between patients with Type 1 osteoporosis and those with Type 2 osteoporosis, as well as a significant negative correlation between the ratio of T-helper to T-cytotoxic cells and spinal bone mineral density. This "disorder" of the immune system suggests that an immununologic defect occurring in the perimenopausal period may alter the production of local factors, and thus play a role in the pathogenesis of postmenopausal osteoporosis.

Body Weight

There is strong statistical evidence that body weight influences bone density in both pre- and postmenopausal women independently of other variables.[90] This observation led to speculation that obesity may influence bone density either by a mechanical loading stress or because of an associated increase in estrogen secretion. A decrease in SHBG, possibly resulting in an increased bioavailability of sex steroids, has been observed in obese women.[93] Davidson et al.[93] also reported that postmenopausal patients with osteoporotic fractures were significantly more slender than age-matched controls. Osteoporosis is also more common in slender smokers.[99,187]

Genetic Factors

Studies in twins have demonstrated that both environmental and genetic factors influence bone mass in the adult and, although the problem has not been adressed specifically, it is conceivable that genetic factors influence perimenopausal bone metabolism as well.[188,189] Ethnic and racial background also appear to be determinants of perimenopausal bone mass,[190,191] as blacks compared to whites have a higher bone mass and a lower risk of fracture,[192] as well as lower postmenopausal biochemical indices of bone formation and resorption.[193] Differences in the vitamin D system in different races may be the reason,[194] although this notion has recently been disputed.[195]

Conclusion

This chapter has discussed the major determinants and biochemical consequences of the menopause with regard to bone mineral metabolism. Although estrogen deficiency is the major predisposing factor for postmenopausal osteoporosis, progesterone deficiency as well as environmental and genetic factors also play roles in its pathogenesis. Cortisol, IGF, and the thyroid hormones may contribute in selected cases.

Epidemiologically, postmenopausal, premenopausal, and osteoporotic postmenopausal women differ in numerous biochemical parameters of bone mineral metabolism. However, no single serum or urinary biochemical marker is available for the diagnosis of postmenopausal osteoporosis in an individual patient. Unsuccessful attempts have been made to combine and analyze several markers together. Despite intensive investigation, bone histomorphometric analysis at present remains the gold standard for describing the features of bone turnover in individual postmenopausal women, although bone density measurements using newer techniques such as dual x-ray absorptiometry may further help define those at risk. There is an urgent need for comprehensive longitudinal studies of determinants and markers of bone mineral metabolism in the perimenopausal period.

References

1. Riggs BL, Melton LJ III. Clinical heterogeneity of involutional osteoporosis: Implications for preventive therapy. *J Clin Endocrinol Metab*. 1990;70: 1229–1232.
2. Frost HM. Tetracycline-based histological analysis of bone remodelling. *Calcif Tissue Res*. 1969;3:211–237.
3. Stevenson JC, Whitehead MI. Postmenopausal osteoporosis. *Br Med J*. 1982;285:585–588.
4. Editorial. Osteoporosis. *Lancet*. 1987;ii:833–835.
5. Gallagher JC. The pathogenesis of osteoporosis. *Bone*. 1990;9:215–227.
6. Albright F, Smith PH, Richardson AM. Postmenopausal osteoporosis—its clinical features. *JAMA*. 1941;116:2465–2473.
7. Arlot M, Edouard C, Meunier PJ, Neer RM, Reeve J. Impaired osteoblast function in osteoporosis: comparison between calcium intake and dynamic histomorphometry. *Br Med J*. 1984;289:517–520.
8. Ettinger B, Genant HK, Cann CE. Long-term estrogen replacement therapy prevents bone loss and fractures. *Ann Intern Med*. 1985;102:319–324.
9. Christiansen C, Riis BJ, Rodbro P. Prediction of rapid bone loss in postmenopausal women. *Lancet*. 1987;1:1105–1108.
10. Tsai K-S, Heath H III, Kumar R, Riggs BL. Impaired vitamin D metabolism with aging in women. Possible role in pathogenesis of senile osteoporosis. *J Clin Invest*. 1984;73:1668–1672.
11. Gallagher JC, Riggs BL, Eisman J, Hamstra A, Arnaud SB, DeLuca HF. Intestinal calcium absorption and serum vitamin D metabolites in normal

subjects and osteoporotic patients: effect of age and dietary calcium. *J Clin Invest.* 1979;64:729–736.

12. Fujisawa Y, Kida K, Matsuda H. Role of change in vitamin D metabolism with age in calcium and phosphorus metabolism in normal human subjects. *J Clin Endocrinol Metab.* 1984;59:719–726.

13. Epstein S, Bryce G, Hinman JW, Miller ON, Riggs BL, Hui SL, Johnston CC Jr. The influence of age on bone mineral regulating hormones. *Bone.* 1986;7:421–425.

14. Orwoll ES, Meier DE. Alterations in calcium, vitamin D, and parathyroid hormone physiology in normal men with aging: relationship to the development of senile osteopenia. *J Clin Endocrinol Metab.* 1986;63:1262–1269.

15. Sherman SS, Hollis BW, Tobin JD. Vitamin D status in a healthy population: the effects of age, sex, and season. *J Clin Endocrinol Metab.* 1990;71:405–413.

16. Lamberg-Allardt C. The relationship between serum 25-hydroxyvitamin D levels and other variables related to calcium and phosphorus metabolism in the elderly. *Acta Endocrinol.* 1984;105:139–144.

17. Prince RL, Dick I, Garcia-Webb P, Retallack RW. The effects of the menopause on calcitriol and parathyroid hormone: response to a low dietary calcium stress test. *J Clin Endocrinol Metab.* 1990;70:1119–1123.

18. McPherson K, Healy MJR, Flynn FV, Piper KAJ, Garcia-Webb P. The effect of age, sex, and other factors on blood chemistry in health. *Clin Chim Acta.* 1978;84:373–397.

19. Prince R, Dick I, Boyd F, Kent N, Garcia-Webb P. The effects of dietary calcium deprivation on serum calcitriol levels in premenopausal and postmenopausal women. *Metabolism.* 1988;37:727–731.

20. Falch JA, Oftebro H, Haug E. Early postmenopausal bone loss is not associated with a decrease in circulating levels of 25-hydroxyvitamin D, 1,25-dihydroxyvitamin D, or vitamin D-binding protein. *J Clin Endocrinol Metab.* 1987;64:836–841.

21. Hartwell D, Riis BJ, Christiansen C. Changes in vitamin D metabolism during natural and medical menopause. *J Clin Endocrinol Metab.* 1990;71:127–132.

22. Scharla SH, Minne HW, Waibel-Treber S, Schaible A, Lempert UG, Wuster C, Leyendecker G, Ziegler R. Bone mass reduction after estrogen deprivation by long-acting gonadotropin-releasing hormone agonists and its relation to pretreatment serum concentrations of 1,25-dihydroxyvitamin D_3. *J Clin Endocrinol Metab.* 1990;70:1055–1061.

23. Lore F, Di Cairano G, Periti P, Caniggia A. Effect of the administration of 1,25-dihydroxyvitamin D_3 on serum levels of 25-hydroxyvitamin D in postmenopausal osteoporosis. *Calcif Tissue Int.* 1982;34:539–541.

24. Ambrecht HJ, Zenser TV, Davis BB. Effect of age on the conversion of 25-hydroxyvitamin D_3 to 1,25-dihydroxyvitamin D_3 by kidney of rat. *J Clin Invest.* 1980;66:1118–1123.

25. Lore F, Di Cairano G, Signorini AM, Caniggia A. Serum levels of 25-hydroxyvitamin D in postmenopausal osteoporosis. *Calcif Tissue Int.* 1981;33:467–470.

26. Buchanan JR, Cauffman S. Serum 25(OH)D and 24,25(OH)$_2$ concentrations in patients with postmenopausal osteoporosis. *Am Coll Surg*. 1981;22: 531–533.

27. Cheema C, Grant BF, Marcus R. Effects of estrogen on circulating "free" and total 1,25-dihydroxyvitamin D and the parathyroid-vitamin D axis in postmenopausal women. *J Clin Invest*. 1989;83:537–542.

28. Bouillon R, Van Assche FA, Van Baelen H, Heyns W, De Moor P. Influence of the vitamin D-binding protein on the serum concentration of 1,25-dihydroxyvitamin D$_3$: significance of the free 1,25-dihydroxyvitamin D$_3$ concentration. *J Clin Invest*. 1981;67:589–596.

29. Riggs BL, Melton LJ III. Involutional osteoporosis. *N Engl J Med*. 1986;314:1676–1686.

30. Stock JL, Coderre JA, Mallette LE. Effects of a short course of estrogen on mineral metabolism in postmenopausal women. *J Clin Endocrinol Metab*. 1985;61:595–600.

31. Goltzman D, Henderson B, Loveridge N. Cytochemical bioassay of parathyroid hormone. Characteristics of the assay and analysis of circulating hormonal forms. *J Clin Invest*. 1980;65:1309–1317.

32. Wiske PS, Epstein S, Bell NH, Queener SF, Edmonson J, Johnston CC Jr. Increases in immunoreactive parathyroid hormone with age. *N Engl J Med*. 1979;300:1419–1421.

33. Marcus R, Madvig P, Young G. Age-related changes in parathyroid hormone and parathyroid hormone action in normal humans. *J Clin Endocrinol Metab*. 1981;58:223–230.

34. Gallagher JC, Riggs BL, Jerpbak CM, Arnaud CD. The effect of age on serum immunoreactive parathyroid hormone in normal and osteoporotic women. *J Lab Clin Med*. 1980;95:373–385.

35. Boucher A, D'Amour P, Hamel L, Fugere P, Gascon-Barre M, Lepage R, Ste-Marie LG. Estrogen replacement decreases the set point of parathyroid hormone stimulation by calcium in normal postmenopausal women. *J Clin Endocrinol Metab*. 1989;68:831–836.

36. Silverberg SJ, Shane E, de la Cruz L, Segre GV, Clemens TL, Bilezikian JP. Abnormalities in parathyroid hormone secretion and 1,25-dihydroxyvitamin D$_3$ formation in women with osteoporosis. *N Engl J Med*. 1989;320: 277–281.

37. Gallagher JC, Riggs BL, DeLuca HF. Effect of estrogen on calcium absorption and serum vitamin D metabolites in postmenopausal osteoporosis. *J Clin Endocrinol Metab*. 1980;51:1359–1364.

38. Copp DH, Cameron EC, Cheney BA, Davidson AFG, Henze KG. Evidence of calcitonin: a new hormone from the parathyroid that lowers blood calcium. *Endocrinology*. 1962;70:638–649.

39. McDermott MT, Kidd GS. The role of calcitonin in the development and treatment of osteoporosis. *Endocr Rev*. 1987;8:377–390.

40. Heath H III, Sizemore GW. Plasma calcitonin in normal man: differences between men and women. *J Clin Invest*. 1977;69:1135–1140.

41. Deftos LJ, Weisman MH, Williams GW, Karpf DB, Frumar AM, Davidson BJ, Parthemore JG, Judd HL. Influence of age and sex on plasma calcitonin in human beings. *N Engl J Med*. 1980;302:1351–1353.

42. Body JJ, Struelens M, Borkowski A, Mandart G. Effect of estrogens and calcium on calcitonin secretion in postmenopausal women. *J Clin Endocrinol Metab*. 1989;68:223–226.

43. Greenberg C, Kukreja SC, Bowser EN, Hargis GK, Henderson WJ, Williams GA. Effects of estradiol and progesterone on calcitonin secretion. *Endocrinology*. 1986;118:2594–2598.

44. Body JJ, Heath H III. Estimates of circulating monomeric calcitonin: physiological studies in normal and thyroidectomized man. *J Clin Endocrinol Metab*. 1983;57:897–903.

45. Tiegs RD, Body JJ, Barta JM, Heath H III. Secretion and metabolism of monomeric human calcitonin: effects of age, sex, and thyroid damage. *J Bone Min Res*. 1986;1:339–349.

46. Hillyard CJ, Stevenson JC, MacIntyre I. Relative deficiency of plasma-calcitonin in normal women. *Lancet*. 1978;I:961–962.

47. Perez Cano R, Montoya MJ, Moruno R, Vazquez A, Galan F, Garrido M. Calcitonin reserve in healthy women and patients with postmenopausal osteoporosis. *Calcif Tissue Int*. 1989;45:203–208.

48. Prince RL, Dick IM, Price RI. Plasma calcitonin levels are not lower than normal in osteoporotic women. *J Clin Endocrinol Metab*. 1989;68:684–687.

49. Chestnut CH III, Baylink DJ, Sisom K, Nelp WB, Roos BA. Basal plasma immunoreactive calcitonin in postmenopausal osteoporosis. *Metabolism*. 1980;29:559–562.

50. Lore F, Galli M, Franci B, Martorelli MT. Calcitonin levels in normal subjects according to age and sex. *Biomed Pharmacother*. 1984;38:261–263.

51. Stevenson JC, Abeyasekera G, Hillyard CJ, Phang KG, MacIntyre I, Campbell S, Townsend PT, Young O, Whitehead MI. Calcitonin and the calcium-regulating hormones in postmenopausal women: effects of oestrogens. *Lancet*. 1981;1:693–695.

52. Parfitt AM. Calcitonin in the pathogenesis and treatment of osteoporosis. *Triangle*. 1983;22:91–102.

53. Leggate J, Farish E, Fletcher CD, McIntosh W, Hart DM, Sommerville JM. Calcitonin and postmenopausal osteoporosis. *Clin Endocrinol*. 1984;20:85–92.

54. Zseli J, Szucs J, Steczek K, Szathmari M, Kollin E, Horvath Cs, Guoth M, Hollo I. Decreased calcitonin reserve in accelerated postmenopausal osteoporosis. *Horm Metab Res*. 1985;17:696–697.

55. Reginster J-Y, Deroisy R, Denis D, Lecart MP, Sarlet N, Franchimont P. Is there any place for salmon calcitonin in prevention of postmenopausal bone loss? *Gynecol Endocrinol*. 1988;2:195–204.

56. Avioli LV. Rationale for the use of calcitonin in postmenopausal osteoporosis. *Ann Chir Gynaecol*. 1988;77:224–228.

57. Hurley DL, Tiegs RD, Barta J, Laakso K, Heath H III. Effects of oral contraceptive and estrogen administration on plasma calcitonin in pre- and postmenopausal women. *J Bone Min Res*. 1989;4:89–95.

58. Reginster JY, Deroisy R, Albert A, Denis D, Lecart MP, Collette J, Franchimont P. Relationship between whole plasma calcitonin levels, calcitonin secretory capacity, and plasma levels of estrone in healthy women and postmenopausal osteoporotics. *J Clin Invest*. 1989;83:1073–1077.

59. Tiegs RD, Body JJ, Wahner HW, Barta J, Riggs BL, Heath H III. Calcitonin secretion in postmenopausal osteoporosis. *N Engl J Med.* 1985;312:1097–1100.

60. Overgaard K, Riis BJ, Christiansen C, Podenphant J, Johansen JS. Nasal calcitonin for treatment of established osteoporosis. *Clin Endocrinol.* 1990;30:435–442.

61. Rudman D, Kutner MH, Rogers CM, Lubin MF, Fleming GA, Bain RP. Impaired growth hormone secretion in the adult population. Relation to age and adiposity. *J Clin Invest.* 1981;67:1361–1369.

62. Rudman D, Feller AG, Nagraj HS, Gergans GA, Lalitha PY, Goldberg AF, Schlenker RA, Cohn L, Rudman IW, Mattson DE. Effects of human growth hormone in men over 60 years old. *N Engl J Med.* 1990;323:1–6.

63. Bennet AE, Wahner HW, Riggs BL, Hintz RL. Insulin-like growth factors I and II: aging and bone density in women. *J Clin Endocrinol Metab.* 1984; 59:701–704.

64. Duursma SA, Biljlsma JWJ, Van Paassen HC, Van Buul-Offers SC, Skottner-Lundin A. Changes in serum somatomedin and growth hormone concentrations after 3 weeks' oestrogen sustitution in postmenopausal women; a pilot study. *Acta Endocrinol.* 1974;75:233–242.

65. Mosekilde L, Eriksen EF, Charles P. Effects of thyroid hormones on bone and mineral metabolism. In: Tiegs RD, ed. *Endocrinology and metabolism clinics of North America: metabolic bone disease, part II.* Philadelphia: Saunders; 1990;1:35–63.

66. Custro N, Scaffidi A, Gangi M, Pepe S. Functional manners of hypothalamus-pituitary-thyroid axis in post-menopause. *Minerva Endocrinol.* 1982;7: 269–276.

67. Custro N, Scafidi V. Mild hyperthyroidism with inappropriate secretion of TSH in postmenopausal women. *Acta Endocrinol.* 1986;111:204–208.

68. Bottiglioni F, de Aloysio D, Nicoletti G, Mauloni M, Mantuano R, Capelli M. A Study of thyroid function in the pre- and post-menopause. *Maturitas.* 1983;5:105–114.

69. Ballinger CB, Browning MCK, Smith AHW. Hormone profiles and psychological symptoms in peri-menopausal women. *Maturitas.* 1987;9: 235–251.

70. Rubenstein HA, Butler VP Jr., Werner SC. Progressive decrease in serum triiodothyronine concentration with human aging: radioimmunoassay following extraction of serum. *J Clin Endocrinol Metab.* 1973;37:247–253.

71. Rico H, Charro A, Depablos I, Bordiu E, Hernandez ER, Espinos D. Lack of hormonal changes in postmenopausal women of equal weight with and without osteoporosis, including relation to time of menopause. *Clin Rheumatol.* 1984;3:337–343.

72. Abdalla HI, Beastall G, Fletcher D, Hawthorn JS, Smith J, McK Hart D. Sex steroid replacement in postmenopausal women: effects on thyroid hormone status. *Maturitas.* 1987;9:49–54.

73. Smythe HA. Osteoarthritis, insulin and bone density. *J. Rheumatol.* 1987;14:91–93.

74. Sambrook PN, Eisman JA, Pocock NA, Jenkins AB. Serum insulin and bone density in normal subjects. *J Rheumatol.* 1988;15:1415–1417.

75. Knapen MH, Hamulyak K, Vermeer C. The effect of vitamin K supplementation on circulating osteocalcin (bone Gla protein) and urinary calcium excretion. *Ann Intern Med.* 1989;111:1001–1005.
76. Manwol SE, Menan KMJ. Changes in reproductive hormone secretion during the climateric and postmenopausal periods. *Clin Obstet Gynecol.* 1977;20:113–122.
77. Yen SSC. The biology of the menopause. *J Reprod Med.* 1977;18:287–296.
78. Lindsay R. Sex steroids in the pathogenesis and prevention of osteoporosis. In: Riggs BL, Melton LJ, eds. *Osteoporosis: etiology, diagnosis, and management.* New York: Raven Press; 1988:333–357.
79. Cann CE, Martin MC, Genant HK, Jaffe RB. Decreased spinal mineral content in amenorrheic women. *JAMA.* 1984;251:626–629.
80. Eriksen EF, Colvard DS, Berg NJ, Graham ML, Mann KG, Spelsberg TC, Riggs BL. Evidence of estrogen receptors in normal human osteoblast-like cells. *Science.* 1988;241:84–86.
81. Komm BS, Terpening CM, Benz DJ, Graeme KA, Gallegos A, Korc M, Greene GL, O'Malley BW, Haussler MR. Estrogen binding, receptor mRNA, and biologic response in osteoblast-like osteosarcoma cells. *Science.* 1988;241:81–84.
82. Sherman BW, Korenman SC. Hormonal characteristics of the human menstrual cycle throughout reproductive life. *J Clin Invest.* 1975;55:699–706.
83. Treloar AE, Boynton RE, Behn BG, Brown BW. Variation of the human menstrual cycle through reproductive life. *Int J Fertil.* 1967;12:77–126.
84. Monroe SE, Menon KMJ. Changes in reproductive hormone secretion during the climacteric and postmenopausal periods. *Clin Obstet Gynaecol.* 1977;20:1:113–122.
85. Metcalf MG, Donald RA, Livesey JH. Pituitary-ovarian function in normal women during the menopausal transition. *Clin Endocrinol.* 1981;14:245–255.
86. Elders PJM, Netelenbos JC, Lip P, Khoe E, van Ginkel FC, Hulshof KFAM, van der Stelt PF. Perimenopausal bone loss and risk factors. *Bone.* 1989;7:289–299.
87. Riggs BL, Wahner HW, Melton LJ III, Richelson LS, Judd HL, Offord KP. Rates of bone loss in the appendicular and axial skeletons of women: Evidence of substantial vertebral bone loss before menopause. *J Clin Invest.* 1986;77:1487–1491.
88. Cutler AB, Garcia R. The menopausal transition and beyond. In: Biello L, ed. *The medical management of the menopause and premenopause.* Philadelphia: Lippencott; 1984:1–48.
89. Richelson LS, Wahner HW, Melton LJ III, Riggs BL. Relative contributions of aging and estrogen deficiency to postmenopausal bone loss. *N Engl J Med.* 1984;311:1273–1275.
90. Steinberg KK, Freni-Titulaer LW, DePuey EG, Miller DT, Sgoutas DS, Coralli CH, Phillips DL, Rogers TN, Clark RV. Sex steroids and bone density in premenopausal and perimenopausal women. *J Clin Endocrinol Metab.* 1989;69:533–539.
91. Nilas L, Christiansen C. The pathophysiology of peri- and postmenopausal bone loss. *Br J Ostet Gynaecol.* 1989;96:580–587.

92. Riggs BL, Ryan RJ, Wahner HW, Jiang N-S, Mattox VR. Serum concentrations of estrogen, testosterone and gonadotropins in osteoporotic and non-osteoporotic postmenopausal women. *J Clin Endocrinol Metab.* 1973;36:1097–1099.

93. Davidson BJ, Ross RK, Paganini-Hill A, Hammond GD, Siiteri PK, Judd HL. Total and free estrogens and androgens in postmenopausal women with hip fractures. *J Clin Endocrinol Metab.* 1982;54:115–120.

94. Marshall DH, Crilly RG, Nordin BEC. Plasma androstenedione and oestrone levels in normal and osteoporotic postmenopausal women. *Br Med J.* 1977;2:1177–1179.

95. Slemenda C, Hui SL, Johnston CC. Sex steroids and bone mass. A study of changes around the time of the menopause. *J Clin Invest.* 1987;80: 1261–1269.

96. Dunn JF, Nisula BC, Rodbard D. Transport of steroid hormones: binding of 21 endogenous steroids to both testosterone-binding globulin and corticosteroid-binding globulin in human plasma. *J Clin Endocrinol Metab.* 1981;53:58–68.

97. Judd HL. Hormonal dynamics associated with the menopause. *Clin Obstet Gynecol.* 1976;19:775–788.

98. Wild RA, Buchanan JR, Myers C, Lloyd T, Demers LM. Adrenal androgens, sex-hormone binding globulin and bone density in osteoporotic menopausal women: is there a relationship. *Maturitas.* 1987;9:55–61.

99. Daniell HW. Osteoporosis of the slender smoker. Verterbral compression fractures and loss of metacarpal cortex in relation to postmenopausal cigarette smoking and lack of obesity. *Arch Intern Med.* 1976;136:298–304.

100. Anderson FS, Transbol I, Christiansen C. Is cigarette smoking a promoter of the menopause? *Acta Med Scand.* 1982;212:137–139.

101. MacMahon B, Trichopoulos D, Cole P, Brown J. Cigarette smoking and urinary estrogens. *N Engl J Med.* 1982;307:1062–1065.

102. Jensen J, Christiansen C, Rodbro P. Cigarette smoking, serum estrogens, and bone loss during hormone-replacement therapy early after menopause. *N Engl J Med.* 1985;313:973–975.

103. Longcope C, Johnston CC. Androgen and estrogen dynamics in pre- and postmenopausal women: a comparison between smokers and nonsmokers. *J Clin Endocrinol Metab.* 1988;67:379–383.

104. Slemenda CW, Hui SL, Longcope C, Johnston CC Jr. Cigarette smoking, obesity, and bone mass. *J Bone Miner Res.* 1989;737–741.

104a. Kill DP, Baron JA, Andemon JJ, Hannan MT, Felson DT. Smoking eliminates the protective effect of oral estrogens on the risk for hip fracture in women. *Ann Intern Med.* 1992;116:716–721.

105. Lindsay R. The menopause: Sex steroids and osteoporosis. *Clin Obstet Gynecol.* 1987;30:847–859.

106. Prior JC. Progesterone as a bone-tropic hormone. *Endocr Rev.* 1990;11:386–398.

107. Feldman D, Dziak R, Koehler R, Stern P. Cytoplasmic glucocorticoid binding proteins in bone cells. *Endocrinology.* 1975;96:29–36.

108. Chen TL, Aronow L, Feldman D. Glucocorticoid receptors and inhibition of bone cell growth in primary culture. *Endocrinology.* 1977;100:619–628.

109. Johnston CC, Norton JA, Khairi RA, Longcope C. Age related bone loss. In: Barzel E, ed. *Osteoporosis II*. New York: Grune & Stratton; 1980: 91–100.

110. Stepan JJ, Pospichal J, Schreiber V, Kanka J, Mensik J, Presl J, Pacovsky V. The application of plasma tartrate-resistant acid phosphatase to assess changes in bone resorption in response to artificial menopause and its treatment with estrogen and norethisterone. *Calcif Tissue Int*. 1989;45: 273–280.

111. Ettinger B. Prevention of osteoporosis: treatment of estradiol deficiency. *Obstet Gynecol*. 1988;72:12S–17S.

112. Christiansen C, Riis BJ, Nilas L, Rodbro P, Deftos P. Uncoupling of bone formation and resorption by combined oestrogen and progesterone therapy in postmenopausal osteoporosis. *Lancet*. 1985;ii:800–801.

113. Johansen JS, Riss BJ, Hassager C, Moen M, Jacobsen J, Christiansen C. The effect of a gonadotropin-releasing hormone agonist analog (nafarelin) on bone metabolism. *J Clin Endocrinol Metab*. 1988;67:701–706.

114. Manologas SC, Anderson DC, Lindsay R. Adrenal steroids and the development of osteoporosis in oophorectomised women. *Lancet*. 1979;ii:597–600.

115. Brody S, Carlstrom K, Lagrelius A, Lunell NO, Rosenberg L. Adrenocortical steroids, bone mineral content and endometrial conditions in postmenopausal women. *Maturitas*. 1982;4:113–122.

116. Nordin BEC, Robertson A, Seamark RF, Bridges A, Philcox JC, Need AG, Horowitz M, Morris HA, Deam S. The relation between calcium absorption, serum dehydroepiandrostenedione, and vertebral mineral density in postmenopausal women. *J Clin Endocrinol Metab*. 1985;60:651–657.

117. Chestnut CH, Ivey JL, Gruber HE, Mathews M, Nelp WB, Sisom K, Baylink DJ. Stanazol in postmenopausal osteoporosis: therapeutic efficacy and possible mechanisms of action. *Metabolism*. 1983;32:571–580.

118. Finklestein JS, Klibanski A, Neer RM, Greenspan SL, Rosenthal DI, Crowley WF. Osteoporosis in men with hypogonadotropic hypogonadism. *Ann Intern Med*. 1987;106:354–361.

118a. Finklestein JS, Neer RM, Beverley MK, Biller MD, Crawford JD, Klibanski A. Osteopenia in men with a history of delayed puberty. *New Engl J Med*. 1992;326:600–610.

119. Duke CJS. The multifactorial nature of osteoporosis: the potential of corticosteroid-binding globulin as a unifying regulator. *Med Hypotheses*. 1986;21:431–439.

120. Epstein S. Serum and urinary markers of bone remodeling: assessment of bone turnover. *Endocr Rev*. 1988;9:437–449.

121. Delmas PD. Biochemical markers of bone turnover for the clinical assessment of metabolic bone disease. In: Tiegs RD, ed. *Endocrinology and metabolism clinics of North America: metabolic bone disease, part II*. Philadelphia: Saunders; 1990;1:1–18.

122. Price PA. Vitamin K-dependent formation of bone Gla protein (osteocalcin) and its function. *Vit Horm*. 1985;42:65–108.

123. Hauschka PV. Osteocalcin: the vitamin K-dependent Ca^{2+}-binding protein of bone matrix. *Haemostasis*. 1986;16:258–272.

124. Lian JB, Gundberg CM. Osteocalcin: biochemical considerations and clinical applications. *Clin Orthop*. 1988;226:267–291.
125. Epstein S. Bone derived proteins. *Trends Endocrinol Metab*. 1989;Sept/Oct:9–14.
126. Price PA, Otsuka AS, Poser JW, Kristaponis J, Raman N. Characterization of a γ-carboxyglutamic acid containing protein from bone. *Proc Natl Acad Sci USA*. 1976;73:1447–1451.
127. Gallop PM, Lian JB, Hauschka PV. Carboxylated calcium binding proteins and vitamin K. *N Engl J Med*. 1980;302:1460–1466.
128. Poser JW, Esch FS, Ling NC, Price PA. Isolation and sequence of the vitamin K-dependent protein from human bone: under decarboxylation of the first glutamic acid residue. *J Biol Chem*. 1980;255:8685–8691.
129. Nishimoto SK, Price PA. Secretion of the vitamin K-dependent protein of bone by rat osteosarcoma cells: Evidence for an intracellular precursor. *J Biol Chem*. 1980;255:18574–18577.
130. Lian JB, Coutts M, Canalis E. Studies of hormonal regulation of osteocalcin synthesis in cultured rat calvariae. *J Biol Chem*. 1985;8706–8710.
131. Lian JB, Tassinari M, Glowacki J. Resorption of implanted bone prepared from normal and warfarin treated rats. *J Clin Invest*. 1984;73:1223–1226.
132. Rodan GA, Martin TJ. Role of osteoblasts in hormonal control of bone resorption—a hypothesis. *Calcif Tissue Int*. 1981;33:349–351.
133. Price PA, Parthemore JG, Deftos LJ. New biochemical marker for bone metabolism. Measurement by radioimmunoassay of bone Gla protein in the plasma of normal subjects and patients with bone disease. *J Clin Invest*. 1980;66:878–883.
134. Price PA, Williamson MK, Lothringer JW. Origin of the vitamin K-dependent bone protein found in plasma and its clearance by kidney and bone. *J Biol Chem*. 1981;256:12760–12766.
135. Garcia-Carrasco M, Gruson M, de Vernejoul MC, Denne MA, Miravet L. Osteocalcin and bone morphometric parameters in adults without bone disease. *Calcif Tissue Int*. 1988;42:13–17.
136. Epstein S, Poser J, McClintock R, Johnston CC Jr., Bryce G, Hui S. Differences in serum bone GLA-protein with age and sex. *Lancet*. 1984;i:307–310.
137. Delmas PD, Stenner D, Wahner HW, Mann KG, Riggs BL. Increase in serum bone γ-carboxyglutamic acid protein with aging in women: implications for the mechanisms of age-related bone loss. *J Clin Invest*. 1983;71:1316–1321.
138. Duda RJ, O'Brien JF, Katzman JA, Peterson JM, Mann KG, Riggs BL. Concurrent assays of circulating bone Gla-protein and bone alkaline phosphatase: effects of sex, age, and metabolic bone disease. *J Clin Endocrinol Metab*. 1988;66:951–957.
139. Galli M, Caniggia M. Osteocalcin in normal adult humans of different sex and age. *Horm Metab Res*. 1985;17:165–166.
140. Kotowicz MA, Melton LJ III, Cedel SL, O'Fallon WM, Riggs BL. Effect of age on variables relating to calcium and phosphorus metabolism in women. *J Bone Min Res*. 1990;5:345–352.
141. Melick RA, Farrugia W, Quelch KJ. Plasma osteocalcin in man. *Aust N Z J Med*. 1985;15:410–415.

142. Vanderschueren D, Gevers G, Raymaekers G, Devos P, Dequeker J. Sex- and age-related changes in bone and serum osteocalcin. *Calcif Tissue Int.* 1990;46:179–182.

143. Johansen JS, Riis BJ, Delmas PD, Christiansen C. Plasma BGP: an indicator of spontaneous bone loss and of effect of treatment in post-menopausal women. *Eur J Clin Invest.* 1988;18:191–195.

144. Ismail F, Epstein S, Fallon MD, Thomas SB, Reinhardt TA. Serum bone Gla protein and the vitamin D endocrine system in the oophorectomized rat. *Endocrinology.* 1988;122:624–630.

145. Brown JP, Malaval L, Chapuy MC, Delmas PD, Edouard C, Meunier PJ. Serum bone Gla protein: a specific marker for bone formation in post-menopausal osteoporosis. *Lancet* 1984;i:1091–1093.

146. Yasumura S, Aloia JF, Gunberg CM, Yeh J, Vaswani AN, Yuen K, Lo Monte AF, Ellis KJ, Cohn SH. Serum osteocalcin and total body calcium in normal pre- and postmenopausal women and postmenopausal osteoporotic patients. *J Clin Endocrinol Metab.* 1987;64:681–685.

147. Ismail F, Epstein S, Pacifici R, Droke D, Thomas SB, Avioli LV. Serum bone Gla protein (BGP) and other markers of bone mineral metabolism in postmenopausal osteoporosis. *Calcif Tissue Int.* 1986;39:230–233.

148. Delmas PD, Wahner HW, Mann KG, Riggs BL. Assessment of bone turn-over in postmenopausal osteoporosis by measurement of serum bone Gla-protein. *J Lab Clin Med.* 1983;102:470–476.

149. Podenphant J, Christiansen C, Catherwood BD, Deftos LJ. Serum bone Gla protein variations during estrogen and calcium prophylaxis of post-menopausal women. *Calcif Tissue Int.* 1984;36:536–540.

150. Moss DW. Alkaline phosphatase isoenzymes. *Clin Chem.* 1982;28: 2007–2016.

151. Crilly RG, Jones MM, Horsman A, Nordin BEC. Rise in plasma alkaline phosphatase at the menopause. *Clin Sci.* 1980;53:341–342.

152. Christiansen C, Christiansen ME, Larsen N-E, Transbol I. Patho-physiological mechanisms of estrogen effect on bone metabolism. Dose-response relationships in early postmenopausal women. *J Clin Endocrinol Metab.* 1982;55:1124–1130.

153. Nordin BEC, Polley K. Metabolic consequences of the menopause: a cross-sectional, longitudinal, and intervention study on 557 normal post-menopausal women. *Calcif Tissue Int.* 1987;41(Suppl 1):s1–s59.153.

154. Brown JP, Delmas PD, Arlot M, Meunier PJ. Active bone turnover at the cortico-endosteal envelope in postmenopausal osteoporosis. *J Clin Endocrino! Metab.* 1987;64:954–959.

155. Shifrin LZ. Correlation of serum alkaline phosphatase with bone formation rates. *Clin Orthop.* 1970;70:212–215.

156. Brixen K, Nielson HK, Eriksen EF, Charles P, Mosekilde L. Efficacy of wheat germ lectin-precipitated alkaline phosphatase in serum as an estimator of bone mineralization rate: comparison to serum total alkaline phosphatase and serum bone Gla-protein. *Calcif Tissue Int.* 1989;44:93–98.

157. Simon LS, Krane SMK. Procollagen extension peptides as markers of collagen synthesis. In: Frame B, Potts JT Jr., eds. *Clinical disorders of bone and mineral metabolism.* Amsterdam: Excerpta Medica; 1983:108–111.

158. Parfitt AM, Simon LS, Villanueva AR, Krane SM. Procollagen type I carboxy-terminal extension peptide in serum as a marker of collagen biosynthesis in bone. Correlation with iliac bone formation rates and comparison with total alkaline phosphatase. *J Bone Min Res*. 1987;2: 427–436.
159. Li CY, Chuda RA, Lam WKW, Yam LT. Acid phosphatases in human plasma. *J Lab Clin Med*. 1973;82:446–460.
160. Minkin C. Bone acid phosphatase: tartrate-resistant acid phosphatase as a marker of osteoclast function. *Calcif Tissue Int*. 1982;34:285–290.
161. Kraenzlin ME, Lau K-H, Liang L, Freeman TK, Singer FR, Stepan J, Baylink DJ. Development of an immunoassay for human osteoclast tartrate-resistant acid phophatase. *J Clin Endocrinol Metab*. 1990;71:442–451.
162. Marshall RW, Francis RW, Hodgkinson A. Plasma total and ionised calcium, albumin and globulin concentrations in pre- and post-menopausal women and the effects of estrogen administration. *Clin Chim Acta*. 1982;122:283–287.
163. Sokoll LJ, Dawson-Hughes B. Effect of menopause and aging on serum total ionized calcium and protein concentrations. *Calcif Tissue Int*. 1989;44:181–185.
164. Nordin BEC, Need AG, Hartley TF, Philcox JC, Wilcox M, Thomas DW. Improved method for calculating calcium fractions in plasma: reference values and effect of menopause. *Clin Chem*. 1989;35:14–17.
165. Young MM, Nordin BEC. Effects of natural and artificial menopause on plasma and urinary calcium and phosphorus. *Lancet*. 1967:i:118–120.
166. Gallager JC, Young MM, Nordin BEC. Effects of artificial menopause on plasma and urine calcium and phosphate. *Clin Endocrinol*. 1972;1:57–64.
167. Nordin BEC, Morris HA. The calcium deficiency model for osteoporosis. *Nutr Rev*. 1989;47:65–72.
168. Bichara M, Mercier O, Borensztein P, Paillard M. Acute metabolic acidosis enhances circulating parathyroid hormone, which contributes to the renal response against acidosis in the rat. *J Clin Invest*. 1990;86:430–443.
169. Nordin BEC. Diagnostic procedures in disorders of calcium metabolism. *Clin Endocrinol*. 1978;8:55–67.
170. Prockop DJ, Udenfriend S. A specific method for the analysis of hydroxyproline in tissues and urine. *Anal Biochem*. 1960;1:228–239.
171. Prockop DJ, Kivirikko KI, Tuderman L, Guzman NA. The biosynthesis of collagen and its disorders (first of two parts). *N Engl J Med*. 1979;301: 13–23.
172. Moro L, Mucelli RSP, Gazzarrini C, Modricky C, Marotti F, de Bernard B. Urinary β-1-galactosyl-O-hydroxylysine (GH) as a marker of collagen turnover of bone. *Calcif Tissue Int*. 1988;42:87–90.
173. Moro L, Modricky C, Rovis L, de Bernard B. Determination of galactosyl hydroxylysine in urine as a means of identification of osteoporotic women. *Bone*. 1988;3:271–276.
174. Robins SP, Stewart P, Astbury C, Bird HA. Measurement of the cross linking compound, pyridinoline, in urine as an index of collagen degradation in joint disease. *Ann Rheum Dis*. 1986;45:969–973.

242   Ian A. Katz, Sol Epstein

175. Black D, Duncan A, Robins SP. Quantitative analyses of the pyridinium crosslinks of collagen in urine using ion-paired reversed-phase high-performance liquid chromatography. *Anal Biochem*. 1988;169:197–203.
176. Black D, Farquharson C, Robins SP. Excretion of pyridinium cross-links of collagen in ovariectomized rats as urinary markers for increased bone resorption. *Calcif Tissue Int*. 1989;44:343–347.
177. Uebelhart D, Gineyts E, Chapuy M-C, Delmas PD. Urinary excretion of pyridinium crosslinks: a new marker of bone resorption in metabolic bone disease. *Bone*. 1990;8:87–96.
177a. Eyre D. Editorial: New biomarkers of bone resorption. *J Clin Endocrinol Metab*. 1992;74:470a–470c.
178. Taylor AK, Linkhart S, Mohan S, Christenson RA, Singer FR, Baylink DJ. Multiple osteocalcin fragments in human urine and serum as detected by a midmolecule osteocalcin radioimmunoassay. *J Clin Endocrinol Metab*. 1990;70:467–472.
179. Gundberg CM, Lian JB, Gallop PM, Steinberg JJ. Urinary γ-carboxyglutamic acid and serum osteocalcin as bone marker. Studies in osteoporosis and Paget's disease. *J Clin Endocrinol Metab*. 1983;57:1221–1225.
180. Canalis E, McCarthy T, Centrella M. Growth factors and the regulation of bone remodeling. *J Clin Invest*. 1988;81:277–281.
181. Canalis E, McCarthy TL, Centrella M. The role of growth factors in skeletal remodeling. In: Tiegs RD, ed. *Endocrinology and metabolism clinics of North America: metabolic bone disease, part I*. Philadelphia: Saunders; 1989:903–918.
181a. Manolagas SC, Jilka R. Cytokines, hematopoeisis osteoclastogenesis, and estrogens. *Calcif Tissue Int*. 1992;50:199–202.
182. Pacifici R, Rifas L, Teitelbaum S, Slatopolsky E, McCracken R, Bergfeld, M, Lee W, Avioli LV, Peck WA. Spontaneous release of interleukin 1 from human blood monocytes reflects bone formation in idiopathic osteoporosis. *Proc Natl Acad Sci USA*. 1987;84:4616–4620.
183. Lorenzo JA, Sousa SL, Alander C, Raisz LG, Dinarello CA. Comparison of the bone-resorbing activity in the supernatants from phytohemagglutinin-stimulated human peripheral blood mononuclear cells with that of cytokines through the use of an antiserum to interleukin 1. *Endocrinology*. 1987;121:1164–1170.
184. Pacifici R, Rifas L, McCracken R, Vered I, McMurty C, Avioli LV, Peck WA. Ovarian steroid treatment blocks a postmenopausal increase in blood monocyte interleukin 1 release. *Proc Natl Acad Sci USA*. 1989;86:2398–2402.
185. Hustmyer FG, Benninger L, Girasole G, Sakagami Y, Yu XP, Walker EB, Peacock M, Manolagas SC. Cytokine production and cell-surface marker analysis in blood mononuclear cells in osteoporosis. *J Bone Min Res*. 1990;5:s109 no. 141 (abstr.).
186. Rosen CJ, Usisken K, Owens M, Barlascini CO, Belsky M, Adler RA. T lymphocyte surface antigen markers in osteoporosis. *J Bone Min Res*. 1990;5:851–855.
187. Slemenda CW, Hui SL, Longcope C, Wellman H, Johnston CC Jr. Predictors of bone mass in perimenopausal women: a prospective study of

clinical data using photon absorbtiometry. *Ann Intern Med.* 1990;112: 96–101.

188. Pocock NA, Eisman JA, Hopper JL, Yeates MG, Sambrook PN, Eberl S. Genetic determinants of bone mass in adults. A twin study. *J Clin Invest.* 1987;80:706–710.

189. Christian JC, Yu P-L, Slemenda CW, Johnston CC. Heritability of bone mass: a longitidinal study in aging male twins. *Am J Hum Genet.* 1989;44:429–433.

190. Lappe JM, Stegman MR, Heaney RP, Recker RR. Ethnic background as a predictor of osteoporotic fracture syndrome. *J Bone Min Res.* 1990;5:S118 no. 177 (abstr.).

191. Liel Y, Edwards J, Shary J, Spicer KM, Gordon L, Bell NH. The effects of race and body habitus on bone mineral density of the radius, hip, and spine in premenopausal women. *J Clin Endocrinol Metab.* 1988;66:1247–1250.

192. Farmer ME, White LR, Brody JA, Baily KR. Race and sex differences in hip fracture incidence. *Am J Public Health* 1984;74:1374–1380.

193. Meier DE, Luckey S, Wallenstein S, Lapinski R, Catherwood BD. Biochemical evidence of racial differences in postemenopausal bone homeostasis in healthy white and black women. *J Bone Min Res.* 1990;5:s174 no. 402 (abstr.).

194. Bell NH, Greene A, Epstein S, Oexmann MJ, Shary J. Evidence for alteration of the vitamin D-endocrine system in blacks. *J Clin Invest.* 1985;76:470–473.

195. Reid IR, Cullen S, Schooler BA, Livingston NE, Evans MC. Calcitropic hormone levels in polynesians: evidence against their role in interracial differences in bone mass. *J Clin Endocrinol Metab.* 1990;70:1452–1456.

11
Influence of Hypokinesia and Weightlessness on Jaw Bones and Teeth

BERND-MICHAEL KLEBER and KARL HECHT

Hypokinesia has been used to simulate some effects of weightlessness on the organism in preparation for manned space flight on the assumption that weightlessness and hypokinesia would cause similar stress reduction on bones and muscles and hence similar effects on their structure and function. This assumption proved correct. Thus, research on hypokinesia has become an integral part of the space sciences and has expanded to include structural and functional studies of all organ systems and of the human adaptation mechanisms. Indeed, antiorthostatic hypokinesia is now part of all cosmonaut and astronaut training programs. These observations began more than 20 years ago, when Kakurin[1] began to describe the effects of long-term hypokinesia and antiorthostatic positions at various angles in relation to the horizontal on physical and psychologic training. The subjective feelings and objective changes during space flight and hypokinesia were similar: under both sets of conditions the subjects felt an increased pressure in the head, difficulty in breathing through the nose, nausea upon rapid movement of the head and eyes, and erroneous preception of their actual body position. Meanwhile, the pathogenesis of the phenomena induced by hypokinesia was being investigated[2] and the first papers on the physiologic and biochemical changes during hypokinesia were published.[3] More recently, new noninvasive methods for the examination of human bone tissue became available, allowing for better studies of the changes in mineral metabolism under hypokinetic conditions and their possible avoidance through suitable physical exercise and pharmacologic treatment.[4] Different components of the immune system have also been examined during 20 days of antiorthostatic hypokinesia in 30 men, 20 to 38 years of age.[5] Unspecific suppressor activities of T lymphocytes were found to increase whereas activities of T-helper cells showed no change under such extreme long-term conditions. Cytotoxic activities of killer lymphocytes remained normal at first, but later decreased drastically, a condition that continued for about a month after the end of the experiment, indicating that long-term hypokinesia reduces the resistance to stress and to infection. During the antiorthostatic experi-

TABLE 11.1. Chronologic comparison of some physiologic parameters in humans and rats.[52]

Parameter	Rat (mos)	Human (mos)	Ratio
Gestation	0.7	9	1:13
Puberty	2.7	153	1:57
Growth period	7.3	294	1:34
Average life span	30.7	849	1:28

ment, blood volume, plasma volume, total erythrocyte mass, and extra- and intracellular fluid volume decreased as well.[6] However, all these changes were reversible. Generally, the effects of chronic hypokinesia on the organism are similar to those of stress. Emotional and physical stress generate unspecific and specific reactions that can influence metabolism, especially that of the bones and muscles, all of which, as far as they have been investigated, have an adverse effect on the working capacity of the human organism. Nevertheless, hypokinesia should not be considered a static condition; rather, it should be viewed as an evolving process, with periods of greater deviation from the individual homeostatic values and periods during which functions remain within the range of normalcy. Thus, hypokinesia is useful to detect how far an organism can adapt to conditions of external stress and to study the relations between normal and pathologic states. For long-term studies on the effects of hypokinesia and weightlessness on bone tissues, the rat has proven to be a useful experimental animal, especially when the chronology of some major biologic events, relative to man, is kept in mind (Table 11.1). However, so far no systematic investigations on the specific impact of hypokinesia on jaw bones and teeth have been carried out.[3] We describe here the results of such investigations.

Hypokinesia

Several weeks of hypokinesia constitute a powerful stress for the mammalian organism.[7] The effects of stress, such as disturbances of circadian rhythms, noise, lack of oxygen, and social and emotional stress may be divided into three stages:[8,9] a stage of inhibition (protection), a stage of adaptation (restoration of health or setting the stage for pathogenesis), and, when the duration of the stress finally weakens the adaptive potential, a premorbid stage (etiology of diseases)[10] (Fig. 11.1). Chronic hypokinesia stresses the organism by the very lack of stress,[11] a condition comparable to denervation. Two components may be identified in hypokinesic stress: an emotional one and a metabolic one. Some well known effects of chronic hypokinesia on the mammalian organism are summarized in Table 11.2.

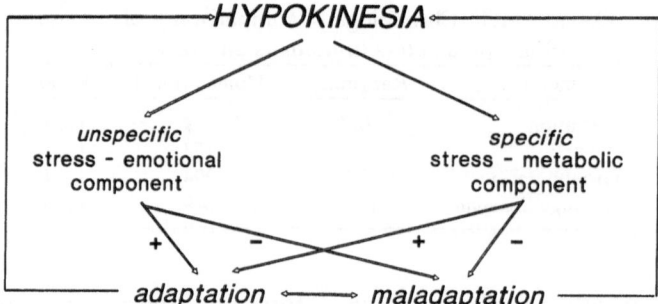

FIGURE 11.1. Diagram of hypokinesia as a stressor.

TABLE 11.2. Effect of chronic hypokinesia on the mammalian organism.[3]

Event	Consequences
Reduced energy requirements	Change in gas exchange
	Reduced muscle strength
	Reduced capacity for work
Reduced use of muscles	Reduced muscle mass
	Reduced body mass
	Reduced muscle strength
Reduced afferent impulses	Disturbed coordination of body movements
Redistribution of circulating blood, reduced	Deconditioning of cardiovascular system
demand on cardiovascular system	Loss of cardiovascular function under stress
Changes in skeletal loading	Structural changes in bones
	Excretion of calcium
	Reduced bone strength
Changes in central nervous system activity	Disturbed trophic reactions
	Reduced reaction capacity
Reduced tonus in central nervous system	General weakness of organism

Many workers have investigated the influence of hypokinesia on bone metabolism. The most important findings have been a reduction of cortical tissue[12,13] and of spongy substance (decreased amount and thickness of trabeculae, destruction of trabeculae, irregular trabecular positioning),[13,14] an increased amount of fat in the bone marrow,[2,13] osteoporosis,[2,13] demineralization and changes in bone density,[13,15] a decreased number of osteoblasts and osteocytes,[2] and an increased number of osteoclasts.[14] Although most of these results were obtained from the study of weight-bearing bones under terrestrial conditions, their similarity to the effects of chronic hypokinesia and long-term space conditions (up to 3 weeks) on the bone tissue of rats is impressive[9] (Table 11.3).

TABLE 11.3. Macroscopic, microscopic, and mechanical bone changes caused by chronic hypokinesia and long-term orbital flights in rats.[9]

Chronic hypokinesia	Long-term weightlessness
Reduced periosteal tissue	Reduced periosteal growth
Reduced cortical thickness	Reduced cortical thickness
Reduced volume of spongy substance (fewer trabeculae, reduced thickness, bone loss, irregularities)	Reduced volume of spongy substance (fewer trabeculae, reduced thickness)
Reduced endosteal bone diameter	Reduced endosteal bone diameter
Widened bone marrow canal; increased fat content of bone marrow	Widened bone marrow canal; increased fat content of bone marrow
Osteoporosis	Osteoporosis
Demineralization, reduced fatty content of bones	Demineralization, redistribution of minerals, reduced bone strength
Disturbed protein metabolism	Reduced collagen content, slower collagen maturation
Reduced vascularization	Reduced vascularization
Reduced number of osteoblasts	Reduced number of osteoblasts
Increased number of osteoclasts	Increased number of osteoclasts

Influence of Hypokinesia on Mandibular Bone and Teeth: Personal Investigations

The first phase in our program of biomedical space research included model experiments about the influence of hypokinesia on mandibular bone and teeth. The second (and current) phase is dedicated to the impact of weightlessness on these structures.

Hypokinesia

After 1 week of acclimatization, 34 Wistar rats, with a body weight of about 250 g each, but otherwise unselected, were separated into a 10-rat control group and an experimental group, consisting of 24 animals. Hypokinesia was achieved by immobilization in a specially constructed cell that prevented all movements except those of the head and the jaw. The animals were taken from the cells for 1 hr at different times of each day. This allowed the animals to groom themselves while the cells were being cleaned. The control animals were kept in standard cages, five animals in each. Both groups received standard food pellets and piped-in water ad libitum. Weights were measured twice daily. Ater 1 month, six animals from the experimental group were exposed to additional stress by tying them up for 2 hr to test for the exhaustion of the general system of adaptation.[16] After this, the six rats, six additional experimental rats, and five controls were killed by cutting the carotid artery; 5 ml of blood per animal was collected for plasma analysis. Immediately thereafter, the adrenal glands were extirpated and weighed and the mandibles and the right tibia were dissected and stored in 5% buffered formalin. After an additional month of hypokinesia, the remaining six experimental rats and five controls were killed using the same procedure. The blood plasma of the animals was used to determine corticosterone concentrations according to Stahl et al.[17] Dopamine-hydroxylase (DBH) activity was measured photometrically according to Kato et al.[18] and phenyl-ethanolamine-*N*-methyltransferase (PNMT) was determined radioenzymatically using the method of Axelrod,[19] as modified by Molinoff et al.[20] The distances between the distal surface of the third molars and the margin of the incisors were measured on both sides of the upper and lower jaws, as was the distance between the condylus medialis and the malleolus lateralis of the right tibia. For this purpose, we used a ruler and dividers.[9,21,22]

The bones were evaluated histomorphometrically after decalcification (in 5% nitrohydrochloric acid for 36 hr) and embedded in paraffin, using 5-μm sagittal sections stained with hematoxylin-eosin or PAS/van Giesen. At least five preparations per animal and jaw were used. Qualitative and quantitative determinations of the osteocytes, cementocytes, and osteoclasts in the central root region were made. In addition, we examined the quality and quantity of the nuclei of fibroblasts and the capillaries in the periodontal ligament of the molars in the upper jaw. These studies were carried out according to the criteria of Weibel[23] employing the method of counting points at a 1000× magnification. All data obtained were examined statistically (with a $p < .05$ threshold of significance) using international computer programs for Student, Mann-Whitney, F-, and chi-square tests.

Results

General Findings

During the 2 months of the experiment, the two groups of rats did not differ in their food and water intake. Nevertheless, the hypokinetic animals lost weight during the first 5 days and, during the the first month, put on less weight than those of the control group. This trend continued during the second month, although the difference in weight gain between hypokinetic and control animals was somewhat less than in the first month (Table 11.4).

Endocrinologic Findings

After 1 and 2 months of experimentation, the plasma levels of corticosterone of the hypokinetic animals were markedly lower than those of

TABLE 11.4. Changes of body mass after 2 months of hypokinesia.

Day of experiment	Exp. animals (n = 24) ($\bar{x}$)	SD	Controls (n = 10) $\bar{x}$	SD
1	239.6	6.7	255.5	9.2
5	224.9	6.0	259.3	9.6[a]
10	229.4	6.6	270.5	10.4[a]
15	232.7	5.9	274.7	6.6[a]
20	233.8	6.3	278.8	8.8[a]
25	236.1	6.6	287.0	9.0[a]
30	242.7	11.0	271.6	12.0
35	242.9	10.8	272.0	12.0
40	242.8	10.8	273.2	11.4
45	248.9	10.7	277.8	14.4
50	248.2	11.6	285.0	15.9
55	249.0	11.4	289.8	16.9

[a] $p < .05$.

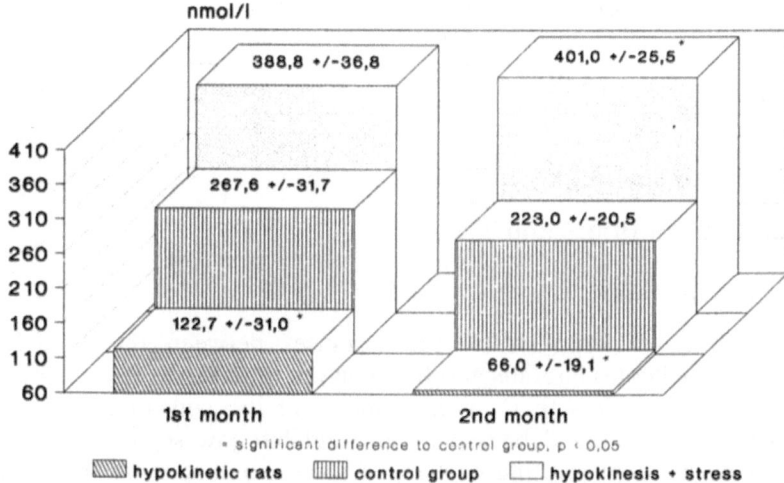

FIGURE 11.2. Concentrations of corticosterone (nmol/l) in plasma of hypokinetic (n = 24) and control (n = 10) rats after 1 and 2 months of hypokinesia (*$p \leq .05$).

the control group, whereas after 2 hr of total mechanical restraint, these levels increased drastically (Fig. 11.2). The weight of the adrenal glands did not change after either 1 or 2 months of hypokinesia. DBH and PNMT enzyme activities showed a significant decrease after 1 month of hypokinesia. After 2 months, the DBH activity was found to be even lower whereas the PNMT activity had returned to control levels (Tables 11.5 and 11.6).

TABLE 11.5. Dopamine-β-hydroxylase activity ($M\,mol/g \cdot h^{-1}$) in adrenal glands of hypokinetic (n = 24) and control (n = 10) rats after 1 or 2 months of hypokinesia.

		1st month	2nd month
Exp. animals	x̄	1.25	0.88
	SD	0.25	0.14
Exp. animals + stress (2 hr)	x̄	2.82	1.41
	SD	0.35	0.48
Controls	x̄	3.14	3.51
	SD	0.56	0.64

TABLE 11.6. Phenylethanolamine-N-methyltransferase activity ($\mu mol/g \cdot h^{-1}$) in the adrenal glands of hypokinetic (n = 24) and control (n = 10) rats after 1 or 2 months of hypokinesia.

		1st month	2nd month
Exp. animals	x̄	61.8	72.5
	SD	2.5	2.2
Exp. animals + stress (2 hr)	x̄	74.9	79.6
	SD	2.7	4.8
Controls	x̄	69.9	72.1
	SD	4.4	4.3

Observations on Bones and Teeth

MACROSCOPIC FINDINGS

After 1 month of hypokinesia, the distances between the condylus medialis and malleolus lateralis and between the third molar and the edge of the incisor were less than in the animals of the control group. No differences were observed between right and left jaw segments. After 2 months of hypokinesia, these changes could no longer be detected. (Table 11.7).

MICROSCOPIC FINDINGS

After 1 month, quantitative and qualitative changes in the osteocytes of the upper jaw could be shown in the hypokinetic rats, whereas no such changes occurred in the osteocytes of the control animals (Fig. 11.3). In particular, decreases were noted in the number of osteocytes, the density of cytoplasm, the nuclear volume, the numerical density of nuclei and cytoplasm, and the surface density of the cytoplasm. The findings for the lower jaw were similar. Further changes were noted after the additional stress of fettering, mostly in the morphometric parameters of the osteocyte cytoplasma.

TABLE 11.7. Distances (mm) between the condylus medialis and malleolus lateralis of the right tibia and between the distal site of the third molar and the edge of the incisor in the jaw bones of hypokinetic (n = 24) and control (n = 10) rats after 1 or 2 months of hypokinesia.

		1st month		2nd month	
		Exp. animals	Controls	Exp. animals	Controls
Tibia	$\bar{x}$	37.87[a]	39.3	38.58	39.14
	SD	0.3	0.25	0.27	0.18
Upper jaw, right	$\bar{x}$	20.9[a]	21.45	20.68	20.84
	SD	0.21	0.12	0.2	0.2
Upper jaw, left	$\bar{x}$	20.97[a]	21.37	20.7	20.78
	SD	0.24	0.02	0.2	0.21
Lower jaw, right	$\bar{x}$	19.47	19.82	19.04	19.24
	SD	0.15	0.07	0.18	0.2
Lower jaw, left	$\bar{x}$	19.5	19.74	19.05	19.22
	SD	0.18	0.16	0.19	0.21

[a] $p < .05$.

After 2 months of hypokinesia, the situation was reversed and the following changes were observed: an increase in the number of osteocytes in the upper jaw with no further decrease of osteocytes showing decreased amounts of cytoplasm in the lower jaw (Fig. 11.3); increases in the density of the nuclear surface, in the nuclear volume, and in the numerical nuclear density; a decrease in individual cell cytoplasmic volume and in the median cut surface area of the cytoplasm. There was also a reduction in the number of cementocytes per unit area. After 2 months of hypokinesia, however, this trend had reversed and the number of cementocytes had increased (Fig. 11.4). In the lower jaw, no statistically significant changes were noted. The average number of osteoclasts per root in the upper jaw was increased after 1 month of hypokinesia, but tended to decrease after 2 months.

The following changes in the fibroblasts could be seen after 2 months of hypokinesia: an increase in the number of fibroblast nuclei per unit area, an increase in the numerical nuclear density and nuclear surface density, and a decrease of nuclear volume, median nuclear cut surface, and density of nuclear volume (Fig. 11.5). Similarly, a reduced number and size of the capillaries could be shown (Fig. 11.6). This trend was found to be reversed after 2 months of hypokinesia.

Discussion and Conclusions

Our experiments were motivated by the well known stress effect of hypokinesia,[7,10,24] by the recognition that chronic hypokinesia may be used as a model for some of the effects of weightlessness,[3,14] and by our

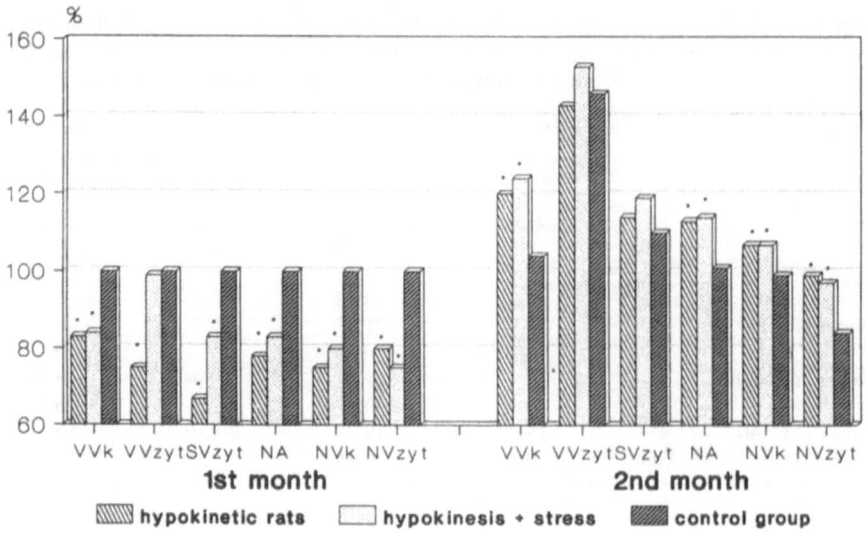

FIGURE 11.3. Morphometry of osteocytes of hypokinetic and control rats. Upper jaw, region of the first molar. Effects of 1 or 2 months of hypokinesia followed by tying for 2 hr. VVk, density of nuclear volume; VVzyt, volume density of cytoplasm; SVzyt, surface density of cytoplasm; NA, number of cells per unit area; NVk, numerical density of nuclei; NVzyt, numerical density of cytoplasm ($*p \leq .05$).

previous experience with hypokinesia in rats. To evaluate the effect of stress we studied corticosterone levels, the weight of the adrenal glands, and the activities of DBH and PNMT. Our corticosterone findings must be interpreted with caution as we worked with a cortisol assay with a relatively low specificity. We have interpreted the decreases in body weight of the animals in the hypokinetic group during the first month of the experiment as an expression of unspecific[16] and specific (reduced muscle and bone mass, loss of fluid)[3] stress responses to hypokinesia. On the other hand, the slight weight increase in the second month of the experiment suggests adaptation. Woloschin[25] observed a 15% to 20% reduction in the mass of femur and pelvis associated with a marked increase in calcium excretion after 30 days of hypokinesia in rats. Poppei[26] found that chronic intermittent hypokinesia had no effect on the food consumption of rats. This was confirmed by our observations. The lower plasma levels of corticoids after 1 and 2 months of hypokinesia is in full agreement with the findings of other investigators,[3,27,28] who also noticed a decrease of corticoid concentrátions in situations of chronic stress. However, in agreement with Kvetnansky et al.,[29] we believe that this finding is evidence of an adaptive reaction of the organism to hypokinesia

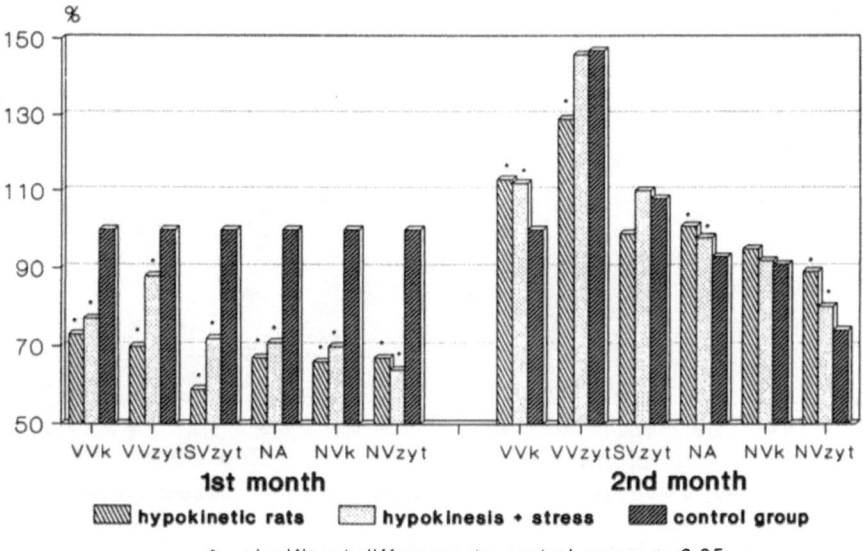

FIGURE 11.4. Morphometry of cementocytes of hypokinetic (n = 24) and control (n = 10) rats. Upper jaw, region of the first molar. Effects of 1 or 2 months of hypokinesia followed by tying for 2 hr. VVk, density of nuclear volume; VVzyt, volume density of cytoplasm; SVzyt, surface density of cytoplasm; NA, number of cells per unit area; NVk, numerical density of nuclei; NVzyt, numerical density of cytoplasm (*$p \leq .05$).

rather than a sign of exhaustion. Indeed, after acute stress our hypokinetic animals showed marked increases in corticoid concentrations. DBH and PNMT play an essential role in the biosynthesis of noradrenalin and adrenalin in the adrenal gland. We observed an initial decrease in enzyme activity after the onset of chronic hypokinesia. After additional acute stress, however, this activity increased markedly, in agreement with the findings of Van Loon.[30] Generally, the unspecific reactions of stress result in loss of body weight, in changes of hypothalamo–pituitary–adrenal axis activity, and in changes in the activity of enzymes involved in the synthesis of catecholamines.[31,32]

Our findings regarding changes in the length of the tibiae and of both jaw bones confirm the observations of Woloschin[25] in rabbits, who also described an inhibition of osteogenesis and an increase of bone absorption during long-term hypokinesia. Previous observations in our laboratory on hypokinesia-induced changes in jaw bones in rats[21] were also confirmed. We attribute the fact that the changes observed after 1 month of hypokinesia had disappeared after 2 months to "overadaptation" or to a "catching up" effect. The less conspicuous changes observed in the

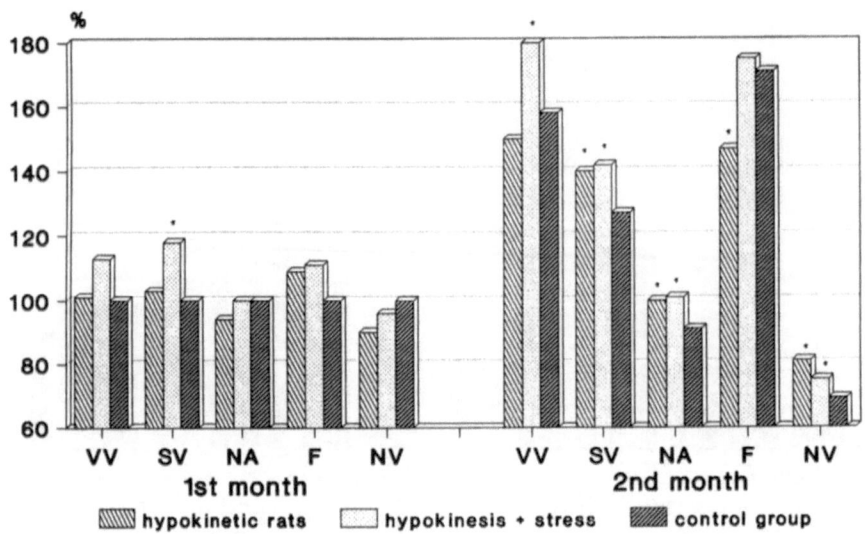

FIGURE 11.5. Morphometry of fibroblast nuclei in the periodontal ligament from the first molar of the upper jaw. Effects of 1 or 2 months of hypokinesia followed by tying for 2 hr. VV, volume density; SV, surface density; NA, number per unit area; F, median cut surface area; NV, numerical density (*$p < .05$).

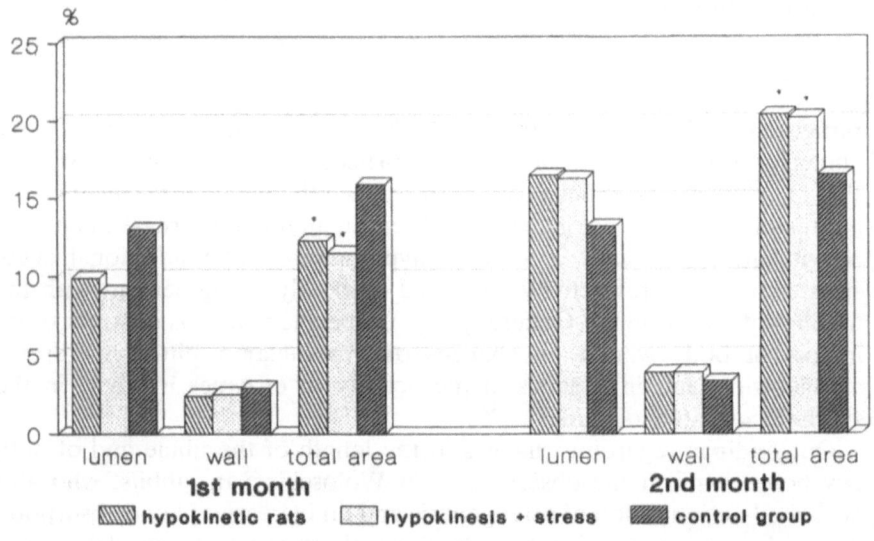

FIGURE 11.6. Relative areas of capillaries in the periodontal ligament of the first molar in hypokinetic (n = 24) and control (n = 10) rats. Effects of 1 or 2 months of hypokinesia followed by tying for 2 hr (*$p \leq .05$).

mandibles must be interpreted in light of the more stable structure of this bone.[33]

The demonstrated decrease in the number of osteocytes in the root area of the upper and lower jaw after 1 month of hypokinesia corresponds to the reduced number of osteocytes in the hip bone after chronic hypokinesia described by Kowalenko and Gurowski.[3] Simmons et al.[34] noted a similar effect in the root areas of animals after experiments in space. The reduced size of the osteocyte nuclei and cytoplasm, added to a drop in cell number, is an additional indicator of stress. The cementocytes reacted in the same way, suggesting a similarity between the two cell types. At the same time we noticed an increase of osteoclasts in the root area of the upper jaw, which indicates an increased absorptive activity during chronic hypokinesia.[3,35] The reduced number of vessels with a corresponding decrease in the volume of the periodontal ligament of the upper jaws of hypokinetic animals confirms previous observations, whereas the decreased number of capillaries, also previously observed, may represent a prerequisite for the quantitative and qualitative changes in cell structure.[25]

After 2 months of hypokinesia, the animals showed adaptive reactions that nearly abolished most differences between experimental and control groups. A possible indication of this overreaction may be the reduced cytoplasmic content of the cells involved in the process. According to Roberts et al.,[36] the periodontal ligament constitutes a limited but good model of bone response to stress.

In summary, one can say that during chronic hypokinesia adaptive and maladaptive reactions occur simultaneously and in succession. Some of them result in significant changes of the orofacial system.

Weightlessness

The impact of weightlessness on jaw bones and teeth was examined in Soviet biosatellites. Such satellites are unmanned automatic space vehicles, specially constructed for biologic experiments in perigee orbits (Fig. 11.7). They are equipped, launched, and remain in orbit in accordance with the objectives of the experiment. Besides the scientific investigations in such biosatellites, control experiments are made on earth. These include experiments that almost completely imitate the space conditions except for weightlessness and vivarium observations under the usual conditions existing in terrestrial laboratories. In these experiments, clinical and physiologic examination of the animals started 3 to 5 hr after the soft landing of the biosatellite. The animals most frequently used in such experiments were Wistar rats, of the same strain used for the experiments described in the previous section. The animals were kept in special cages with 5 to 10 rats per cage. Food, water,

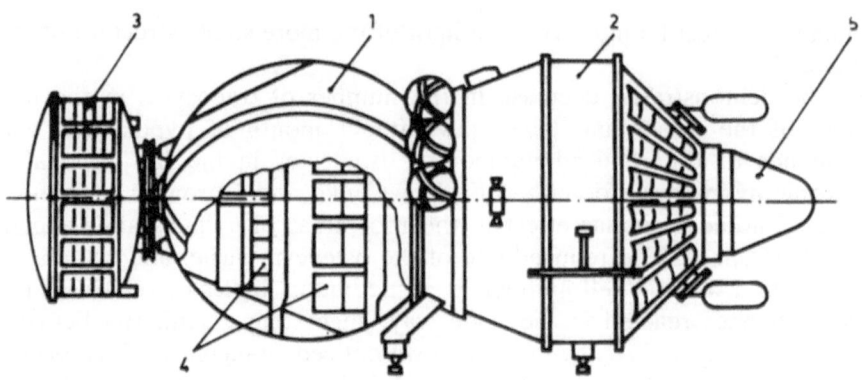

FIGURE 11.7. Schematic plan of COSMOS biosatellite: 1, landing apparatus instrument section; 3, hermetically sealed container; 4, scientific apparatus braking system.

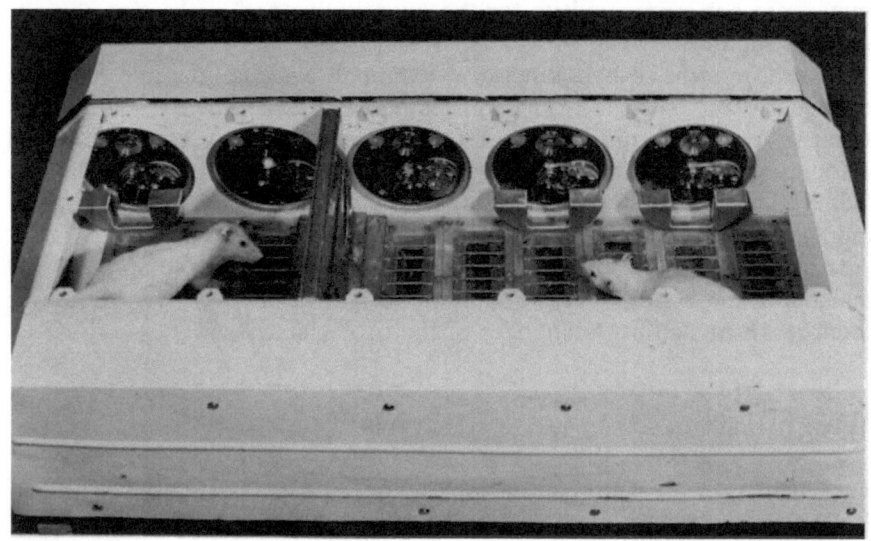

FIGURE 11.8. Cages for five rats on board Soviet biosatellite, COSMOS series, providing food, water, and light automatically. The lattice floor allows for collection of feces.

ventilation, and light were provided automatically. The rats were exposed to light for 16 hr and to darkness for 8 hr. Food consisted of a special paste, offered four times daily at 6-hr intervals and providing 70 to 80 kcal per day.[37] Water was constantly available. The hermetically closed boxes used in terrestrial experiments imitated the microclimate and living conditions of the satellite (Fig. 11.8). All rats underwent a comprehensive

TABLE 11.8. Selected changes in shoulder bones and tibiae of rats after 19 days of weightlessness.[42]

	Bone density (g/cm^2)	Rate of maturation of periosteal osteoids (%/hr)	Rate of periosteal mineralization (%/hr)
5–11 hr after space flight			
Exp. animals	1.475 ± 0.032	0.62 ± 0.24	0.43 ± 0.13
Controls	1.490 ± 0.022	1.11 ± 0.11	0.64 ± 0.06
25 days after space flight			
Exp. animals	1.562 ± 0.017	2.18 ± 0.42	0.76 ± 0.17
Controls	1.562 ± 0.047	1.41 ± 0.28	0.56 ± 0.09

adaptation period (learning to feed, adjusting to the boxes, and to the implanted sensors) lasting about 1 to 2 months.

Some Effects of Weightlessness on Behavior and Metabolic Processes

The animals were exposed to weightlessness for periods varying between 7 and 22 days, depending on the type of experiment. During this period, they ate normally, had normal body temperatures, and showed normal myokinetic activities.[37] All in all, changes of physiologic parameters are comparable to those produced by chronic stress.[38] In particular, during weightlessness greater amounts of sodium, potassium, and calcium were excreted, whereas after the flight the excretion of these ions decreased. Indeed, the excretion of calcium and of sodium did not return to normal until 3 weeks and 11 to 12 days after the flight, respectively. The thyroid gland of rats exposed to several weeks of space flight showed reduced amounts of calcitonin-secreting C cells. Focal cellular hyperplasia was detected in the parathyroid gland, suggesting an increased output of parathyroid hormone.[37] Atrophic changes in skeletal muscles, such as decreased muscle mass and diameter of muscle fibers, were also noted after space flight. Such changes are believed to be the result of inactivity. The most conspicuous changes induced by weightlessness on the skeletal system occurred in the bones, which under gravitational conditions are weight-bearing; that is, the bones of the extremities and the spinal column. The following changes were observed: a 37% to 56% decrease in periostal bone formation,[39] reduced linear growth, demineralization of bone tissues, generalized osteoporosis, and a 30% reduction of mechanical strength[40–42]; (Table 11.8). Clear signs of osteoporosis and reduced bone strength could be observed as early as after 7 days of weightlessness. Less drastic changes were observed on non–weight-bearing bones: the maturing process of bone minerals was delayed and less differentiated

TABLE 11.9. Number and distribution of Wistar rats used for examination of alveolar bones and dental tissue.

Satellite, flight number	Sex	No. of animals		Material	Days of flight
		Hypokinetic controls	Vivarium		
1514, 10	Female, pregnant	10	10	Dentine Bones	13–18 days of gestation
1514, 5	90-day-old offspring	5	5	Dentine Bones	Same as their mother's
1667, 10	Male adult	5	5	Dentine Bones	7
1887, 10	Male adult	5	5	Dentine Bones	13

cells were transformed into preosteoblasts, indicating changes in the system of hormonal regulation. In fact, reduced concentrations of somatotropic hormone, a decreased ouput of calcitonin, and an increased production of parathormone were observed.[37] Changes in the skeletal system induced by weightlessness normalized within 25 days of readaptation. A group of animals placed in a centrifuge producing an artificial gravity of 1 g was put into orbit on board the biosatellite Cosmos 936 (Aug. 8–21, 1977; apogee 419 km, perigee 224 km). This artificial gravity, sustained for the entire flight of 18.5 days, prevented the bone tissue changes usually observed under conditions of weightlessness. An embryologic experiment in biosatellite 1514 was designed to study the influence of weightlessness on the concentration of 10 elements (calcium, magnesium, strontium, iron, phosphorus, copper, manganese, potassium, sodium) in the hair and tails of pregnant rats and their offspring.[43] Wistar rats, aged 4 months and weighing 280 to 310 g, were exposed to weightlessness between day 13 and 18 of their pregnancy. Measurements carried out in mothers and offspring demonstrated that the effects of weightlessness on mineral metabolism were comparable to those of psychic stress.[43] In particular, as the result of staying in space, the 15-day-old offspring showed reduced concentrations of potassium, sodium, calcium, magnesium, strontium, iron, and phosphorus, suggesting a reduced mineralization of the bones.

Effects on Jaw Bones and Teeth

We measured apatite content of the dentine of 750 molars, 80 incisors, and parts of the corpus mandibulae of 40 adult male and 40 adult female Wistar rats and 15 of their 90-day-old offspring and of control materials derived from animals kept under space-simulated (except for weightlessness) or vivarium conditions (Table 11.9). The male animals were in orbit

TABLE 11.10. Some flight characteristics of biosatellites used to examine the influence of weightlessness on jaw bones and teeth.

Biosatellite "Cosmos" No.	Dates of launch and landing	Duration of flight (days)	Apogee (km)	Perigee (km)
1514	Dec. 14–19, 1983	5	288	226
1667	July 10–17, 1985	7	297	222
1887	Sept. 29–Oct. 12, 1985	13	224	406

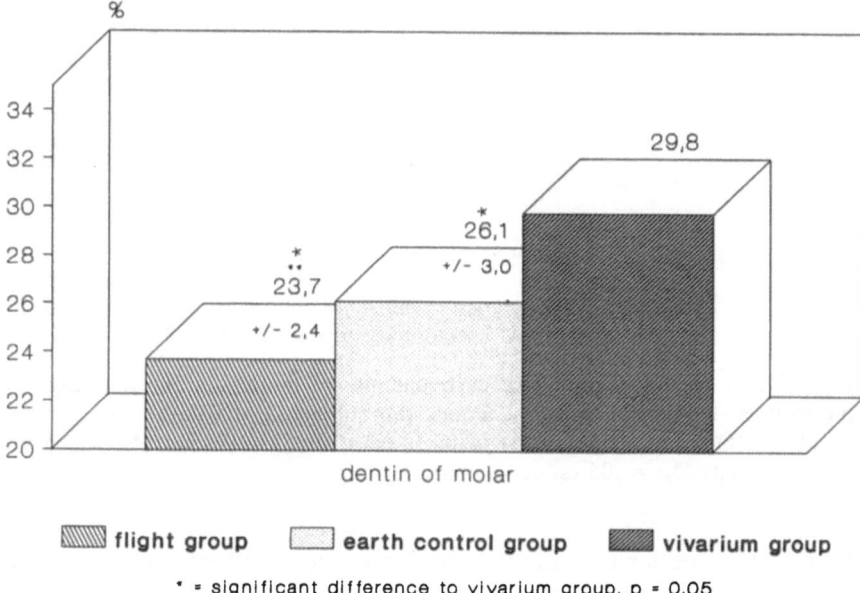

* = significant difference to vivarium group, p = 0,05
** = significant difference to earth control group, p = 0,05

FIGURE 11.9. Relative content of carbonapatite of molar dentine of pregnant rats after 5 days of weightlessness (13th–18th day of gestation), biosatellite COSMOS 1514 (*p = .05 vs. vivarium group; **p = .05 vs. earth control group).

for 7 days (biosatellite "Cosmos 1667") or 13 days (biosatellite "Cosmos 1887"); the female animals, which had been inseminated before the flight, were put into orbit between day 13 and 18 of their pregnancy (biosatellite "Cosmos 1514", see Table 11.10). The dentine was separated from the enamel according to Rozeik et al.[44] or to Kleber et al.[45] Apatite content was measured by infrared spectroscopy (Specord IR 75, Carl Zeiss, Jena), using potassium bromide pellets.[46–48] Weightlessness decreased the carbonate apatite fraction in various degrees, although it appeared that longer periods of weightlessness (13 days for adult males) resulted in a smaller effect. The female animals showed the greatest drop in apatite (see Fig. 11.9). In contrast, their 90-day-old offspring, which underwent

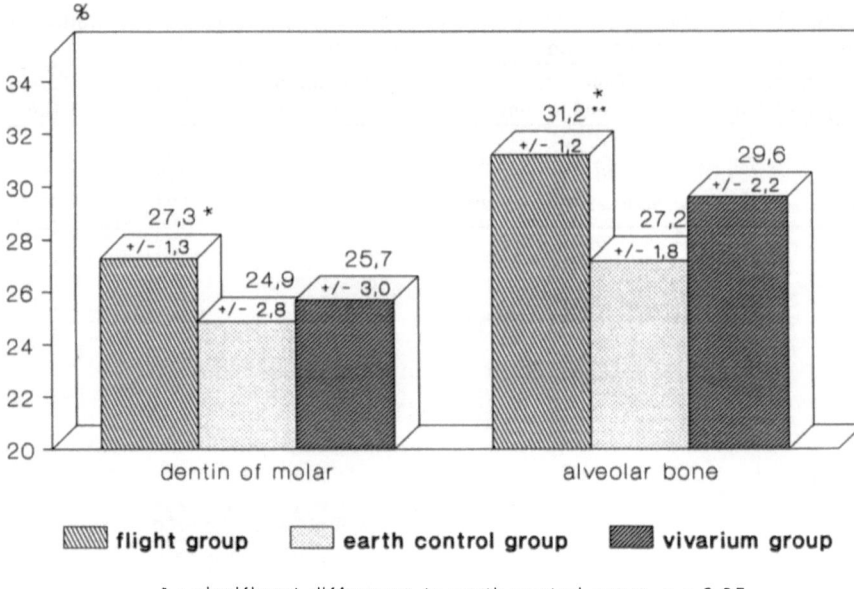

FIGURE 11.10. Relative content of carbonapatite of molar dentine and alveolar bone from 90-day-old offspring of female rats subjected to 5 days of weightlessness (13th–18th day of gestation) in biosatellite COSMOS 1514 (*p = .05 vs. vivarium group; **p = .05 vs. earth control group).

odontogenesis under conditions of weightlessness and later continued to grow under normal terrestrial conditions, showed a roughly 6% higher apatite content in their molar dentine than the animals of the control group (Fig. 11.10). An analysis of the alveolar bones of male rats showed that after 7 days of space flight there was a 20% reduction in apatite content whereas after 13 days of weightlessness there was a 7% increase (Figs. 11.11 and 11.12). Similarly, the 90-day-old offspring showed higher contents of apatite than those of the control group (Fig. 11.10). Clearly, short-term weightlessness appears to reduce apatite whereas longer periods of reduced gravity resulted in counterregulatory adaptation. Dentine, a slow turnover tissue, appeared to react more slowly to external stress; indeed, the decreased apatite content had not compensated after 13 days of weightlessness in the adult animals, but was evident in their offspring. Investigations on mineral metabolism carried out on the same animals[43] confirm this interpretation of the findings. After the flight, less sodium, potassium, and calcium were excreted whereas more were excreted during the period of weightlessness. The calcium balance returned to normal homeostatic values only 3 weeks after the space flight. Similarly, weightlessness caused texture changes in teeth and alveolar bones. The

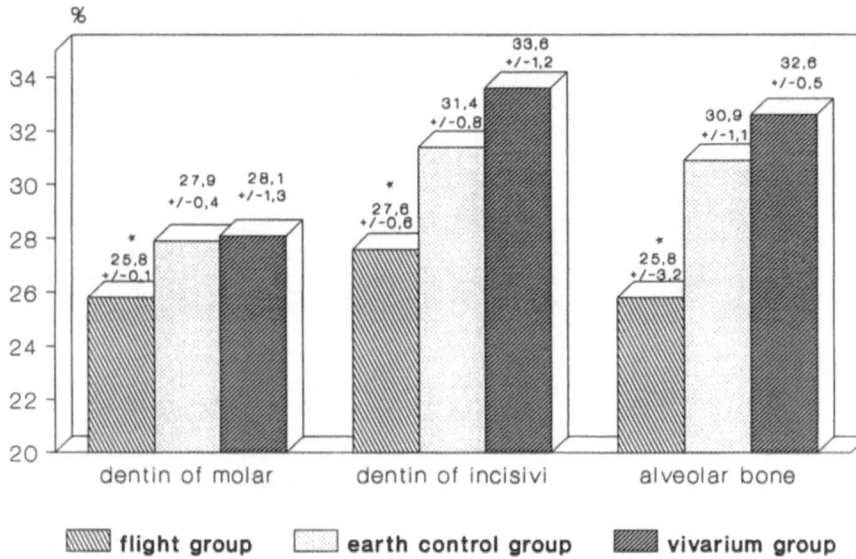

FIGURE 11.11. Relative content of carbonapatite of molar dentine and alveolar bone from adult male rats after 7 days of weightlessness in biosatellit COSMOS 1667 (*p = .05 vs. vivarium group; **p = .05 vs. earth control group).

intensity of these changes depended on the duration of exposure and the metabolic activity of the tissue. Similar observations were made in weight-bearing and non–weight-bearing bones under terrestrial conditions. The onset of counterregulation compensated for the initial loss of apatite. Thus, an orofacial adaptation to the stresses of space flight seems to be possible.

Final Comments

The changes in jaw bones and teeth under conditions of hypokinesia or weightlessness described in this chapter are important for the understanding of physiologic mechanisms of adaptation in the orofacial system.[49] The results of the two types of experiment (hypokinesia and weightlessness) supplement and confirm each other, obviating any criticism that may be raised against the exclusive use of infrared spectroscopy for the examination of mineralized tissues. No histologic and histomorphologic investigations could be carried out, so far, owing to the scarcity of materials from animals kept under extraterrestrial conditions. Such experiments, however, are planned for the next flights of Cosmos biosatellites. Nevertheless, the results of hypokinetic experiments allow us to reach

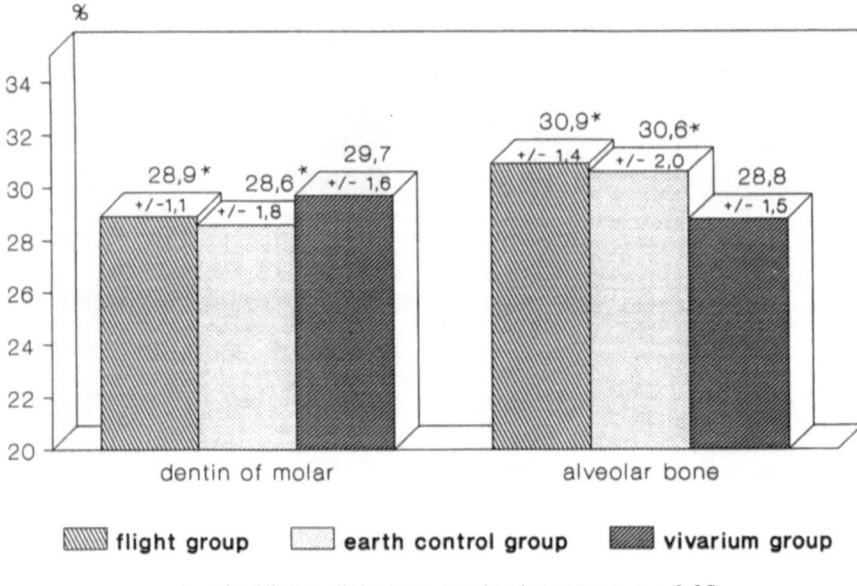

FIGURE 11.12. Relative content of carbonapatite in molar dentine and alveolar bone from adult male rats after 13 days of weightlessness in biosatellite COSMOS 1887 (*p = .05 vs. vivarium group).

some tentative generalizations. The jaw bones react to the external stress of hypokinesia and weightlessness in the same way as other bones.[50] After initial osteoporosislike morphologic changes, these bones adapt to the stress by an occasional excessive counterregulation. Depending on species and metabolic properties of the tissue, this soon results in the restoration of the tissue's morphologic integrity. The reactions in the dentine occur in a similar, although somewhat delayed manner, owing to the slow metabolic turnover of this tissue. That dentine reacts at all is surprising considering that changes in dentine caused by external influences (except for microbial injury and caries) rarely have been observed. The findings gathered from the offspring of pregnant rats are of special relevance to the prevention of caries and an early loss of teeth. Even after a longer period of normal terrestrial conditions (90 days), the reactions to stress on the dentine texture had not been fully compensated. We observed similar changes in the texture and morphology of dentine in the offspring of Wistar rats exposed to a slight, nutritionally induced magnesium deficiency. This resulted in a markedly increased tendency to caries and dental erosions.[51] These findings suggest that further studies of odontogenesis may lead to the detection of factors that may increase the resistance of dentine to microbial attack. Thus, the results obtained under conditions of hypokinesia and temporary weightlessness acquire greater and more general pathogenetic relevance.

References

1. Kakurin LI. Vlijanie dlitelnogo ogranicenija mysechoj dejatelnosti na organizm celoveka i gipokineticeskij komponent nevesomosti. *Kosmic Biol Med.* 1968;2:59–63.
2. Kovalenko EA. Patofiziologija dlitelnoj gipokinezii. *Kosmic Biol Aviakosm Med.* 1977;11:3–9.
3. Kovalenko EA, Gurowski NN. Gipokinesija. *Medizina*; Moscow: 1980.
4. Oganov V, Rachmanow A, Morukow B, Janson C, Dzenis B, Zaitschik W. Untersuchungen des Knochengewebes des Menschen mit nichtinvasiven Methoden im Prozeß einer 120tägigen Hypokinese. In: Grigorjew A, Hecht, K, Haase H. (eds.) *Hypokinese*. Humboldt-Universität; Berlin: 1989.
5. Konstantiv I, Lesnjak A, Antropowa E, Rykowa M, Boschkow N. Immunreaktivität des Menschen bei einer 120tägigen antiorthostatischen Hypokinese. In: Grigorjew A, Hecht, K, Haase H. (eds.) *Hypokinese*. Humboldt-Universität; Berlin: 1989.
6. Lobatshik W. Zustand von Flüssigkeiten und der Hämozirkulation unter einer 120tägigen antiorthostatischen Hypokinese. In: Grigorjew A, Hecht K, Haase H. (eds.) *Hypokinese*. Humboldt-Universität; Berlin: 1989.
7. Poppei M, Hecht K. Stressoreinwirkung und chronischer Streß in Abhängigkeit von Zuständen des Organismus und von Umweltkonstellationen. *Wiss Zschrf HUB Math-Nat R.* 1980;29:667–680.
8. Hecht K, Alpatow AM, Gazenko OG et al. Biosatelliten in der medikobiologischen Kosmosforschung—Entwicklung und einige Ergebnisse. *Kosmische Biol Med.* 1981;10:7–69.
9. Erben A. *Einfluß von chronischem und akutem Streß in Form einer langandauernden Hypokinese und anschließender Fesselung als Modell zur Simulatation von Teileffekten der Schwerelosigkeit auf Wachstumsvorgänge im Alveolarknochen der Wistarratte.* Med. Diss. A. Humboldt-Universität; Berlin: 1987.
10. Fjodorow IW. Obmenwetschestw pre gipodinamii. *Nautsch Dokl Wys Skol Biol Nauk.* 1972;12:24–36.
11. Fjodorow IW, Tscherni A, Fjodorow AJ. Die Intensität der Synthese und des Zerfalls des Gewebeeiweisses bei Hypokinese und erhöhter Muskelaktivität. *Physiol J SSSR.* 1977;63:1128.
12. Niklowitz WJ, Bunch TE, Young DR, Nemestrina M. The effects of immobilization on cortical bone in monkeys. Proceedings of Fifth annual meeting of the IUPS Commission on gravitational physiology. *Physiologist.* 1983;26(Suppl):115–116.
13. Szilagyi T, Rapcsak N, Szöör A, Földes I, Gyarmati Jr J. The effect of immobilization on the rat's bone. Proceedings of Fifth annual meeting of the IUPS Commission on gravitational physiology. *Physiologist.* 1983;26(Suppl): 94–95.
14. Holton EM, Wronski TJ. Animal models for simulating weightlessness. Proceedings of Third annual meeting of the IUPS Commission on gravitational physiology. *Physiologist.* 1981;24(Suppl):45–48.

15. Ushakov AS, Spirichev VB, Belakovsky MS, Sergeev IN, Kondratyev YL. Calcium-phoshporus metabolism and prevention of its disorders in hypokinetic rats. Proceedings of Fifth annual meeting of the IUPS Commission on gravitational physiology. *Physiologist.* 1982;25(Suppl):45–48.
16. Selye H. Stress beherrscht unser Leben. *Econ*; Düsseldorf: 1957.
17. Stahl F, Amendt G, Dörner G. Total and free cortisol plasma levels in pre- und postnatal life. *Endokrinologie.* 1979;74:243–246.
18. Kato T, Wakui Y, Nagatsu T, Onishi T. An improved dual wavelength spectrophotometric assay for dopamine-β-hydroxylase. *Biochem Pharmacol.* 1978;27:829–831.
19. Axelrod J. Purification and properties of Phenylethanolamine-N-methyl-transferase. *Biol Chem.* 1962;237:1657–1660.
20. Molinoff PB, Weinshilbaum R, Axelrod J. A sensitive enzymatic assay for dopamine-β-hydroxylase. *J Pharmacol Exp Ther.* 1971;178:425–431.
21. Erben A. *Einfluß von langandauernder Hypokinese als ein Modell der Simulation von Teileffekten der Schwerelosigkeit auf Wachstumsvorgänge im Alveolarknochen der Wistarratte.* Med. Diplom. Humboldt-Universität; Berlin: 1984.
22. Pott F, Jantzen VV. Röntgenologische Untersuchungen zur Knochenentwicklung der Ratte. *Z Versuchstierkd.* 1972;14:35–47.
23. Weibel ER. *Stereological methods. Vol. 1. Practical methods for biological morphometry.* Academic Press; New York: 1979.
24. Kvetnansky R, Weise VK, Kopin IJ. Elevation of adrenal tyrosine hydroxylase and phenylethanolamine-N-methyltransferase by repeated immobilization of rats. *Endocrinology.* 1970;87:744–749.
25. Woloschin IA. cited In: Kowalenko EA, Gurowski NN. *Hipokinesija.* Medizina; Moscow: 1980;25:279–281. Hipokinesia.
26. Poppei M. *Chronischer Stress in Abhängigkeit von Zuständen des Organismus und von Umweltkonstellationen (Tierexperimentelle Studien).* Diss. B. Humboldt-Universität; Berlin: 1980.
27. Mikulaj L, Kvetnansky R, Murgas K, Parizkova I, Vencel P. Catecholamines and corticosteroids in acute and repeated stress. In: Usdin E, Kvetnansky R, Kopin JJ (eds.): *Catecholamines and stress.* Proceedings of the International Symposium on Catecholamines and Stress. Bratislava, Czechoslovakia, July 27–30, 1975. Pergamon Press; Oxford: 1976.
28. Piesche L, Oehme P, Hilse H, Hecht K. Zur Wirkung von Substanz P auf Dopamin-β-hydroxylase und Phenylethanolamin-N-methyltransferase in der Nebenniere der Ratte. *Pharmazie.* 1982;37:591–593.
29. Kvetnansky R, Mitro A, Palkovits M et al. Catecholamines in individual hypothalamic nuclei in stressed rats. In: Usdin E, Kvetnansky R, Kopin JJ. *Catecholamines and Stress.* Proceedings of the International Symposium on Catecholamines and Stress. Bratislava, Czechoslovakia, July 27–30, 1975. Pergamon Press; Oxford: 1976.
30. Van Loon GR. Brain dopamine-beta-hydroxylase activity: Response to stress, tyrosine hydroxylase inhibition, hypophysectomy and ACTH administration. In: Usdin E, Kvetnansky R, Kopin JJ. *Catecholamines and stress.* Proceedings of the International Symposium on Catecholamines and Stress. Bratislava, Czechoslovakia, July 27–30, 1975. Pergamon Press; Oxford: 1976.

31. Hilse H, Oehme P, Hecht K. Asymmetrisches Verhalten von Dopamine-β-hydroxylase und Phenylethanolamine-N-methyltransferase in den Nebennieren spontan—hypertensiver und normotensiver Wistar-Kyoto-Ratten. *Biomed Biochim Acta*. 1983;42:745–750.
32. Bhagal BD, Horenstein S. Modulation of adrenal medullary enzymes by stress. In: Usdin E, Kvetnansky R, Kopin JJ. *Catecholamines and stress*. Proceedings of the International Symposium on Catecholamines and Stress. Bratislava, Czechoslovakia, July 27–30, 1975. Pergamon Press; Oxford: 1976.
33. Festin M. Mouse strain identification. *Nature*. 1972;283:351–352.
34. Simmons DJ, Russel JE, Winter F et al. Bone growth in the rat mandible during spaceflight. *Physiologist*. 1980;23:87–90.
35. Simmons DJ. Adaptation of the rat skeleton to weightlessness and its physiological mechanisms. Result of animal experiments aboard the Cosmos-1129 biosatellite. Proceedings of Third Annual Meeting of the IUPS Commission on Gravitational Physiology. *Physiologist*. 1981;24(Suppl):65–68.
36. Roberts WE, Mozary PG, Morey ER. Suppression of osteoclast differentiation during weightlessness. Proceedings of the Third Annual Meeting of the IUPS Commission on gravitational physiology. *Physiologist*. 1981;24(Suppl):75–76.
37. Ilyin EA, Hecht K. *Biosatelliten*. Berichte. Humboldt-Universität; Berlin: 1987;7.
38. Simonov PV. Psychophysiological stress of spaceflight. In: Calvin M, Gazenko DG (eds.): *Foundations of space biology and medicine*. Vol. 2. NASA Washington 1975.
39. Wronski TJ, Morey-Holton E, Jee WSS. Cosmos 1129. Spaceflight and bone changes. Proceedings 28th Congress IUSP; Budapest: 1980.
40. Ilyin EA. Issledovanya na biosputnikah "Kosmos". *Kosm Biol Aviakosm Med*. 1984;18:57–66.
41. Gazenko OG, Ilyin EA, Oganov VS, Serowa LV. Eksperimenty s gyvotnymi na biosputnikah serii "Kosmos" (itogi i perspektivy). *Kosm Biol Aviakosm Med*. 1981;15:60–66.
42. Prochonchukov AA, Komissarova HA, Ilyina NA, Volojin AL. Sravnitelnoje izuchenije vlijanija nevesomosti i iskusstvennoy sily tjajesti na plotnost, soderjanije zoly, kalcija i fosfora v obizvestvlennych tkanjah. *Kosm Biol Aviakosm Med*. 1980;14:23–26.
43. Lüderitz P, Serowa LW, Marquardt D et al. Mineralstoffwechsel von Ratten und deren Nachkommen, die im Biosputnik "Kosmos 1514" der Schwerelosigkeit ausgesetzt waren. 6. *Gemeinsch Ges Exp Med*. 1985; Abstracts p. 538.
44. Rozeik F, Hannover R, Hermann M, Cremer HD. Die Trennung von Schmelz und Dentin bei Kaninchen- und Rattenzähnen. *Dtsch Zahnärztl Z*. 1956;11:254–258.
45. Kleber BM, Weigt D, Weingart H, Wachtel E, Denisowa LA, Serowa LS. Strukturveränderungen an tierischen Zahnhartsubstanzen und Kieferknochen durch kurzdauernde Schwerelosigkeit. *Zahn- Mund-Kieferheilkd*. 1989;77:668–673.
46. Brügel W. *Einführung in die Ultraspektroskopie*. Steinkopff Verlag; Darmstadt: 1962.
47. Münzenberg KJ, Gebhardt M. Kristallographische Untersuchungen der Knochenminerale. *Dtsch Med Wschr*. 1969;94:1325.

48. Termine JD, Posner AS. Infrared determination of the percentage of crystallinity in apatite calciumphosphates. *Nature*. 1966;211:268–270.
49. Hummel H, Kleber BM, Hecht K, Geissler K, Serowa L. Der Einfluß eines kosmischen Fluges auf tierische Zahnhartsubstanzen (Biosputnikexperiment Kosmos 1514). *6. Gemeinsch Ges Exp Med*. 1985; Abstracts p. 541.
50. Weingart H, Kleber BM, Geissler H et al. The influence of oneweek space flight on teeth and jaw bones of Wistar rats (Cosmos 1514 and Cosmos 1667). *Physiologist*. 1988;31(Suppl):34–35.
51. Kleber BM, Fehlinger R. Dental and periodontal disturbances due to mangnesium deficit. *Magnesium Res*. 1989;2:235–237.
52. Asdell SA. Comparative chronologic age in man and other mammals. *J Gerontol*. 1946:224–236.

12
Humoral Factors in the Pathogenesis of Osteoarthritis

DAVID HAMERMAN and STEPHEN TAYLOR

Osteoarthritis is the outcome of a number of conditions that affect certain synovial joints, and is characterized histologically by loss of articular cartilage and remodeling of the joint surface by new cartilage and bone. Age is the most frequently associated condition; by 80 years, osteoarthritic changes in the distal interphalangeal joints (Heberden's nodes) are almost universal.[1] Clinically symptomatic osteoarthritis of the major weight-bearing joints, the hips and knees, is less prevalent but still results in disability in the elderly second only to cardiovascular diseases.[2]

The association of osteoarthritis with age and with acquired injury of the knee[3] or with congenital abnormalities of the hip[4] indicates the importance of altered mechanical forces over time.[5] Additional contributory factors include a genetic predisposition to certain types of osteoarthritis,[6] obesity (at least for the knee[7]), and, possibly, crystal deposition in the joint.[8] At a more fundamental level, investigators in many disciplines are exploring the ways in which altered cellular functions in joint components contribute to the mechanisms of this disease.[9,10]

This chapter is organized into two major parts. In the first, we consider the structure of joint components, their interdependence, and the humoral factors that are known to influence cellular functions. In the second part, we discuss how humoral factors may modify cellular functions and contribute to the characteristic features of cartilage breakdown and repair that occur in osteoarthritis.

Structure–Function Relationships of Joint Components

Figure 12.1 is a diagram of a composite normal synovial joint. Enclosing the joint is a fibrous capsule, the inner surface of which is the synovial membrane responsible for the elaboration of synovial fluid. The articulating surface is hyaline cartilage, beneath which is a backing of sub-chondral bone. Within the joint, ligaments, tendons, and fibrocartilages

TABLE 12.1. Structure-function relationships and interdependence of joint components.

Components	Structure	Functions
Ligaments and tendons	Dense, fibrous, connective tissue	Prevents over-extension of joints, provides stability and strength
Synovial membrane	Areolar, vascular and cellular	Secretes synovial fluid. Phagocytoses particulate material in synovial fluid.
Synovial fluid	Viscous fluid	Provides nutrients for articular cartilage, lubricates cartilage during joint motion.
Articular cartilage	Firm hyaline cartilage	Constitutes the articular surface, bears weight, responds elastically to compression.
Tidemark	Calcified cartilage	Separates articular cartilage from underlying bone.
Subchondral bone	Hard bone with marrow spaces	Provides backing for articular surface. Marrow cavity provides nutrients to base of cartilage and is the source of stromal cells with osteogenic potential.

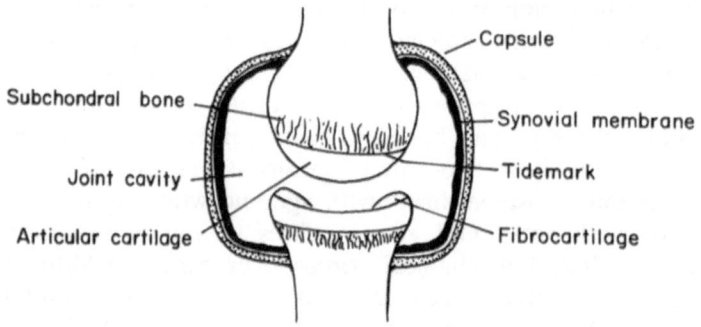

FIGURE 12.1. Diagram of a composite normal synovial joint.

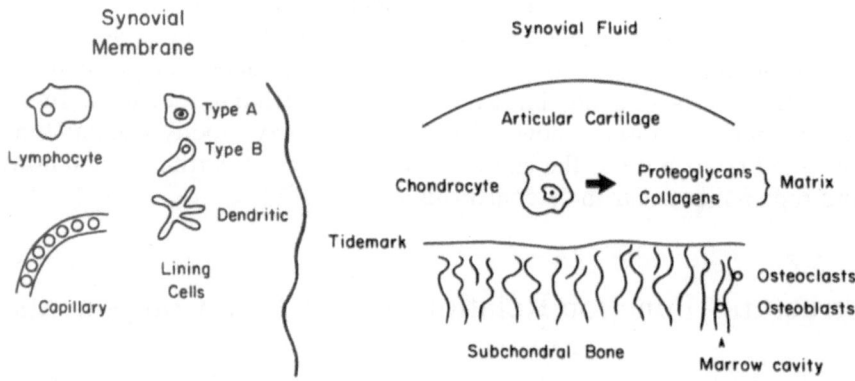

FIGURE 12.2. Diagram of cells with specialized functions within the synovial membrane, articular cartilage, and subchondral bone, to emphasize the interaction of the components of a normal synovial joint.

as represented in the knee provide stability and a close fit between articulating surfaces.

The specialized cells of the joint components (Fig. 12.2) synthesize and maintain the macromolecules whose interactions are responsible for tissue integrity and function: the tensile strength of ligaments and tendons, the loose connective tissue that supports the vascular and cellular elements of the synovial membrane, the viscous synovial fluid, the elasticity of hyaline articular cartilage, and the rigid strength of subchondral bone. The interdependence of these joint components is presented in Table 12.1.

Synovial Membrane

The term "synovial cells" refers to a group of cells at the surface of the synovial membrane. These "lining cells" are distinctive connective tissue cells in terms of their fine structure and metabolic functions. The structure of the lining cells has been characterized by electron microscopy in the intact membrane.[11] Type A cells contain many Golgi vesicles and mitochondria, whereas the type B cells are characterized by abundant rough endoplasmic reticulum. Cytochemical studies of the synovial membrane also showed cells in the lining layer with many interdigitating cytoplasmic processes.[12] Collagenase digestion disrupts the synovial membrane matrix and liberates the lining cells, which can then be maintained in cell culture in vitro. A heterogeneous cell population can be cloned; three cell types that correspond to those that have been described in the intact membrane can be identified in vitro. These are an oval- or rhomboid-shaped macrophage-like cell that is similar to the type A cell and is phagocytic in vitro, an elongated, spindle-shaped fibroblast-like cell similar to the type B cell, and a cell described as stellate or dendritic with extended, multiple interdigitating cytoplasmic processes.[13] These dendritic cells are nonphagocytic and their cell surface membrane stains for Ia antigen,[14] a marker for tissue macrophages. The Ia-positive cells belong to a lineage similar to that of the epidermal Langerhans cells and may be the principal accessory cell for antigen presentation to lymphocytes.[15] Reversion between cell types in vitro occurs in the presence of prostaglandins (PGE_2), with fibroblast-like cells taking on the appearance and properties of dendritic cells.[16]

The components of the synovial fluid are derived from the synovial lining cell layer as well as from the vasculature in the synovial membrane. The lining cells secrete principally hyaluronic acid, which accounts for the viscosity of the synovial fluid, smaller amounts of sulfated proteoglycans,[17] and "lubricin," a glycoprotein that appears to lubricate the articular cartilage.[18] The vasculature in the synovial membrane is the source of synovial fluid proteins and solutes.

Effects of Mediators on Synovial Cells

Synovial cells are target cells in the sense that they are stimulated by a number of soluble mediators (endocrine), and effector cells in the sense that their secretory products are capable of acting on other cells in their local environment (paracrine) or on themselves (autocrine). The influence of humoral factors on the metabolic properties of synovial cells was perhaps first demonstrated when conditioned medium from blood-derived monocytes-macrophages was added to synovial cell cultures[19]: the synovial cells thereupon elaborated a variety of products that included PGE_2, collagenase, interleukin-1α (IL-1α), interleukin-1β (IL-1β), interferon, and tumor necrosis factor (TNF).

The proteins IL-1α and IL-1β have 26% amino acid homology and bind to the same high-affinity receptors on synovial cells.[20] Although many of its effects on cells are similar to those of IL-1, TNF has a different amino acid composition and different receptors.[21] IL-1 appears capable of regulating its own production in synovial cells by activating IL-1 gene expression (up-regulation) and stimulating PGE_2 formation, which in turn suppresses IL-1 synthesis (down-regulation).[16,22] Indomethacin, a prostaglandin synthesis inhibitor, can act in this cycle by reversing the down-regulation of PGE_2 on IL-1 formation. IL-1 stimulates the synthesis of types I and III collagens and of fibronectin by synovial cells. However, neither the IL-1–mediated suppression of type II collagen synthesis nor the enhancement of hyaluronic acid synthesis by synovial cells[23] is suppressed by indomethacin. IL-1 also stimulates synthesis of proteinases in latent (inactive) form by synovial cells. One mechanism for this action may be the formation of "autocrine proteins" similar to serum amyloid A and B_2 microglobulins, which feed back on the same cell to enhance latent collagen formation.[24] The procollagenase induced by IL-1 is similar to two other proteinases that potentially have deleterious effects on cartilage matrix components: gelatinase, which attacks denatured collagens, and stromelysin (metalloproteinase-3), which acts on a similar substrate as well as on proteoglycan core protein, type IV collagen, and fibronectin.[25] Latent matrix-degrading proteinases may be converted to their active form by the plasminogen activator formed in response to the stimulation of synovial cells by IL-1.

Other mediators studied include interferon, basic fibroblast growth factor (bFGF), and transforming growth factor-B (TGF-B). Interferon stimulates the synthesis of new Ia surface proteins on synovial cells and inhibits the synthesis of types I, II, and III collagens and fibronectin, effects not modulated by indomethacin.[26,27] Synovial cells have been shown to have receptors for and to secrete bFGF,[28] and also to respond to TGF-B and epidermal growth' factor by forming foci "resembling cultures of primary rheumatoid synovial cells," with a two- to fivefold increase in ^{3}H-thymidine incorporation into DNA.[29]

Cartilage

Chondrocytes in the hyaline articular cartilage elaborate large aggregating proteoglycans and collagens whose interactions endow the matrix with its unique physical property of elasticity in response to compressive forces.[30] The major large proteoglycan of cartilage is composed of a protein core to which are attached glycosaminoglycan chains (chondroitin sulfate and keratan sulfate), and short chain N- and O-linked oligosaccharides. Link protein joins a number of proteoglycan monomers to a long filament of hyaluronic acid by way of a binding site on the core protein, thus forming proteoglycan aggregates of enormous molecular size ($40-100 \times 10^6$) in the cartilage matrix.[9,30] Nonaggregating proteoglycan monomers are also found in relatively high proportion in articular cartilages.[30]

Effects of Mediators: Catabolic

Since cartilage is avascular it may at first appear as a contradiction to consider a role for humoral factors in mediating chondrocyte metabolic responses. In fact, joint motion with alternate cartilage compression and relaxation promotes flow through the matrix: products of matrix degradation and chondrocyte metabolism are extruded into the joint fluid, and solutes from the joint fluid and cartilage surface are imbibed. The integrity of the cartilage matrix is maintained under conditions where an intact surface (lamina splendens) and the physical barrier of the matrix macromolecules exclude penetration of large molecules potentially capable of degrading the matrix, but permit access of those solutes needed for metabolic functions of the chondrocytes. The conditions that render cartilage vulnerable have been studied in vitro using organ slices or chondrocyte cultures. How factors gaining access to the cartilage or produced by the chondrocytes themselves degrade the matrix can then be extrapolated to in vivo conditions in the joint.

Reducing the amount of serum in the ambient medium of cartilage slices in vitro results in a reversible state of matrix proteoglycan loss.[31,32] In the absence of serum growth factors (somatomedins), chondrocytes stop making proteoglycans while continuing to produce proteases responsible for proteoglycan turnover.[33,34] Matrix proteoglycan depletion can be documented by the augmented content of matrix-derived chondroitin sulfate present in the medium; the cellular basis for these responses can be demonstrated by the ability of cycloheximide, a metabolic inhibitor, to halt matrix proteoglycan loss.[33] More relevant to humoral conditions that may prevail in osteoarthritis is the observation that cytokines added to organ slices in vitro gain access to the cartilage and promote similar chondrocyte responses.[33-36] Chondrocytes have specific receptors for IL-1 with high binding affinities, and respond by producing PGE_2 and proteases such as collagenase and stromelysin in latent forms.[37] Cartilage-specific type II collagen synthesis is suppressed by IL-1 as well as by TNF, an effect unrelated to prostoglandin levels.[16,38] IL-1 suppresses proteoglycan synthesis and, through activation of protease release, brings about a limited proteolytic degradation of the core protein and link protein; the result is reduced proteoglycan aggregate formation. These effects result in loss of matrix integrity and focal matrix depletion.[35]

Effects of Mediators: Anabolic

Interest in the effects of growth factors on cartilage was spurred by the demonstration of Salmon and Daughaday that a growth hormone–dependent serum factor enhanced $^{35}SO_4$ uptake into chondroitin sulfate by cartilage chondrocytes.[39] Originally termed "sulfation factor," this serum factor is now known to represent a family of somatomedins or insulinlike growth factors (IGF).[40,41] Some of the literature on the effects of insulinlike growth factors on cartilage slices or chondrocyte cultures has been reviewed elsewhere.[42-44]

In cultured chondrocytes derived from rib or knee cartilages, IGF-1 enhances thymidine incorporation into DNA, and $^{35}SO_4$ and proline incorporation into proteoglycans and collagen, respectively.[42] Growth hormone does not increase thymidine incorporation into DNA or $^{35}SO_4$ into proteoglycan in bovine articular chondrocytes,[43,44] but does so in rat growth plate chondrocytes.[45] Growth hormone may promote a selective response of growth plate chondrocytes to IGF-1, with clonal expansion through paracrine or autocrine mechanisms.[45]

Polypeptide growth factors are produced locally in the cartilage itself. One such factor, called cartilage-derived factor, appears to be similar to the somatomedins.[46] Sullivan and Klagsbrun[47] isolated a cartilage-derived growth factor that proved identical to basic fibroblast growth factor (bFGF). A great advance in both the chemical characterization and the isolation of growth factors from many tissues was made possible by the use of heparin affinity chromatography. This led to the identification of two distinct classes of polypeptide growth factors that shared an ability to stimulate endothelial cell growth in vitro.[48] One class of heparin-binding growth factors can be eluted with 1.0 M NaCl; these factors, with isoelectric points of 5 to 7 and molecular weights of 15,000 to 18,000 daltons, are the acidic fibroblast growth factors (aFGF). Another class of heparin-binding growth factors, which are also mitogenic for endothelial cells and can be eluted from heparin-Sepharose with 1.5 M NaCl, have isoelectric points of 8 to 10, molecular weights between 16,000 and 18,500 daltons, and are the basic fibroblast growth factors (bFGF). The bFGFs have a much wider tissue distribution than the aFGF, with which they share 55% of their amino acid sequences.[49] Heparin[50] and heparan sulfate[51,52] potentiate the mitogenicity of aFGF when tested on endothelial cells. An important distinction exists between growth factors such as PDGF, which are secreted into the ambient environment[53] to act on cells at a distance (endocrine) or locally (paracrine, autocrine), and bFGF, which appear to be bound to the cell membrane or to the extracellular matrix. The novel term "matricine" designates matrix-bound bFGF, which acts on contiguous cells.[54] The sequestering of bFGF in the matrix may stabilize and protect this growth factor from proteolytic cleavage.[55] The cell surface or matrix component responsible for binding bFGF appears to be heparan sulfate proteoglycan.[56,57] The relevance of matrix-bound bFGF to the joint is not clear, since cartilage matrix does not appear to contain heparan sulfate; however, this proteoglycan is likely to be part of the chondrocyte cell surface. It is possible that bFGF in the cartilage matrix is "bound" by ionic forces to the anionic groups of the glycosaminoglycans and "sequestered" within the domains of the proteoglycan aggregates. The concept that proteoglycans may play a role in the control of cell proliferation either directly or by mediation through FGF is an exciting one, recently reviewed by Ruoslahti.[58]

There has been an "exponential increase"[59] in knowledge relating to TGF-B. Recent work led to the identification of at least four forms of this molecule and of a large gene family comprising many other structurally related, but functionally distinct, regulatory proteins.[59,60] The effects of TGF-B in vitro are variable depending on cells of origin, serum content, and concentration of the growth factor. TGF-B enhanced proteoglycan and hyaluronate synthesis in cartilage slices,[61] but other studies suggest destabilization of the chondrocyte phenotype. "Terminal differentiation" of growth plate chondrocytes, which involves the formation of a calcified matrix when the cells are sedimented in a centrifuge tube, was inhibited by addition of TGF-B.[62] A similar inhibitory effect was demonstrated more recently for bFGF.[63] TGF-B is mitogenic for chondrocyte cultures in vitro[64] but appears to decrease type II collagen synthesis while enhancing synthesis of type I collagen and fibronectin.[65] TGF-B may exert its effects at the genetic level by modifying cell attachment proteins, or integrins, which transduce signals from the extracellular matrix to the cytoskeleton.[66,67]

Although this chapter has discussed the effects of cytokines (catabolic) and growth factors (anabolic) separately, their potential synergism should be appreciated. Both aFGF and bFGF appear to enhance the number of IL-1 receptors on chondrocytes, with potentiation of protease secretion,[37] whereas TGF-B induces the synthesis of IL-1.[68]

Subchondral Bone

The subchondral bone bears a key structural and functional relationship to the overlying articular cartilage. Stresses on the articular cartilage impact on the subchondral bone and the latter's apparent increasing stiffness with aging has been considered one factor promoting cartilage degradation as a prelude to osteoarthritis.[5] The subchondral bone's marrow cavity is the source of nutrients for the basal chondrocytes in the articular cartilage. Cellular and vascular elements from the subchondral bone violate the tidemark in osteoarthritis, overcome the antiangiogenic properties of the cartilage,[69] migrate to the articular margin, and form new cartilage and bone (the osteophyte) by the process of endochondral ossification.

A brief account of the evolution of the long bone and its articular surface tells us a good deal about the interrelations between the articular cartilage and subchondral bone. The mesenchymal cells in the developing long bone elaborate a variety of matrix proteins sequentially over time.[70] These mesenchymal cells in their prechondrogenic phase have large coats of hyaluronic acid that the cells appear to synthesize in response to bFGF.[71,72] As the cells differentiate and chondrogenesis takes place, a process that may be promoted by TGF-B from the ectodermal ridge,[73] the cells lose their hyaluronate coat, hypertrophy, condense, and synthesize chondroitin sulfate proteoglycan.[71,72] Chondrocytes in the primary and secondary centers of ossification progress from resting to proliferative to hypertrophic cells[74,75] with different phenotypic properties at each stage, as elucidated by immunohistochemical stains[76] and in situ hybridization.[77,78] Thus, the expression of type II collagen diminishes from resting cells to hypertrophic cells, which synthesize predominantly type X collagen and alkaline phosphatase as part of the matrix vesicles[79] in preparation for calcification. The calcified septa formed by the most hypertrophic cells are invaded by vascular buds; osteoid deposition by osteoblasts, or even by chondrocytes in transition to osteoblasts,[80,81] results in new bone formation by the process termed endochondral ossification.[74]

The articular surface at the end of the bone is composed of hyaline cartilage. Chondrocytes at the base of the cartilage appear more oval in shape and synthesize more keratan sulfate than chondrocytes at the surface,[82,83] but articular cartilage chondrocytes do not evolve into hypertrophic cells, and under normal circumstances do not make type X collagen or alkaline phosphatase. Repair of small defects in articular cartilage is generally, limited and not associated with new bone formation.

When growth ceases in the developing long bone, the tidemark of calcified cartilage separates the articular cartilage from the subchondral bone. Within the marrow cavity of the bone a diversified cell population of hematopoietic and mesenchymal cells develops along two lines: blood elements and the osteoclasts are derived from the hematopoietic line, whereas osteogenic, fibroblastic, reticular, and adipocytic cells arise[84,85] from the stromal–fibroblastic system. Macrophages may be derived from stromal cells, but they appear to be more related histogenetically to the hematopoietic stem cells. The precise origin of the endothelial cell is unknown.

Differentiating and differentiated cells can be identified by their ability to express cell type–specific molecules that may be components of the cell or structural molecules of the tissue matrix.[85] Owen and Friedenstein note that the best characterized cells of the marrow stromal system are the osteogenic cells. This term implies that the cells have chondrogenic as well as osteogenic potential, with the common precursors synthesizing type III collagen, laminin, and fibronectin. Subsequent expression of type I or II collagen identifies the osteoblastic or the chondroblastic phenotype, respectively.[86] However, "despite considerable effort being expended in this area, specific molecules have yet to be identified unambiguously for osteoblastic cells."[85] Osteogenic capacity expressed by bone tissue formation in vitro provides the best evidence for osteoblastic activity,[85] but other markers include (1) high levels of alkaline phosphatase activity localized largely in the plasma membrane, (2) synthesis of type I collagen, which forms the structural framework of the mineralized matrix, (3) formation of a number of noncollagenous proteins present in mineralized tissue, especially the bone-specific osteocalcin or bone carboxyglutamic acid, and perhaps the phosphorylated glycoprotein osteopontin,[87–89] and (4) the presence of receptors for parathyroid hormone.

Effects of Mediators on Osteoblasts

In the coupling of bone formation and resorption,[90,91] the osteogenic role of osteoblasts is balanced by their aparent capacity to remove osteoid and thereby expose the mineralized bone surfaces to osteoclasts.[85] In this capacity, osteoblasts produce collagenolytic enzymes and respond to parathyroid hormone, glucocorticoids, PGE_2, prostacyclin, vitamin D_3, IL-1, and retinoic acid, factors that may stimulate bone resorption.[91] Parathyroid hormone appears to have a strong synergistic effect on IL-1–stimulated PGE_2 formation in rat osteoblastic cells.[92]

The rich vasculature of bone accounts for the local presence of more than 26 circulating proteins[88]; in addition, regulators of skeletal growth, as well as immune cell products, may be made in situ. Among the growth factors present in bone are IGF-I, IGF-II, aFGF, bFGF, PDGF, TGF-B, B_2 microglobulin, and osteoinductive factors.[91,93–96]

IGF-1, as well as the FGFs and TGF-B, act on intact rat calvaria to increase the replication of bone cells, particularly that of the osteoprogenitor cells, and promote the synthesis of noncollagenous proteins.[94–98] In osteoblast-enriched cultures, bFGF is mitogenic when the serum content of the medium is low;[97] neither IGF-I or II have a pronounced mitogenic effect,[40] and the mitogenic action of TGF-B is bifunctional: DNA synthesis rates increase, peak, and decline in response to increasing TGF-B levels.[98] The influence of TGF-B in bone itself may depend on the extent to which local factors convert the latent form to the

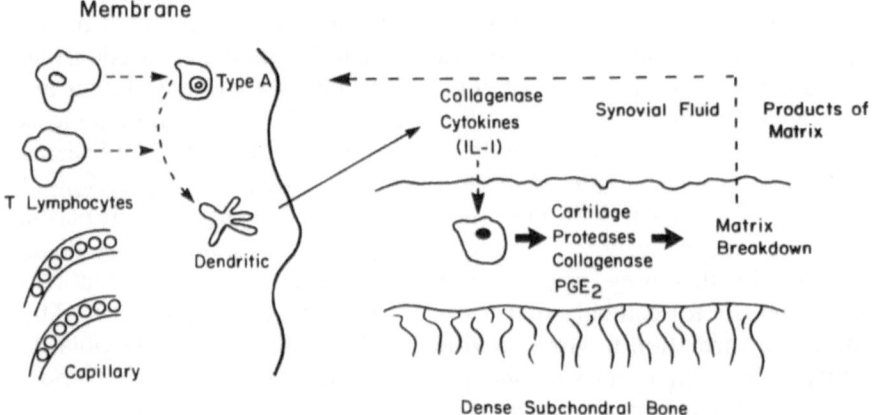

FIGURE 12.3. Diagram of cellular interactions that may contribute to cartilage matrix breakdown in the osteoarthritic joint.

active one.[91,99] "While growth factors are not specific in origin or effects for bone tissue, they probably play a role in bone remodeling"[94] through processes that couple resorption and formation.

Interplay of Humoral Factors in the Pathogenesis of Osteoarthritis

There are many "humoral" components, some circulating, some made in situ, that are capable of modifying the metabolic functions of the cells of the joint tissues; do these humoral factors contribute to the evolution of osteoarthritis? Our present understanding of osteoarthritis indicates that it is a disease of multifactorial origin, probably evolving over decades, with an uneven balance of cartilage matrix degradation and joint remodeling. The articular cartilage shows splintering and fissuring with focal loss of matrix proteoglycan and cell death in the superficial areas, with chondrocyte proliferation in clusters in the more basal areas where histologic stains and biochemical studies suggest locally enhanced synthesis of proteoglycans, at least early in the disease.[9] A favored hypothesis to account for the irreversible loss of cartilage during the evolution of osteoarthritis is that humoral factors promote the release of matrix-degrading proteoglycanases from chondrocytes, particularly in the more superficial zone.[100] Among the cytokines, IL-1 is a prime candidate, based on the in vitro studies discussed above and depicted in Figure 12.3, as well as on the following evidence:

1. IL-1 promotes cartilage matrix degradation when injected into the intact rabbit joint.[101]
2. IL-1 has been isolated directly from lining cells in the synovial membrane obtained from human osteoarthritic joints.[102]
3. IL-1 has been identified in synovial fluids obtained from human osteoarthritic joints.[103]

However, concern has been expressed about the role of IL-1 in promoting cartilage matrix degradation in osteoarthritis because a number of carboxyl, cysteine, serine, and metallo proteinase inhibitors fail to inhibit IL-1 effects on cartilage slices in vitro.[33]

Chondrocytes remaining in the more basal zone of articular cartilage in osteoarthritis appear unable to repair the ongoing matrix degradation. This is in contrast to matrix restoration by chondrocytes in cartilage organ slices when serum or IGF-1 is reintroduced into the medium, as noted above. The healing of cartilage lesions has also been studied experimentally in the rabbit joint. When a defect strictly limited to the cartilage is made with a drill, repair does not occur. However in two circumstances this defect can be repaired. One is to infuse bFGF into the joint.[104] The other is to have the drill penetrate through the cartilage into the subchondral bone. In this event blood welling up into the defect clots and provides a scaffold for cellular elements from the marrow to migrate into the clot and form a fibrocartilage. However, this "repair" cartilage has limited physical integrity compared with hyaline cartilage, and breaks down within a year.[105] This process does not resemble articular surface "repair" in osteoarthritis where matrix formation by chondrocytes appears unable to keep pace with matrix loss, and tidemark violation by vascular and cellular elements from the marrow produce new cartilage and bone at the articular margin.

The conditions that promote cellular and vascular violation of the tidemark are not known, but growth factors within the joint fluid, cartilage, and bone may play a role (Fig. 12.4). Growth factors are known to be involved in the complex events that prepare cells for division: competence factors, such as PDGF and FGF, and progression factors, such as IGF-1, promote cell proliferation, migration, and differentiation. These processes could result in chondrogenesis and osteogenesis, and in the case of vascular cells, growth factors such as FGF and TGF-B could promote angiogenesis in vivo[106] (although TGF-B appears not to do so in vitro[107]). The presumed influence of these and other growth factors on cartilage repair and osteophyte formation in osteoarthritis[108,109] may be comparable to their demonstrated roles in the healing of skin wounds or fracture repair,[110] events more accessible for study.

Bone formation using animal models involves subcutaneous implantation of demineralized bone and depends on the cellular response of the host. The inductive components in the bone include osteogenin and a

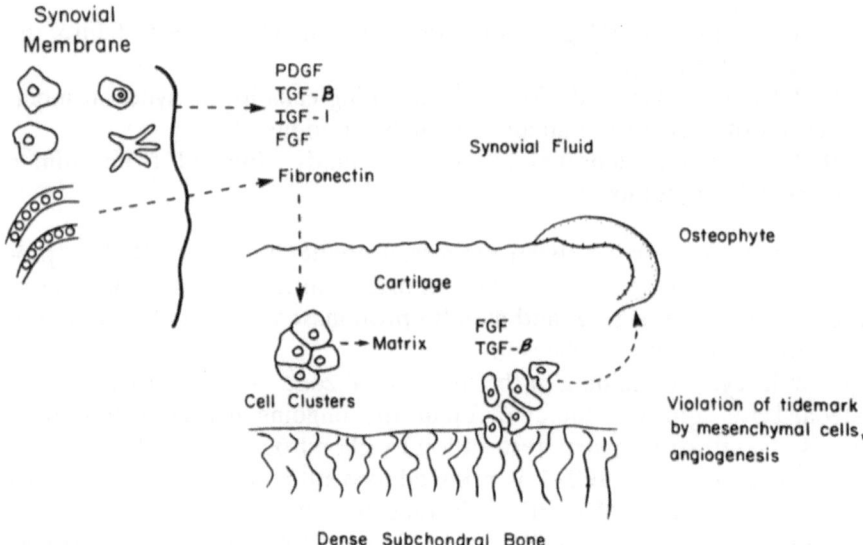

FIGURE 12.4. Diagram of cellular interactions associated with attempts at repair in the articular cartilage, and osteophyte formation at the articular margin in the osteoarthritic joint. The possible role of growth factors in these processes is discussed in the text.

variety of bone morphogenetic proteins.[111] Bone formation can also be explored in vitro. In one experiment, it was noted that bone marrow from "young" human patients (ages 4 and 7 years) possessed osteogenic potential, whereas bone marrow from "older" patients (ages 28, 58, and 63 years) produced only unmineralized fibrous tissues.[112] Whether this age-related difference reflects selection of different cell populations from the donors or diminished sensitivity to cell-produced factors on the progenitor cells is not clear. Growth factor receptors appear to be diminished on aging cells.[113] However, in osteoarthritis, osteophytes occur in patients who are almost universally older than 60 (and often in their 80s). As yet we can only speculate about the multiple events that contribute to osteophyte formation: humoral and matrix-bound growth factors and other mediators, the availability of osteoprogenitor cells, and mechanical tension that increases chondrocyte and osteoblast division and metabolic functions.[114]

We anticipate that further insights will be gained about the factors that contribute to osteoarthritis from studies on cartilage samples obtained during arthroscopy or total joint replacement, and that new therapeutic modalities will result from approaches aimed at modifying the catabolic effects of cytokines, or promoting cartilage matrix formation through local action of growth factors.

References

1. Kelsey JL. Prevalence studies of the epidemiology of osteoarthrosis. In: Lawrence RC, Shulman LE, eds. *Epidemiology of the rheumatic diseases.* New York: Gower Medical Publishing; 1984:282–288.
2. Holbrook TL, Grazier K, Kelsey JL. *The frequency of occurrence, impact and cost of musculoskeletal conditions in the United States.* Chicago: American Academy of Orthopaedic Surgeons; 1984:1–180.
3. Dieppe P, Doherty M, Watt I. Primary and secondary osteoarthritis: inter-relationships. *J Rheumatol.* 1983;10(Suppl 9):16–17.
4. Harris WH. Primary osteoarthritis of the hip: a vanishing diagnosis. *J Rheumatol.* 1983;10(Suppl 9):64.
5. Radin EL, Rose RM. Role of subchondral bone in the initiation and progression of cartilage damage. *Clin Orthop.* 1986;213:34–40.
6. Peyron JG. Osteoarthritis: the epidemiologic viewpoint. *Clin Orthop.* 1986; 213:13–9.
7. Felson DT, Anderson JJ, Naimark A, Walker AM, Meehan AF. Obesity and knee osteoarthritis: the Farmingham study. *Ann Intern Med.* 1988; 109:18–24.
8. Reginato AJ, Schumacher HR Jr. Crystal associated arthropathies. *Clin Geriatr Med.* 1988;4:295–322.
9. Hamerman D. Mechanisms of disease: the biology of osteoarthritis. *N Engl J Med.* 1989;320;1322–1330.
10. Mankin HJ, Brandt KD, Shulman LE. Workshop on etiopathogenesis of osteoarthritis: proceedings and recommendations. *J Rheumatol.* 1986;13: 1130–60.
11. Barland P, Novikoff AB, Hamerman D. Electron microscopy of the human synovial membrane. *J Cell Biol.* 1962;14:207–220.
12. Hamerman D, Stephens M, Barland P. Comparative histology and metabolism of synovial tissue in normal and arthritic joints. In: Mills LC, Mayer JH, eds. *Inflammation and diseases of connective tissue.* Philadelphia: Saunders; 1961:158–180.
13. Goto M, Sasano M, Yamanaka H, Miyasaka N, Kamatani N, Inoue K, Nishioka K, Miyamoto T. Spontaneous production of an interleukin l-like factor by cloned rheumatoid synovial cells in long-term culture. *J Clin Invest.* 1987;80:786–796.
14. Winchester RJ, Burmester GR. Demonstration of Ia antigens on certain dendritic cells and on a novel elongate cell found in human synovial tissue. *Scand J Immunol.* 1981;14:439–443.
15. Harding B, Knight SC. The distribution of dendritic cells in the synovial fluids of patients with arthritis. *Clin Exp Immunol.* 1986;63:594–600.
16. Krane SM, Goldring MB, Goldring SR. Cytokines. In: Evered D. Harnett S, eds. *Cell and molecular biology of vertebrate hard tissues.* Ciba Foundation symposium series. No. 136. New York: John Wiley; 1988:239–256.
17. Hamerman D, Smith C, Keiser HD, Craig R. Glycosaminoglycans produced by human synovial cell cultures. *Coll Rel Res.* 1982;2:313–329.
18. Swan DA, Bloch KJ, Swindell D, Shore E. The lubricating activity of human synovial fluids. *Arthritis Rheum.* 1984;27:552–556.

19. Dayer J-M, Seckinger P. Interleukin and interleukin-1 inhibitors and their relation to connective tissue distruction. In: Glauert A, ed. *The control of tissue damage*. New York: Elsvier; 1988:151–163.

20. Chin J, Rupp E, Cameron PM, MacNaul KL, Lotke PA, Tocci MJ, Schmidt JA, Bayne EK. Identification of a high-affinity receptor for interleukin lα and interleukin lβ on cultured human rheumatoid synovial cells. *J Clin Invest*. 1988;82:420–6.

21. Sherry B, Cerami A. Cachectin/tumor necrosis factor exerts endocrine, paracrine, and autocrine control of inflammatory responses. *J Cell Biol*. 1988;107:1269–77.

22. Dalton BJ, Connor JR, Johnson WJ. Interleukin-1 induces interleukin-1α and interleukin-1β gene expression in synovial fibroblasts and peripheral blood monocytes. *Arthritis Rheum*. 1989;32:279–287.

23. Hamerman D, Wood DD. Interleukin-1 enhances synovial cell hyaluronate synthesis. *Proc Soc Exp Biol Med*. 1984;177:205–210.

24. Brinckerhoff CE, Mitchell TI, Karmilowicz MJ, Kluve-Beckerman B, Benson MD. Autocrine induction of collagenase by serum amyloid A-like and B_2-microglobulin-like proteins. *Science*. 1989;243:655–657.

25. Okada Y, Takeuch N, Tomita K, Nakanishi I, Nagase H. Immunolocalization of matrix metalloproteinase 3 (stromelysin) in rheumatoid synovioblasts (B cells): correlation with rheumatoid arthritis. *Ann Rheum Dis*. 1989;48:645–653.

26. Krane SM, Amento EP, Goldring MB, Goldring SR, Stephenson ML. Modulation of matrix synthesis and degradation in joint inflammation. In: Glauert AM, ed. *The control of tissue damage*. New York: Elsevier; 1988:179–195.

27. Brinckerhoff CE, Guyre PM. Increased proliferation of human synovial fibroblasts treated with recombinant immune interferon. *J Immunol*. 1985;134:3142–3146.

28. Melnyk VO, Shipley GD, Sternfeld MD, Sherman L, Rosenbaum JT. Synoviocytes synthesize, bind, and respond to basic fibroblast growth factor. *Arthritis Rheum*. 1990;33:493–500.

29. Brinckerhoff CE. Morphologic and mitogenic responses of rabbit synovial fibroblasts to transforming growth factor-B require transforming growth factor- or epidermal growth factor. *Arthritis Rheum*. 1983;26:1370–1379.

30. Heinegard D, Oldberg A. Structure and biology of cartilage and bone matrix noncollagenous macromolecules. *FASEB J*. 1989;2:2042–2051.

31. Handley CJ, McQuillan DJ, Campbell MA, Bolis, S. Steady-state metabolism in cartilage explants. In: Kuettner KE, Schleyerbach R, Hascall VC, eds. *Articular cartilage biochemistry*. New York: Raven Press; 1986:163–176.

32. Hascall VC, Morales TI, Hascall GK, Handley CJ, McQuillan DJ. Biosynthesis and turnover of proteoglycans in organ culture of bovine articular cartilage. *J Rheumatol* 1983;11(Suppl):45–52.

33. Saklatvala J, Sarsfield SJ. How do interleukin-1 and tumour necrosis factor induce degradation of proteoglycan in cartilage? In: Glauert AM, ed. *The control of tissue damage*. New York: Elsevier; 1988:97–108.

34. Tyler J. The influence of interleukin-1 and insulin-like growth factor 1 on the integrity of cartilage matrix. In: Glauert AM, ed. *The control of tissue damage*. New York: Elsevier; 1988:197–215.

35. Dingle JT, Tyler JA. Role of intercellular messengers in the control of cartilage matrix dynamics. In: Kuettner KE, Schleyerbach R, Hascall VC, eds. *Articular cartilage biochemistry*. New York: Raven Press; 1986:181–191.
36. Arner EC, Pratta MA. Independent effects of interleukin-1 on proteoglycan breakdown, proteoglycan synthesis, and prostaglandin E_2 release from cartilage in organ culture. *Arthritis Rheum*. 1989;32:288–297.
37. Chandrasekhar S, Harvey AK. Induction of interleukin-1 receptors on chondrocytes by fibroblast growth factor: a possible mechanism for modulation of interleukin-1 activity. *J Cell Physiol*. 1989;138:236–246.
38. Tyler JA, Benton HP. Synthesis of type II collagen is decreased in cartilage cultured with interleukin-1 while the rate of intracellular degradation remains unchanged. *Coll Rel Res*. 1988;8:393–406.
39. Salmon ED Jr, Daughaday WH. A hormonally controlled serum factor which stimulates sulfate incorporation by cartilage in vitro. *J Lab Clin Med*. 1957;49:825–836.
40. McCarthy TL, Centrella M, Canalis E. Insulin-like growth factor (IGF) and bone. *Connect Tissue Res*. 1989;20:277–282.
41. Schoenle E, Zapf J, Humbel RE, Froesch ER. Insulin-like growth factor 1 stimulates growth in hypophysectomized rats. *Nature*. 1982;296:252–253.
42. Osborn KD, Trippel SB, Mankin HJ. Growth factor stimulation of adult articular cartilage. *J Orthop Res*. 1989;7:35–42.
43. Smith RL, Palathumpat MV, Ku CW, Hintz RL. Growth hormone stimulates insulin-like growth factor 1 actions on adult articular chondrocytes. *J Orthop Res*.1989;7:198–207.
44. McQuillan DJ, Handley CJ, Campbell MA, Bolis S, Milway VE, Herington AC. Stimulation of proteoglycan biosynthesis by serum and insulin-like growth factor-1 in cultured bovine articular cells. *Biochem J*. 1986;240:423–430.
45. Nilsson A, Isgaard J, Lindahl A, Dahlstrom A, Skottner A, Isaksson OGP. Regulation by growth hormone of number of chondrocytes containing IGF-1 in rat growth plate. *Science*. 1986;233:571–574.
46. Kato Y, Nomura Y, Tsuju M, Kinoshita M, Ohmae H, Suzuki F. Somatomedin-like peptide(s) isolated from fetal bovine cartilage (cartilage derived factor). *Proc Natl Acad Sci*. 1981;78:6831–6835.
47. Sullivan R, Klagsbrun M. Purification of cartilage-derived growth factor by heparin affinity chromatography. *J Biol Chem*. 1985;260:2399–403.
48. Klagsbrun M, Shing Y. Heparin affinity of anionic and cationic capillary endothelial cell growth factors: analysis of hypothalamus-derived growth factors and fibroblast growth factors. *Proc Natl Acad Sci*. 1985;82:805–809.
49. Gospodarowicz D, Neufeld G, Schweigerer L. Fibroblast growth factor: structural and biological properties. *J Cell Physiol*. 1987;Suppl 5:15–26.
50. Damon DH, Lobb RR, D'Amore PA, Wagner JA. Heparin potentiates the action of acidic fibroblast by prolonging its biological half-life. *J Cell Physiol*. 1989;138:221–226.
51. Mueller SN, Thomas KA, DiSalvo J, Levine EM. Stabilization by heparin of acid fibroblast growth factor mitogenicity for human endothelial in vitro. *J Cell Physiol*. 1989;140:439–448.
52. Gordon PB, Choi HV, Conn G, Ahmed A, Ehrmann B, Rosenberg L, Hatcher VB. Extracellular matrix heparan sulfate proteoglycans modulate

the mitogenic capacity of acidic fibroblast growth factor. *J Cell Physiol.* 1989;140:584–592.

53. Vlodavsky I, Fridman R, Sullivan R, Sasse J, Klagsbrun M. Aortic endothelial cells sunthesize basic fibroblast growth factor which remains cell associated and platelet derived growth factor-like protein which is secreted. *J Cell Physiol.* 1987;131:402–408.

54. Hauschka PV, Chen TL, Mavrakos AE. Polypeptide growth factors in bone matrix. In: Evered D, Harnett S, eds. *Cell and molecular biology of vertebrate hard tissue.* Ciba Foundation Symposium series. No. 136, New York: John Wiley; 1988:207–225.

55. Rifkin DB, Moscatelli D. Mini-review. Recent developments in the cell biology of basic fibroblast growth factor. *J Cell Biol.* 1989;109:1–6.

56. Vladovsky I, Folkman J, Sullivan R, Fridman R, Ishai-Michaeli R, Sasse J, Klagsbrun M. Endothelial cell-derived basic fibroblast growth factor: synthesis and deposition into subendothelial extracellular matrix. *Proc Natl Acad Sci USA.* 1987;84:2292–296.

57. Bashkin P, Doctrow S, Klagsbrun M, Svahn M, Folkman J, Vlodovsky I. Basic fibroblast growth factor binds to subendothelial extracellular matrix and is released by heparitinase and heparin-like molecules. *Biochemistry.* 1989;28:1737–1743.

58. Ruoslahti E. Mini-review. Proteoglycans in cell regulations. *J Biol Chem.* 1989;264:13369–13372.

59. Roberts AB, Sporn MB. The transforming growth factor-Bs. In: Sporn MB, Roberts AB, eds. *Peptide growth factors and their receptors.* New York: Springer Verlag; 1990;1:419–472.

60. Mereola M, Stiles CD. Growth factor superfamilies and mammalian embryogenesis. *Development.* 1988;102:451–460.

61. Morales TI, Roberts AB. Transforming growth factor B regulates the metabolism of proteoglycans in bovine cartilage organ cultures. *J Biol Chem.* 1988;263:12828–12831.

62. Kato Y, Iwamoto M, Koike T, Suzuki F, Takano Y. Terminal differentiation and calcification in rabbit chondrocyte cultures grown in centrifuge tubes: regulation by transforming growth factor B and serum factors. *Proc Natl Acad Sci.* 1988;85:9552–9556.

63. Kato Y, Iwamoto M. Fibroblast growth factor is an inhibitor of chondrocyte terminal differentiation. *J Biol Chem.* 1990;265:5903–5909.

64. Rosier RN, O'Keefe RJ, Crabb ID, Puzas JE. Transforming growth factor beta: an autocrine regulator of chondrocytes. *Connect Tissue Res.* 1989;20:295–301.

65. Rosen DM, Stempien SA, Thompson AY, Seyedin SM. Transforming growth factor beta modulates the expression of osteoblast and chondroblast phenotypes in vitro. *J Cell Physiol.* 1988;134:337–46.

66. Bassols A, Massague J. Transforming growth factor B regulates the expression and structure of extracellular matrix chondroitin/dermatan sulfate proteoglycans. *J Biol Chem.* 1988;263:3039–3045.

67. Roman J, LaChance RM, Broekelmann TJ, Kennedy CJR, Wayner EA, Carter WG, McDonald JA. The fibronectin receptor is organized by extracellular matrix fibronectin: Implications for oncogenic transformation and for cell recognition of fibronectin matrices. *J Cell Biol.* 1989;108:2529–2543.

68. Wahl SM, Hunt DA, Wakefield LM, McCartney-Francis N, Wahl L, Roberts AB, Sporn MB. Transforming growth factor type B induces monocyte chemotaxis and growth factor production. *Proc Natl Acad Sci* 1987;84:5788–5792.
69. Kuettner KE, Pauli BU. Vascularity of cartilage. In: Hall BK, ed. *Cartilage.* New York: Academic Press; 1983;1:281–312.
70. Franzen A, Heinegard D, Solursh M. Evidence for the sequential appearance of cartilage matrix proteins in developing mouse limbs and in cultures of mouse mesenchymal cells. *Differentiation.* 1987;36:199–210.
71. Toole BP, Knudson CB, Munaim SI, Knudson W, Welles S, Chi-Rosso G. Hyaluronate-cell interactions and regulation of hyaluronate synthesis during embryonic limb development. In: Davidovitch Z, ed. *The biological mechanisms of tooth eruption and root resorption.* Birmingham Al.: EBSCO Media; 1989:35–41.
72. Knudson CB, Toole BP. Epithelial-mesenchymal interaction in the regulation of hyaluronate production during limb development. *Biochem Int.* 1988;17:735–745.
73. Solursh M. Ectoderm as a determinant of early tissue pattern in the limb bud. *Cell Diff.* 1984;15:17–24.
74. Cormack DH. Bone. In: *Ham's histology.* 9th ed. Philadelphia: Lippincott; 1987:272–323.
75. Cowell HR, Hunziker EB, Rosenberg L. The role of hypertrophic chondrocytes in endochondral ossification and in the development of secondary centers of ossification. *J Bone Joint Surg.* 1987;69A:159–161.
76. Horton WA, Machado MM. Extracellular matrix alterations during endochondral ossification in humans. *J Orthop Res.* 1988;6:793–803.
77. Oshima O, Leboy PS, McDonald SA, Tuan RS, Shapiro IM. Development expression of genes in chick growth cartilage detected by in situ hybridization. *Calcif Tiss Int.* 1989;45:182–192.
78. Sandberg M, Vuorio E. Localization of types I, II and III collagen mRNAs in developing human skeletal tissues by *in situ* hybridization. *J Cell Biol.* 1987;104:1077–1084.
79. Schwartz Z, Knight G, Swain LD, Boyan BD. Localization of Vitamin D_3-responsive alkaline phosphatase in cultured chondrocytes. *J Biol Chem.* 1988;263:6023–6026.
80. Franzen A, Oldberg A, Solursh M. Possible recruitment of osteoblastic precursor cells from hypertrophic chondrocytes during initial osteogenesis in cartiloginous limbs of young rats. *Matrix.* 1989;9:261–265.
81. Thesingh CW, Scherft JP. Bone matrix formation by transformed chondrocytes in organ culture of stripped embryonic metarsalia. In: Ali S, ed. *Cell Mediated calcification and matrix vesicles.* Amsterdam: Elsevier; 1986:309–314.
82. Mayne R, Irwin MH. Collagen types in cartilage. In: Kuettner KE, Schleyerbach R, Hascall VC, eds. *Articular cartilage biochemistry.* New York: Raven Press; 1986:23–35.
83. Aydelotte MB, Kuettner KE. Differences between subpopulations of cultured bovine articular chondrocytes. I. Morphology and cartilage matrix production. *Connect Tissue Res.* 1988;18:205–22.
84. Beresford JN. Osteogenic stem cells and the stromal system of bone marrow. *Clin Orthop.* 1989;240:270–280.

85. Sodek J, Berkman FA. Bone cell cultures. *Methods Enz*. 1987;145:303–324.
86. Owen M, Friedenstein AJ. Stromal stem cells: marrow-derived osteogenic precursors. In: Evered D, Harnett S, eds. *Cell and molecular biology of vertebrate hard tissues*. Ciba Foundation symposium series. No. 136. New York: John Wiley; 1988:42–53.
87. Butler WT. Mineralized tissues: an overview. *Methods Enz*. 1987;145:255–261.
88. Termine JD. Non-collagen proteins in bone. In: Evered D, Harnett S, eds. *Cell and molecular biology of vertebrate hard tissues*. Ciba Foundation symposium series. No. 136. New York: John Wiley; 1988:178–190.
89. Azria M. The value of biomarkers in detecting alterations in bone metabolism. *Calcif Tissue Int*. 1989;45:7–11.
90. Martin TJ, Na KW, Suda T. Bone cell physiology. *Endocrin Metab Clin North Am*. 1989;18:833–858.
91. Bonewald LF, Mundy GR. Role of transforming growth factor beta in bone remodeling: a review. *Connect Tissue Res*. 1989;23:201–208.
92. Tatakis DN, Schneeberger G, Dziak R. Recombinant interleukin-1 stimulates prostaglandin E_2 production by osteoblastic cells: synergy with parathyroid hormone. *Calcif Tissue Int*. 1988;42:358–362.
93. Mundy GR, Bonewald LF. Effects of immune cell products on bone. In: Sorge C, ed. *Macrophage-derived cell regulatory factors. Cytokines*. Basel: Karger; 1989;1:38–53.
94. Canalis E, McCarthy T, Centrella M. The regulation of bone formation by local growth factors. *Bone Min Res*. 1989;6:27–56.
95. Pfeilschifter J, Bonewald L, Mundy GR. Role of growth factors in cartilage and bone metabolism. In: Sporn MB, Roberts AB eds. *Peptide growth factors and their receptors*. New York: Springer Verlag; 1990;II:371–400.
96. Robey PG, Young MF, Flanders KC, Roche NS, Kondaiah P, Reddi AH, Termine JD, Sporn MB, Roberts AB. Osteoblasts synthesize and respond to transforming growth factor-type B (TGF-B) in vitro. *J Cell Biol*. 1987;105:457–463.
97. Rodan SB, Wesolowski G, Thomas KA, Yoonk Y, Rodan GA. Effects of acidic and basic fibroblast growth factors on osteoblastic cells. *Connect Tissue Res*. 1989;20:283–288.
98. Centrella M, McCarthy TL, Canalis E. Effects of transforming growth factors on bone cells. *Connect Tissue Res*. 1989;20:267–275.
99. Pfeilschifter J, Bonewald L, Mundy GR. Characterization of the latent transforming growth factor B complex in bone. *J Bone Min Res*. 1990;5:49–58.
100. Pelletier JP, Martel-Pelletier J, Howell DS, Ghandur-Mnaymneh L, Ennis JE, Woessner JF Jr. Collagenase and collagenolytic activity in human osteoarthritic cartilage. *Arthritis Rheum*. 1984;27:305–312.
101. Pettipher ER, Higgs GA, Henderson B. Interleukin-1 induces leukocyte infiltration and cartilage proteoglycan degradation in the synovial joint. *Proc Natl Acad Sci*. 1986;83:8749–8753.
102. Wood DD, Ihrie EJ, Hamerman D. Release of interleukin-1 from human synovial tissue in vitro. *Arthritis Rheum*. 1985;28:853–62.
103. Wood DD, Ihrie EJ, Dinarello CA, Cohen PL. Isolation of an interleukin-1-like factor from human joint effusions. *Arthritis Rheum*. 1983;26:975–83.

104. Baird A, Walicke PA. Fibroblast growth factors. *Br Med Bull.* 1989;45: 438–452.
105. Rosenberg L. Biological basis for the imperfect repair of articular cartilage following injury. In: Hunt TK, Heppenstall RB, Pines E, Rovee D, eds. *Soft and hard tissue repair: biological and clinical aspects. Surgical science series.* vol. 2. New York: Praeger; 1984:143–69.
106. Klagsbrun M. Introduction. In: Rifkin DB, Klagsbrun M, eds. *Angiogenesis: mechanisms and pathobiology. Current communications in molecular biology.* Cold Spring Harbor, NY: Cold Spring Harbor Laboratory; 1987: 1–8.
107. Muller G, Behrens J, Nussbaumer U, Bohlen P, Birchmeier W. Inhibitory action of transforming growth factor B on endothelial cells. *Proc Natl Acad Sci* 1987;84:5600–5604.
108. Hamerman D, Klagsbrun M. Osteoarthritis: emerging evidence for cell interactions in the breakdown and remodeling of cartilage. *Am J Med.* 1985;78:495–9.
109. Horton Jr WE, Higinbotham JD, Chandrasekhar S. Transforming growth factor-beta and fibroblast growth factor act synergistically to inhibit collagen II synthesis through a mechanism involving regulatory DNA sequences. *J Cell Physiol.* 1989;141:8–15.
110. Nemeth GG, Bolander ME, Martin GR. Growth factors and their role in wound and fracture healing. In: Barbul A, Pines E, Caldwell M, Hunt TK, eds. *Growth factors and other aspects of wound healing. Biological and clinical implications.* New York: Alan R Liss; 1988:1–17.
111. Reddi AH, Muthakumaran N, Ma S, Carrington J, Luyten FP, Paralkar VM, Cunningham NS. Initiation of bone development by osteogenin and promotion by growth factors. *Connect Tissue Res.* 1989;20:303–312.
112. Bab I, Passi-Even L, Grazit D, Sakeles E, Ashton BA, Peylan-Ramu N, Ziv I, Ulmansky M. Osteogenesis in in vivo diffusion chamber cultures of human marrow cells. *Bone Min.* 1988;4:373–386.
113. Cristofalo VJ, Phillips PD, Sorger T, Gerhard G. Alterations in the responsiveness of senescent cells to growth factors. *J Gerontol.* 1989;44:55–62.
114. Gray ML, Pizzanelli AM, Grodzinsky AJ, Lee RC. Mechanical and physio-chemical determinants of the chondrocyte biosynthetic response. *J Orthop Surg.* 1988;6:777–792.

13
Endocrine Control of Tooth and Periodontal Tissue Growth and Repair

John F. Helfrick

The role of the endocrine system in regulating the normal growth and repair of teeth and periodontal tissues is poorly understood and, although the mandibula and the maxilla are subject to the same controls as other bones, much information has been acquired from the study of disease; its interpretation is obscured by the complex interrelationships existing between humoral, genetic, and nutritional factors.

Pituitary Gland

Current evidence indicates that the anterior pituitary exerts its major effects on tooth development through the action of growth hormone (GH) and through the control of the thyroid by thyrotropin. In turn, most effects of growth hormone depend on two mechanisms: the action of the somatomedins [insulinlike growth factors 1 and 2 (IGF-1 and IGF-2)] secreted by the liver and by peripheral tissues, including bone, in response to GH stimulation and the conversion of cartilage stem cell into cells sensitive to the growth-promoting action of the somatomedins. Indeed, IGF-1 and IGF-2 belong to a large family of growth promoters that include nerve growth factor (NGF), epidermal growth factor (EGF), platelet-derived growth factor (PDGF), bone morphogenic protein (BMP), and others. In addition to their growth-promoting effect, the somatomedins participate in the negative feedback regulation of GH secretion (Fig. 13.1). Schour and Van Dyke,[1] and Baume and associates[2,3] have reported that in rats, hypophysectomy was followed by retardation in the eruption of the incisor tooth. In addition, the tooth attained only about two thirds of normal size and showed a distortion of form, especially at the apical end. Injections of an anterior pituitary extract restored the eruption rate to normal. Becks and associates[4] pointed out that the only constant pathognomonic sign in the hypophysectomized animals was a decrease in the size of the pulp chamber secondary to the thickening of the dentinal walls. Baume and associates[2] also reported that the activity

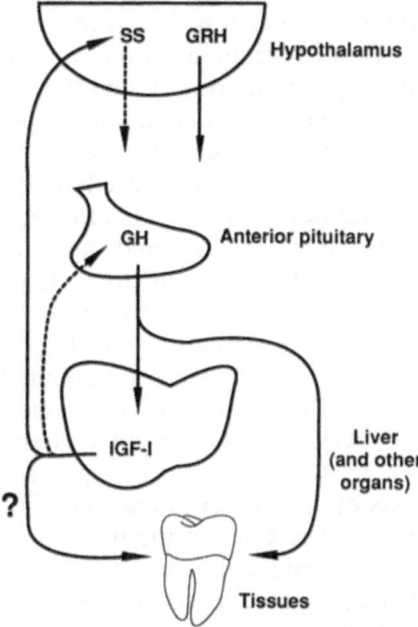

FIGURE 13.1. Diagrammatic illustration of the feedback mechanism for the regulation of grwoth hormone secretion. The exact role of somatomedins (IGF-1 and IGF-2) on dental development are yet to be defined.

of the odontogenic epithelium was entirely anterior pituitary-dependent, whereas dentinogenesis and cementogenesis were able to proceed in the absence of pituitary hormones, even if at a decreased rate. These investigators also noted that the histologic changes in the enamel organ of the hypophysectomized rat incisors were comparable to those of thyroidectomized animals and that the abnormal folding of the incisors at the apical end in these animals was similar to that seen in magnesium-deficient rats. They theorized that these changes may reflect abnormalities in salt and mineral metabolism and in mineralcorticoid activity similar to those implicated in the abnormal root development characteristic of hypophysectomized animals.[1] In another study, Baume and associates[2] investigated the possibility that the pituitary gland may influence tooth eruption not only through the effects of GH, but also by stimulating the release of the thyroid hormones. They noted that the effects of thyroxine and those of GH on dental growth and development differ both qualitatively and quantitatively: thyroxine stimulated eruptive movement and tooth size, but had little effect on alveolar growth, whereas GH influenced dental and alveolar growth, but had little effect on the rate of eruption.

Pituitary gland function also appears to be critical for normal dental development in humans. Tooth formation, eruption, and facial growth are known to be retarded in hypopituitarism. Using the system developed by Demirjian et al.[5] for the evaluation of dental development in relation to height and skeletal development between the ages of 2.5 and 16 years, it has been shown that in patients with isolated GH or with generalized pituitary deficiency the dental development was affected less frequently and to a lesser degree than the skeletal age and the statural growth. In addition, it has been observed that all components of dental development seem equally retarded. Treatment with GH for up to 4 years accelerated dental development in parallel with statural and skeletal development in most cases, whereas when GH therapy was discontinued, the delay in dental eruption again correlated with retardation of skeletal age.

Kosowicz and Rzymski[6] performed radiographic studies of the teeth and maxillofacial skeleton in 48 patients with hypopituitary dwarfism, confirmed by measurements of serum GH in basal conditions and during insulin-induced hypoglycemia. The faces of the patients were "childishly round," the mandibles were delicate and poorly developed, and the chins were recessed. The mental development was normal, but secondary sex characteristics were absent. The bone age was markedly delayed and the maxillofacial skeleton showed a stricking underdevelopment. These radiographic findings were uniform, irrespective of age and sex. The dental radiographic findings were also uniform and consisted of delayed resorption of the roots of the deciduous teeth, delayed exfoliation of deciduous teeth, delayed or absent eruption of the permanent teeth, continued root development and apical closure of the unerupted permanent teeth, displacement of the first permanent molars from the body into the ramus of the mandible, altered axial inclination of a number of the retained deciduous teeth, absence of third molars, underdevelopment of the maxilla and mandible, and lack of chin and gonial angle prominence. In some of these pituitary dwarfs, therapy with anabolic drugs resulted in growth of the facial bones and the mandible. The exfoliation of deciduous teeth was accelerated, the eruption of the permanent ones was stimulated, and the dental age was accelerated. Of interest was the observation that the greater the delay in height age, the greater was the delay in dental development.

These observations are similar to those of Salzmann and Wein,[7] who described a case of pituitary dwarfism with extremely delayed dental development, failure to resorb the roots of several deciduous teeth, and delayed eruption of the permanent teeth. Similar findings have been reported by Hamori and associates,[8] Drews,[9] and Snyder.[10]

Kosowicz and Rzymski[6] summarized their findings by stating, "In primary dwarfism the development of the teeth occurred at the usual time; in Turner's syndrome the teeth were frequently hypoplastic but erupted normally." In congenital hypothyroidism the eruption of the deciduous teeth is markedly delayed, but in contrast to the situation seen in pituitary dwarfs, the roots of retained permanent teeth are not formed even in older patients. Hence, in pituitary dwarfism the roentgenographic appearance of the teeth and jaws is typical.

The mechanism whereby GH affects dental development in humans is not well understood. However, some information may be obtained by comparing patients with isolated GH deficiency and patients with Laron-type dwarfism (LTD), who have normal or high serum levels of GH, but lack hepatic GH receptors and therefore suffer IGF-1 deficiency. Treatment of these patients with exogenous GH has no effect on growth; these patients remain dwarfs with the typical characteristics of GH deficiency. By comparing the dental development in these two groups of patients with and without GH replacement therapy, the role of GH and IGF-1 is better understood.

Such comparison was carried out by Sarnat et al.,[11] who studied 32 patients with either isolated growth hormone deficiency (IGHD) or LTD. Their evaluation included clinical and radiologic examination, study casts, and bone age determinations by standard radiography of the left hand and wrist. Significant differences were noted between the two groups: 92% of the LTD but only 37% of the IGHD patients were missing their third molars. In 75% of the LTD patients, all four third molars were absent. Both groups showed a high prevalence of hypodontia and 10.5% of the patients in the IGHD group had impacted maxillary canine, a rate more than 10 times greater that reported in the general population. The length of the maxillary and mandibular dental arches was well below normal and, whereas no difference was noted in the size of the maxilla between the two groups, the mandible was significantly smaller in the LTD patients than in the IGHD patients. The arch circumferences were also smaller in both groups; the mandibular arch circumference was especially reduced in the LTD patients and the permanent teeth appeared to have smaller mesiodistal diameters. The authors observed that dental development was much less affected than skeletal maturation and that dental age corresponded to chronological age in both the IGHD and LTD groups and was not influenced by GH therapy. This discrepancy between skeletal and dental maturation has been observed in other systemic diseases,[12,13] and supports the hypothesis that tooth development and eruption is an independent growth process. One must be cautious in attributing the absence of third molars, particularly in the LTD group, to a hormone deficiency. LTD is a hereditary disease with a high incidence of consanguineous marriages in the affected families. Since the congenital absence of third molars is also a genetically determined defect, the failure of these teeth to develop in the LTD group could be due to genetic rather than endocrine causes. Although statistical evaluation could not be performed because of the small sample size, the data strongly indicate that pituitary dwarfism is associated with smaller mesiodistal tooth width.

Other investigators have also reported hypodontia in hypopituitarism.[7] The lack of dental crowding in otherwise very small jaws is secondary to the absence of third molars and the small size of the teeth.

In a more recent case report, Shaw and Foster[14] discuss the dental findings in two patients who were found to have pituitary and thyroid dysfunction, respectively. The first patient was a 21-year-old man who at age 9 years was found to have deficient levels of GH and was short of stature. He was referred for dental evaluation for an "unusual state of microdontia." Clinical and panorex radiologic examination disclosed small teeth, advanced root resorption, and marked obliteration of the pulp canals. The second patient was a 12-year-old boy with normal production but deficient release of GH and hypothyroidism. He was short in stature and at age 3 years his pediatrician noted that he had "small teeth." Clinical and radiographic examination disclosed a small face and jaws, an essentially normal dental eruption pattern, absence of the third molars, reduced crown sizes, and shortened conical roots. Despite the small size of the maxilla and the mandible, the small size of the teeth prevented dental crowding. Confirming previous observation, the authors noted that in hypopituitarism there is no constant relationship between the lag in statural and skeletal growth and in dental development and that the former is invariably greater than the latter. Indeed, Edler observed a mean delay in skeletal age of 28% (range 25–32%) and a mean delay in dental age of 9% (range 4–20%).[15]

In summary, GH deficiency results in the following defects of dental development: delayed resorption of the roots and delayed exfoliation of deciduous teeth, delayed and incomplete eruption of permanent teeth, hypodontia with apparent reduced mesiodistal tooth width, reduced maxillary and mandibular dental arch length with or without dental crowding, less delay in dental development than in skeletal development with dental age corresponding to chronologic age, congenitally missing third molars, and obliteration of pulp chambers in permanent teeth.

Hypersecretion of GH is associated with gigantism or acromegaly. If this increase occurs before the epiphyses of the long bones are closed, gigantism results; if the increase occurs later in life, the result is acromegaly. In gigantism the teeth are proportional to the size of the jaw and the roots may be longer than normal. Snyder also noted an increase in the deposition of cementum.[10] The enlarged tongue seen in the acromegalic individual frequently causes buccal and lingual displacement of the teeth, resulting in malocclusion.

Thyroid

The thyroid gland plays a direct role in a host of physiologic functions, including water balance and electrolyte, protein, carbohydrate, and

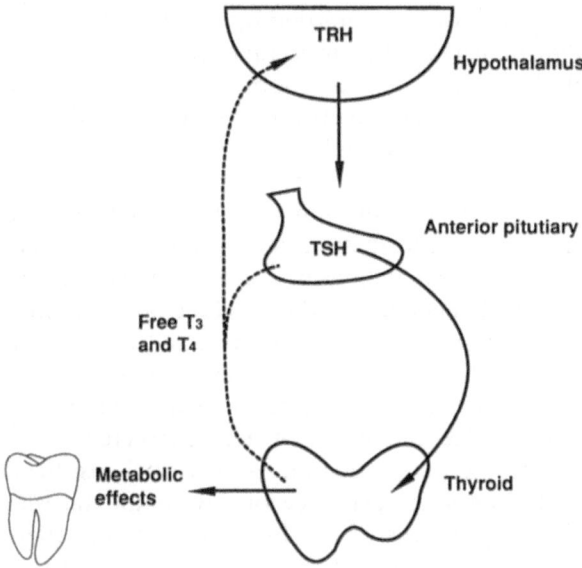

FIGURE 13.2. A normally functioning thyroid gland is necessary for dental development. This figure diagrammatically represents the feedback mechanism involving TRH and TSH secretion.

lipid metabolism, tissue differentiation, growth and maturation of the skeleton, and dental structures (Fig. 13.2). In addition, thyroxine and triiodothyronine potentiate the secretion and the effects of GH. Failure of anterior pituitary thyrotropic function and the atrophy or destruction of the thyroid gland in infancy result in cretinism, whereas in the adult the result is myxedema. The dentofacial changes in cretinism are related to the degree of thyroid insufficiency. In a typical case, the face is wide and fails to develop in the longitudinal direction whereas the base of the skull is shortened, resulting in retraction of the bridge of the nose and alar flaring. The growth of the jaw is disproportionate as the mandible remains underdeveloped and the maxilla overdeveloped. The tongue is enlarged and its chronic protrusion may result in a dental malocclusion.

The earliest experimental findings on the relationship of dental development to thyroid function were reported by Hoskins in 1928,[16] who found that acetyl thyroxin injected into newborn rats accelerated the development of the incisor teeth. The molars were unaffected. The formation of the enamel of the incisors was also accelerated but no more than the other aspects of dental development. Ziskin and Applebaum[17] reported that thyroid ablation with reimplantation of the parathyroid glands in rats resulted in the formation of a line of dentin hypoplasia corresponding to the date of operation. Enamel calcification was defective and dentin apposition was retarded up to 90%. Paynter[18] found that total hypothyroidism caused by propylthiouracil in rats caused retardation of mandibular and tooth development. Garren[19,20] also found that propylthiouracil retarded the eruption of incisor teeth in rats and increased the proportion of dentin to pulp. He also noted that the administration of thyroid hormones accelerated the eruption of the incisor teeth. Baume and Becks[21] noted degeneration of the enamel organ and a significant retardation of enamel formation in thyroidectomized rats.

The reported clinical findings in humans with congenital hypothyroidism are varied, most likely because they include observations made in patients with congenital hypothyroidism and juvenile hypothyroidism and, in most cases, in patients diagnosed and treated at an early age. In general, "cretinism" is associated with a delay in the eruption of teeth and with abnormalities in size and shape of dental structures.[22-24] Cephalometric studies of maxillofacial, dental, and carpal ossification patterns in untreated children show evidence of developmental retardation in comparison with the patterns of children given thyroid therapy.[25] The teeth display delayed eruption but are anatomically normal. Other authors have reported anomalous formation of teeth[26,27] or delayed eruption.[28-30]

Anderson[30] observed an incidence of enamel hypoplasia of 80% in hypothryroid children compared to 15% in the normal population. He also speculated that delayed eruption in juvenile hypothroidism was secondary to the accumulation of mucopolysaccharides in the tissues surrounding the teeth. Noren and Alm[31] found that the deciduous teeth from children with congenital hypothyroidism exhibited areas of increased pore volume distribution in the prenatal and postnatal enamel as compared with the teeth of normal patients. These findings suggest that thyroid hormones may regulate enameloblast activity, in agreement with the experimental results reported previously by Baume. Noren and Alm also reported no abnormalities in the shape or size of deciduous teeth in congenital hypothryoid children and suggested that the increased porosity of the enamel may be ascribed to changes in the stages of matrix deposition and calcification caused by reduced levels of the thyroid hormone. Noren and Alm further suggested that the distribution of abnormalities in the enamel of deciduous teeth in children with congenital hypothyroidism may assist in distinguishing children with prenatal from those with postnatal onset of hypothyroidism.

Bedi and Brook[31] reported a rare case of juvenile hypothyroidism in a girl who remained untreated until the age of 14 years, when she had attained a height age of $5\frac{1}{2}$ years and a dental age of 7 to 8 years. A cephalometric radiograph showed a short facial height and a large sella turcica. She had grossly carious deciduous molars and marked gingival inflammation. Tooth dimensions were within the normal range. Similar observation were made by Hinrich[32] in 36 cases of juvenile hypothyroidism. Again, the degree of abnormality correlated with the length of time the patients had gone before diagnosis and therapy. The most striking finding in patients who had not been treated for prolonged periods was the retardation of exfoliation and eruption. The eruption rate was restored to near normal when the patients were treated with thyroid hormone. Morphologic changes in teeth were not noted.

Since myxedema occurs after dental development and eruption, few abnormalities are noted and are limited to the soft tissues of the face and mouth. The lips, nose, eyelids, and suborbital tissues are myxedematous and swollen. The tongue is large and edematous and speech is frequently altered secondary to tongue size.

Hyperthyroidism is not associated with significant oral pathology. Alveolar atrophy may occur in advanced cases and exfoliation of deciduous teeth in children occurs prematurely. The eruption of permanent teeth may also be accelerated.[33]

Parathyroid Hormone

Idiopathic hypoparathyroidism and hypoparathyroidism secondary to thyroidectomy are associated with a number of clinical dental findings that may vary according to the age of onset. Frensilli and Hinrichs[34] reviewed the literature and described three children with idiopathic hypoparathyroidism. In two of the children who had been hypocalcemic since birth, tooth eruption was not delayed, but enamel hypoplasia was noted. The third patient, who had become hypocalcemic at age 13 years, showed both delayed eruption and enamel hypoplasia. The authors noted that the age of onset did not explain these differences. This is in agreement with previous reports that failed to note a consistent relationship between the rate of tooth eruption and other indices of maturation. In these three patients, however, there was a correlation between the age of onset of hypocalcemia and the occurrence of enamel hypoplasia. The two patients who were hypocalcemic since birth showed not only generalized enamel hypoplasia but also blunting of root tips. The third patient, who became hypocalcemic at the age of 13 years, only displayed enamel hypoplasia of the third molars. This dental finding correlates well with the age of onset of hypoparathyroidism. The delayed eruption, which was not affected by the administration of vitamin D and calcium, may have been caused by a factor other than hypocalcemia. The authors noted that the clear correlation between the enamel hypoplasia and age of onset of the hypoparathyroidism could be a useful tool in helping to determine the age of onset of the disorder.

In primary hyperparathyroidism, the major radiographic findings are a decrease in radiopacity associated with the development of clearly defined radiolucencies in the maxilla and/or mandible (Fig. 13.3). These multilocular cystic lesions are similar to those observed in other bones

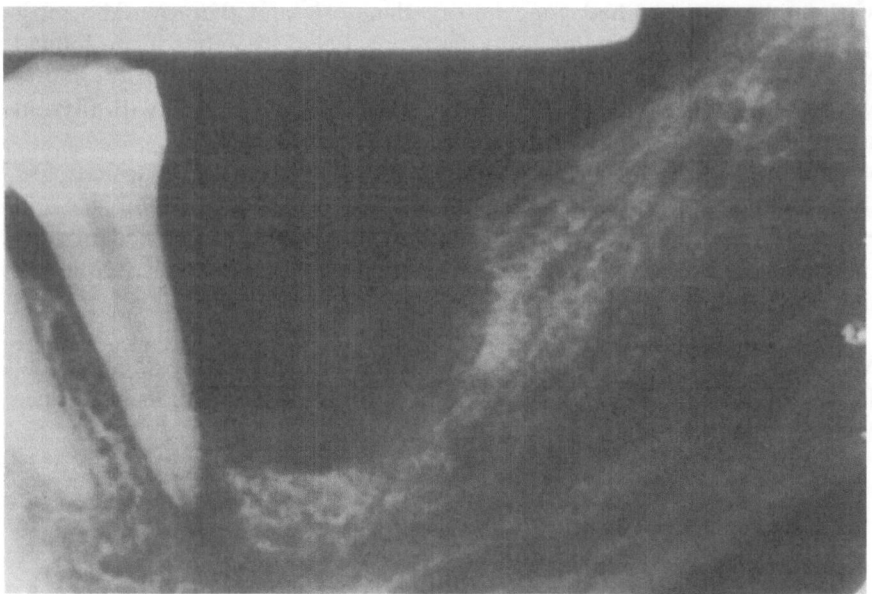

FIGURE 13.3. Radiolucency of the mandible of a patient with hyperthyroidism showing a "Brown tumor." These lesions are histologically identical to a giant cell granuloma. A pathologic diagnosis of central giant cell granuloma following a biopsy requires a laboratory evaluation to rule out the presence of hyperparathyroidism. Courtesy of Dr. Kenneth Abramovitch.

and are characterized by the presence of identical giant cell granulomas. Most striking is the presence of "osteoclastomas," characterized by masses of fibroblasts, numerous capillaries and endothelium-lined blood spaces, red blood cells, hemosiderin, and many multinucleated giant cells. Serum calcium determinations will differentiate the systemic disease from that of the isolated central giant cell ganuloma.[35] No changes in the size, shape, or structure of teeth has been reported in hyperparathyroidism. However, periodontal abnormalities have been described. These consist primarily to a partial loss of the lamina dura adjacent to the roots (Fig. 13.4). Silverman and Ware[35] studied 42 consecutive dentulous patients with hyperparathyroidism. Urinary tract stones were noted in 33 of them: twenty had no radiologic evidence of dental abnormalities, 17 had normal lamina dura but abnormal appearing alveolar bone, and 5 had a partial loss of the lamina dura. None of the patients in this series had a complete loss of lamina dura.

Secondary hyperparathyroidism, which is most commonly a result of end-stage renal disease, can also result in significant alteration in the structure of the jaws and periodontal structures. Decalcification

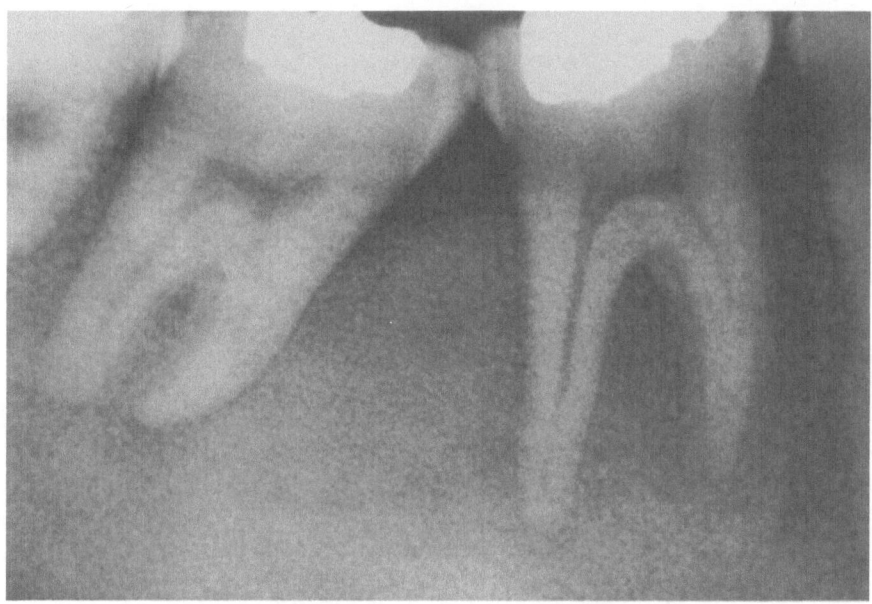

FIGURE 13.4. Radiograph representing the classic round glass appearance of the mandible in a patient with hyperparathyroidism. Also note the loss of the lamina dura. Courtesy of Dr. Benton E. Crawford.

results in a generalized osteoporotic appearance of the mandible and/ or maxilla. "Brown tumors," giant cell lesions similar to those seen in primary hyperparathyroidism, may be present. In addition, loss of the lamina dura may be a prominent finding in patients with secondary hyperparthyroidism.

Adrenal Glands and Gonads

No specific roles in dental growth or development have been ascribed to the hormones of the adrenal medulla. However, the adrenal cortex produces steroids that appear to have an effect on dental development. The only significant oral finding in chronic insufficiency of the adrenal cortex (Addison's disease) is the appearance of darkly pigmented areas on the lips, tongue, gingiva, and buccal mucosa. However, in adreno- cortical hyperplasia associated with precocious puberty,[36] McCune- Albright syndrome, familial male precocious puberty, and in patients with adrenal, gonadal, or human chorionic gonadotropin-secreting tumors, orofacial changes may be observed. For example, the McCune-Albright syndrome characteristically includes fibrous dysplasia of the jaws and skeleton, precocious puberty, and café-au-lait spots (Fig. 13.5).

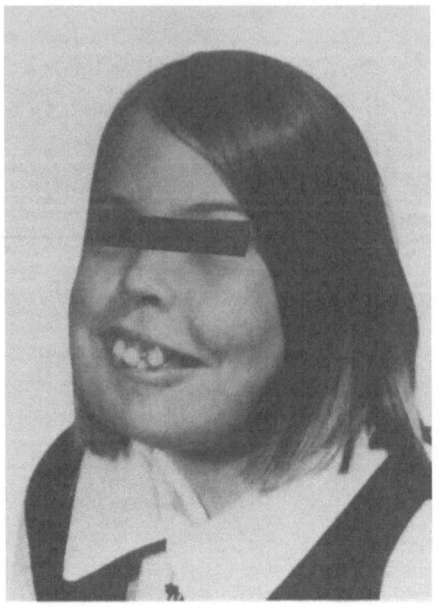

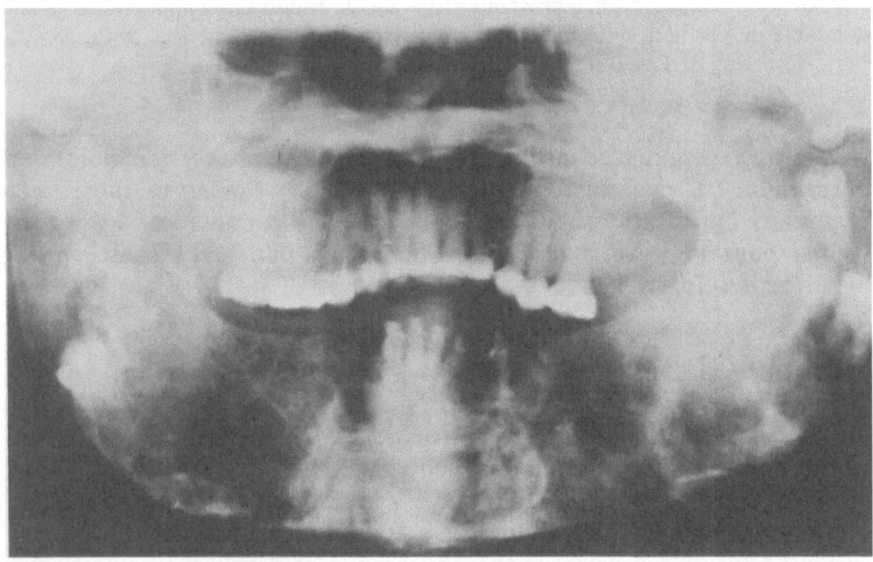

FIGURE 13.5. A: This young girl presented with all of the findings seen in McCune-Albright syndrome. She had polyostotic fibrous dysplasia, endocrine abnormalities, including precocious puberty, and café-au-lait spots.
B: The patient's panorex radiograph disclosed fibrous dysplasia involving the mandible and maxilla. She also had extensive involvement of her ilium and femur.

Roberts and colleagues[37] described the dental findings in 101 children with precocious puberty. Dental root development of canines, premolars, and molars was assessed radiographically. In patients with precocious puberty due to lesions of the central nervous system, the dental age was retarded whereas the skeletal age was advanced relative to the chronologic age; a significant correlation coefficient (0.42) existed between the two abnormalities. In a subsequent publication, Roberts and colleagues[37] reported on 43 additional children with central nervous system (CNS)-related precocious puberty. No abnormal development was noted in the male patients. However, in the female patients there was a significant delay in dental development similar to that observed in female patients with idiopathic precocious puberty. The authors postulated that elevated levels of estrogen associated with precocious puberty may be responsible for advancing skeletal maturation while retarding odontogenesis and suggested that the presence of estrogen may explain why teeth develop and erupt earlier in normal girls than in normal boys. The fact that the development of the third molars, which occurs after puberty, is delayed in girls as compared to boys, was attributed by Moorrees and associates[38] to a dual effect of estrogen, which, at low prepubertal levels, may accelerate tooth development while retarding development at the elevated levels associated with puberty.

Pancreas

The adverse effects of diabetes mellitus on the periodontium are evidence that the endocrine pancreas has a major role in tooth and periodontal growth and repair. The oral and periodontal problems associated with diabetes are many. Xerostomia is a common complaint but neither xerostomia nor the degree of hyperglycemia seems to be associated with an increased incidence of caries. However, the periodontal effects of diabetes are significant. Thus, according to Cianciola and collaborators,[39] gingivitis and periodontal disease, extremely uncommon disorders in normal children, occur frequently in young patients with insulin-dependent diabetes (IDDM). In this study, periodontitis generally first detected after age 12 years reached a prevalence of 9.8% in the 13- to 18-year-old patients, increasing to 39% to those 19 years of age or older. The severity of the disease was greater in the patients who developed periodontitis at an early age, although the occurrence of periodontal disease was related more strongly to chronologic age than to the duration of the disease. Alveolar bone loss affected most frequently the incisor and first molar teeth that erupt at the age of 6 to 7 years. However, in the more advanced cases, bone loss was generalized. Although some patients with advanced disease experienced periodontal abscesses and pathologic migration of teeth, most were asymptomatic. Since the presence of advanced periodontal disease may adversely affect the patient's ability to control their diabetes, the authors urge the periodontal status of these patients be assessed by proper dental consultation. Of importance is the finding that the amounts of supragingival dental plaque was the same in the control and the IDDM group, indicating that the increased prevalence of periodontal disease in IDDM patients cannot be attributed soley to poor oral hygiene and greater plaque accumulation.

Rosenthal, Abrams and Kopczyk[40] studied the relationship between IDDM and periodontal disease in 52 patients between the ages of 11 and 22 years and found an incidence of moderate to advanced periodontitis in 5.8% of the patients. Periodontitis did not correlate significantly with plaque index, age of diabetes onset, duration of diabetes, insulin dosage/weight, or serum glucose levels. However, there was a significant correlation between the presence of periodontal disease and of other complication of diabetes, such as retinopathy and neuropathy. Patients with neurologic complications had a significantly higher gingival index score than those with complications. These findings correlate well with those of Rylander and asociates,[41] who found that although there was no correlation between periodontal variables (the duration of the diabetes and insulin dosage), individuals with both retinopathy and neuropathy had significantly more gingival inflammation than individuals with these complications. Since local factors such as supragingival plaque do not seem to account for the increased susceptibility to periodontal disease in these patients, other possibilities have been investigated. Vascular insufficiency[42] and uncontrolled ketoacidosis and hyperglycemia decrease the resistance of the diabetic patient to infections,[42] and since infection makes the diabetes more difficult to control, a vicious cycle may become established. Of importance in this regard have been the investigations of Lavine et al.[43] and Van Dyke et al.,[44] who demonstrated impairment of the chemotactic response of polymorphonuclear neutrophilic leukocytes (PMN) in young patients with IDDM and rapidly progressing periodontitis. These observations suggest that impaired PMN function is a major factor in the predisposition of patients to periodontal destruction. Manoucheur-Pour et al.[45] also observed a significant impairment of PMN chemotactic response in diabetics with severe periodontitis and suggested that it may contribute to the severity of periodontal disease. Other studies have shown that the susceptibility to gingivitis and periodontitis is not limited to young patients but occurs also in patients between the ages of 20 and 40 years.

The correlation between diabetes and periodontal disease in patients over age 40 years is difficult to evaluate since many of these patients are edentulous and the prevalence of periodontal disease increases with age in the population in general. However, Ainamo et al.[46] reported rapid periodontal destruction in a group of 12 middle-aged or elderly diabetic patients who were either unaware of being diabetic or were unable to control the disease. The bone loss in these patients progressed despite professional care and the patients responded to treatment only after their diabetes was brought under control. Given the relationship between rapid periodontal breakdown and elevated blood glucose levels, Ainamo and his collaborators stressed the importance of ruling out diabetes in patients with periodontitis who are not responding to therapy.

Epidermal Growth Factor

The growth of teeth and the growth and repair of periodontal tissues after injury are also stimulated by the EGF, one of about 30 known peptides capable of stimulating cell division and cell locomotion, and hence of inducing tissue proliferation. In contrast to the hormones discussed above, EGF is generally secreted in the immediate vicinity of its target tissues and exerts its action in a paracrine fashion. Thus, whereas EGF has been isolated from plasma, urine, milk and saliva, its major sources appear to be the salivary glands, the kidney, and possibly some dental and periodontal tissues.[47-54] Indeed, sialoadenectomy is followed by the reduction of salivary EGF to undetectable amounts.[55] Human EGF is a single-chain 53–amino acid peptide with an apparent molecular weight of 6045 daltons derived from the processing of a 1207–amino acid precursor[53,56,57] and appears to be identical to urogastrone, a urinary factor capable of inhibiting hydrochloric acid secretion and of enhancing the repair of experimental gastric ulcers.[58-60] The biologic action of EGF is transduced by specific receptors found in a variety of target cells, including cells of the germinating tooth, such as the epithelium of the tooth bud, the outer and inner enamel, and the mesenchymal cells of the dental papilla.[61,62] The receptor is characterized by an extracellular binding domain, a transmembrane portion, and a cytoplasmic domain with the properties of a protein tyrosine-kinase (PTK.[63]).

Activation of PTK by EGF binding results in the phosphorylation of the receptor itself and of other protein molecules, among them phospholipase C, an enzyme that hydrolyzes phospholipids of the cell membrane, releasing inositol phosphate and diacylglycerol. This, in turn, leads to an increase in the concentration of intracellular calcium, to the activation of calcium-dependent kinases, and eventually to stimulation of DNA synthesis and ornithine decarboxylase activity, two indicators of cell growth and multiplication.[60,64,65] Stimulation of growth by EGF has been observed in renal,[66] gastric,[58] and epidermal[52] epithelia, vascular endothelium,[67] osteoblasts derived from mouse calvaria,[68] tissues undergoing repair after injury, and tissues of the oral cavity. Indeed, EGF, first identified as a "submaxillary protein accelerating incisor eruption" in mice,[52] appears to be critical not only for the development of teeth but for the reabsorption of the roots of human deciduous teeth,[51] for the development of human embryonic palate,[69-71] and possibly for the normal craniofacial morphogenesis during early embryonic development.[69] Finally, salivary EGF accelerates the healing of experimental wounds of the tongue in mice[55] and may be an essential factor in the salivary maintenance of tooth integrity.[71] Perhaps EGF is the reason why animal behavior and popular wisdom teach us to lick our wounds.

References

1. Schour I, Van Dyke HB. Changes in the incisor following hypophysectomy. I. Changes in the incisor of the white rats. *Am J Anat*. 1932;50:397–433.
2. Baume LJ, Becks H, Evans HM. Hormonal control of tooth eruption. III. The response of the incisors of hypophysectomized rats to growth hormone, thyroxin or the combination of both. *J Dent Res*. 1954;33:104–114.
3. Baume LJ, Becks H, Ray JD, Evans HM. Hormonal control to tooth eruption. II. The effects of hypophysectomy on the upper rat incisor following progressively longer intervals. *J Dent Res*. 1954;33:91–103.
4. Becks H, Collins DA, Simpson ME, Evans HM. Changes in the central incisors of hypophysectomized female rats after different postoperative periods. *Arch Pathol*. 1946;41:457–475.
5. Demirjian A, Goldstein H, Tanner JM. A new system of dental age assessment. *Hum Biol*. 1973;42:211–227.
6. Kosowicz J, Rzymski K. Abnormalities of tooth development in pituitary dwarfism. *Oral Surg Oral Med Oral Pathol*. 1977;44:853–863.
7. Salzmann JA, Wein S. Dental correlation in pituitary dwarfism. *Am J Orthodont*. 1952;38:674–686.
8. Hamori J, Gyulavari O, Szabo B. Tooth size in pituitary dwarfs. *J Dent Res*. 1974;53:1302–1426.
9. Drews M. Studies on the organ of mastication in patients with disorders of the growth hormone in cases of pituitary dwarfism and acromegaly. *Poznan Soc. Friends Sci*. 1971;3:61–93.
10. Snyder MB. Endocrine diseases and dysfunction. In: Lynch MA, Brightman VJ, Greenberg MS, eds. *Burket's Oral Medicine, Diagnosis and Treatment*. 8th ed. Philadelphia: Lippincott; 1984:812–841.
11. Sarnat H, Kaplan I, Petzelan A, Laron Z. Comparison of dental findings in patients with isolated growth hormone deficiency treated with human growth hormone (hGH) and in untreated patients with Laron-type dwarfism. *Oral Surg Oral Med Oral Pathol*. 1988;67:581–586.
12. Lozy M, Reed RB, Ken GR, Boutourline E. Nutritional correlates of child development in South Tunisia, IV. The relation of deciduous dental eruption to somatic development. *Growth*. 1975;39:209–211.
13. Wolff A, Stark H, Sarnat H, Binderman I, Eisenstein B, Drukker A. The dental status of children with chronic renal failure. *Int J Pediatr Nephrol*. 1985;6:127–132.
14. Shaw L, Foster TD. Size and development of the dentition in endocrine deficiency. *J Pedodont*. 1989;13:155–159.
15. Edler RJ. Dental and skeletal ages in hypopituitary patients. *J Dent Res*. 1977;56:1145–1153.

16. Hoskins MM. The effect of acetyl thyroxin on the development of the teeth. *J Dent Res*. 1928;8:85–97.

17. Ziskin DE, Applebaum E. Effects of thyroidectomy and thyroid stimulation on growing permanent dentition of rhesus monkeys. *J Dent Res*. 1941;10: 21–27.

18. Paynter KJ. The effect of propylthiouracil on the development of molar teeth of rats. *J Dent Res*. 1954;33:364.

19. Garren LD. The effect of hormones on the eruption rate of the rat incisor. *Harvard Dental Alumni Bull.*, 14 November 1954.

20. Garren LD. Effects of endocrine on the eruption rate of the upper incisor of the rat. *J Dent Res*. 1955;34:687–688.

21. Baume LJ, Beck H. The effect of thyroid hormone on dental and paradental structures. *Paradontologie*. 1952;6:89–106.

22. Wilkins L. *The diagnosis and treatment of endocrine disorders in childhood and adolescence. Adrenal cortex cushing's syndrome*. 3rd ed. Springfield, In: Chartes C Thomas; 1965:382–394.

23. Tiecke RW, Stuteville OH, Calandra JC. *Pathologic physiology of oral disease*. St. Louis: Mosby; 1959:1–480.

24. Engel MB, Bronstein IP, Brodie AG, Wesoke P. A roentgenographic cephalometric appraisal of untreated and treated hypothyroidism. *Am J Dis Child*. 1941;61:1193–1214.

25. Bier SJ. Oral manifestations of hypothyroidism and periodontics. *NY State Dent J*. 1961;27:14–15.

26. Parsons PA. Congenital hypothyroidism. *J Mich. State Dent Assoc*. 1960;42:184–186.

27. Kerley CG: Subthyroidism with defective dental development. *Arch Pediatr*. 1938;55:548–552.

28. Buckman N. Oral manifestations of cretinism. *Oral Surg Oral Med Oral Pathol*. 1957;10:938–947.

29. Gardner AF, Breen LM, Zakarin SL. Oral manifestations of endocrine disturbances. *Aust Dent J*. 1963;8:280–291.

30. Anderson HJ. Studies of hypothyroidism in children. *Acta Paediato Scand*. 1961;50(Suppl 125):1–148.

31. Noren IG, Alm J. Congenital hypothyroidism and changes in the enamel of deciduous teeth. *Acta Paediato Scand*. 1983;72:485–489.

32. Hinrich EH. Dental changes in juvenile hypothyroidism. *J Dent Child*. 1966;33:167–173.

33. Shafer WG, Hine MK, Levy BM, Tomich C. Oral aspects of metabolic diseases. In: *A textbook of oral pathology*. 4th ed. Philadelphia: Saunders; 1983:616–672.

34. Frensilli JA, Hinrich EH. Dental changes of idiopathic hypoparathyroidism: report of three cases. *J Oral Surg*. 1971;29:727–731.

35. Silverman S Jr, Ware WH, Gillooly C Jr. Dental aspects of hyperparathyroidism. *Oral Surg Oral Med Oral Pathol*. 1968;26:184–189.

36. Roberts MW, Comite SHLF, Hench KD, Pescovitz OH, Cutler GB Jr., Loriaux DL. Dental development in precocious puberty. *J Dent Res*. 1985;64:1084–1086.

37. Roberts MW, Li SH, Cutler GB, Hench KD, Loriaux DL. Sex differences in dental development in children with prococious puberty related to central nervous system lesions. *Am Acad Pediato Dent*. 1986;8:276–279.

38. Moorrees CFA, Fanning EA, Hunt E Jr. Age variation of formation stages for ten permanent teeth. *J Dent Res*. 1963;42:1490–1502.
39. Cianciola LJ, Park BH, Bruck E, Mosovich L, Genco RJ. Prevalence of periodontal disease in insulin-dependent diabetes mellitus (juvenile diabetes). *J Am Dent Assoc*. 1982;104:653–660.
40. Rosenthal IM, Abrams H, Kopczyk RA. The relationship of inflammatory periodontal disease to diabetic status in insulin-dependent diabetes mellitus patients. *J Clin Periodontol*. 1988;15:425–429.
41. Rylander H, Ramberg P, Blohme G, Lindhe J. Prevalence of periodontal disease in young diabetics. *J Clin Periodontol*. 1987;14:38–43.
42. Gottsegen R. Dental and oral aspect of diabetes mellitus. In: Ellenberg M, Rifkin H, eds. *Diabetes mellitus: theory and practice*. New York: McGraw-Hill; 1970:760–779.
43. Lavine WS, Maderazo Stolman J, Cogen J, Greenblat RB, Robertson PB. Impaired neutrophil chemotaxis in diabetic patients with severe periodontitis. 1979;14:10–19.
44. Van Dyke TE, Horoszewic HV, Ciomciola LJ, Genco RJ. Neutrophil chemotaxis dysfunction in human periodontitis. *Infect Immun*. 1980;27:124,132.
45. Manoucheu-Pour M, Spagnuolo PJ, Rodman HM, Bissada NF. Impaired neutrophil chemotaxis in diabetic patients with severe periodontitis. *J Dent Res*. 1981;60:729–730.
46. Ainamo J, Lahtinen A, Uitto VJ. Rapid periodontal destruction in adult humans with poor controlled diabetes. *J Clin Periodontol*. 1990;17:22–28.
47. Fisher DA, Salido EC, Baraja L. Epidermal growth factor and the kidney. *Annu Rev Physiol*. 1989;51:67–80.
48. Cohen S, Carpenter G. Human epidermal growth factor: isolation and chemical and biological properties. *Proc Natl Acad Sci USA*. 1975;72:1317–1321.
49. Starkey RH, Cohen S, Orth DN. Epidermal growth factor. Identification of a new hormone in human urine. *Science*. 1975;189:800–802.
50. Miki Y, Narayanan AS, Page RC. Mitogenic activity of cementum components to gingival fibroblasts. *J Dent Res*. 1987;66:1399–1403.
51. Kikuchi K, Miki M, Miyamoto S, Arita K, Nishino M. Effect of epidermal growth factor (EGF) on proliferation of the cells derived from the root resorbing tissue of human deciduous teeth. *Shoni Shikagaku Zasshi*. 1989; 27:92–100.
52. Cohen S. Isolation of a mouse submaxillary gland protein accelerating incisor eruption and eyelid opening in the new-born animal. *J Biol Chem*. 1962;237: 1555–1562.
53. Cohen S, Taylor JM. Recent studies on the chemistry and biology of epidermal growth factor. *Rec Prog Horm Res*. 1974;30:551–574.
54. Bellone C, Barni T, Pagni L, Balboni GC, Vannelli GB. Fattori di crescita nello sviluppo del dente umano. *Boll Soc It Biol Sper*. 1990;66:231–238.
55. Noguchi S, Ohba Y, Oka T. Effect of salivary epidermal growth factor on wound healing of tongue in mice. *Am J Physiol*. 1991;260:E620–E625.
56. Carpenter G, Cohen S. Epidermal growth factor. *Annu Rev Biochem*. 1979; 48:193–216.

57. James R, Bradshaw RA. Polypeptide growth factors. *Annu Rev Biochem.* 1984;53:259–292.
58. Sandweiss DG, Saltzstein HC, Farbman AA. The relation of sex hormones to peptic ulcer. *Am J Dig Dis.* 1939;6:6.
59. Gregory H. The isolation and structure of urogastrone and its relationship to epidermal growth factor. *Nature.* 1975;257:325–327.
60. Feldman EJ, Aures D, Grossman MI. Epidermal growth factor stimulates ornithine decarboylase activity in the digestive tract of mouse. *Proc Soc Exp Biol Med NY.* 1978;159:400–402.
61. Thesleff I, Partanen AM, Rihtniemi L. Localization of epidermal growth factor receptors in mouse incisors and human premolars during eruption. *Eur J Orthodont.* 1987;9:24–32.
62. Carpenter G. Receptors for epidermal growth factor and other polypeptide mitogens. *Annu Rev Biochem.* 1987;56:881–914.
63. Hunter T, Cooper JA. Protein-tyrosine kinases. *Annu Rev Biochem.* 1985; 54:897–930.
64. Deuel TF. Polypeptide growth factors: roles in normal and abnormal cell growth. *Annu Rev Cell Biol.* 1987;3:443–492.
65. Soltoff SP, Cantley LC. Mitogens and ion fluxes. *Annu Rev Physiol.* 1988; 50:207–223.
66. Mendley SR, Toback FG. Autocrine and paracrine regulation of kidney epithelial cell growth. *Annu Rev Physiol.* 1989;51:33–50.
67. Klagsbrun M, D'Amore PA. Regulators of angiogenesis. *Annu Rev Physiol.* 1991;53:217–239.
68. Hiramatsu M, Kumegawa M, Hatakeyama K, Yajima T, Minami N, Kodama H. Effect of epidermal growth factor on collagen synthesis in osteoblastic cells derived from newborn mouse calvaria. *Endocrinology.* 1982;111: 1810–1816.
69. Slavkin HC. Regulatory issues during early craniofacial development: a summary. *Cleft Palate J.* 1990;27:101–109.
70. Yoneda T, Pratt RM. Mesenchymal cells from the human embryonic palate are highly responsive to epidermal growth factor. *Science.* 1981;213:563–565.
71. Mandel ID. The functions of saliva. *J Dent Res.* 1987;66(spec. no):623–627.

Index